HYBRID Electric Vehicle TECHNOLOGY

atp
AMERICAN TECHNICAL PUBLISHERS, INC.
ORLAND PARK, ILLINOIS 60467-5756

Automotive Research and Design

American Technical Publishers, Inc., Editorial Staff

Editor in Chief:
Jonathan F. Gosse
Vice President—Production:
Peter A. Zurlis
Art Manager:
James M. Clarke
Technical Editor:
Eric F. Borreson
Copy Editor:
Talia J. Turner
Cover Design:
Nicole. S. Polak
Illustration/Layout:
Samuel T. Tucker
Jennifer M. Hines
William J. Sinclair
Nick W. Basham
Multimedia Coordinator:
Carl R. Hansen
CD-ROM Development:
Robert E. Stickley
Nicole. S. Polak
Daniel Kundrat

1 2 3 4 5 6 7 8 9 – 10 – 9 8 7 6 5 4 3 2 1

Printed in the United States of America

ISBN 978-0-8269-0066-1

This book is printed on 10% recycled paper.

atp made in USA

® GCIU 3-S

Acknowledgments

The authors and publisher are grateful to the following companies and organizations for providing information, photographs, and technical assistance.

Fluke Corporation

General Motors

Honda

Leeson Electric Co.

Tektronix, Inc.

Toyota Motor Company

HYBRID
Electric Vehicle
TECHNOLOGY

Table of Contents

INTERACTIVE CD-ROM CONTENTS

- Quick Quizzes®
- Illustrated Glossary
- Flash Cards
- Media Clips
- Review Questions
- Link to ATPeResources.com

Introduction

Hybrid Electric Vehicle Technology provides a comprehensive systems overview of the operating principles, maintenance, and service of hybrid electric vehicles. This textbook is designed to develop a basic knowledge of common hybrid systems. The textbook contains 19 chapters, beginning with a review of the types of hybrid vehicles and high-voltage vehicle safety concerns. Subsequent chapters include information on the operation of hybrid engines, electric machines, inverters, converters, and sensing systems. This information is used in the later chapters to cover the topics of hybrid transaxles and battery, climate control, and jump start systems.

At the beginning of each chapter, an Introduction, Objectives, and Key Terms inform the learner about chapter contents. At the end of each chapter, Points to Remember, Review Questions, and Trade Tests help the learner review key concepts and reinforce principles of operation, maintenance, and troubleshooting of hybrid vehicles. Answers to odd-numbered questions are included at the end of the book. Key terms are italicized and defined in the text for additional clarity. An extensive Glossary and Appendix provide useful, easy-to-find information. A comprehensive Table of Contents and Index simplify navigation to desired information.

The *Hybrid Electric Vehicle Technology* CD-ROM located at the back of the book is designed as a study aid to complement information presented in the book and includes Quick Quizzes®, an Illustrated Glossary, Flash Cards, Media Clips, and Review Questions. The Quick Quizzes® provide an interactive review of topics in each chapter. The Illustrated Glossary provides a helpful reference to terms related to hybrid vehicle technology. Flash Cards provide a review of terms and definitions. The Media Clips are a collection of animated graphics that illustrate fundamental concepts. The Review Questions are the questions from the end of each chapter in Microsoft® Word format for efficient completion, review, and printing. Clicking on the American Tech web site button (www.go2atp.com) or the American Tech logo accesses information on related instructional products.

HYBRID Electric Vehicle TECHNOLOGY

Book Features

Introductions present an overview of chapter content.

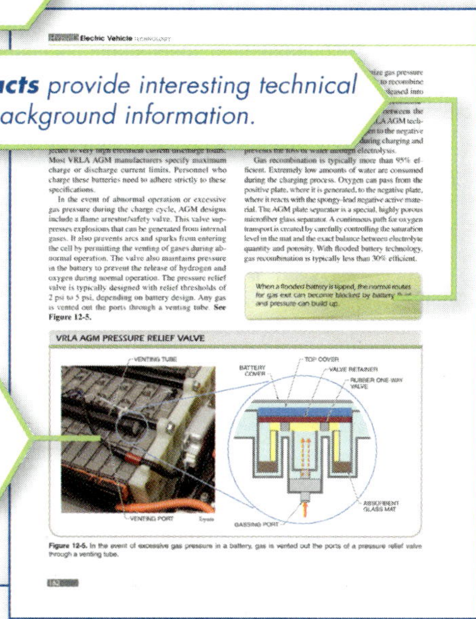

Tech Facts provide interesting technical tips or background information.

Introductory photographs represent chapter content.

Illustrations utilize photographs and detailed drawings to highlight component features and operation.

Objectives list learning goals for the chapter.

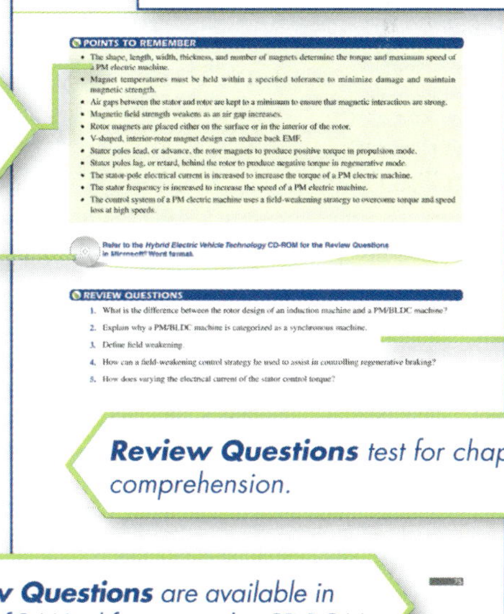

Points to Remember highlight the key concepts of the chapter.

Review Questions test for chapter comprehension.

Review Questions are available in Microsoft® Word format on the CD-ROM.

Photographs show component construction and operation.

Key Terms include common vocabulary terms associated with hybrid vehicles.

HYBRID Electric Vehicle TECHNOLOGY

Interactive CD-ROM Features

The Interactive CD-ROM included with this book is a self-study aid that reinforces the content in *Hybrid Electric Vehicle Technology.* This Windows® compatible CD-ROM includes the following interactive learning tools:

Quick Quizzes® *that reinforce fundamental concepts, with 10 interactive questions for each chapter*

An ***Illustrated Glossary*** *of terms related to hybrid vehicle technology, with links to selected illustrations and animated graphics*

Flash Cards *that provide a self-study review of key terms and definitions*

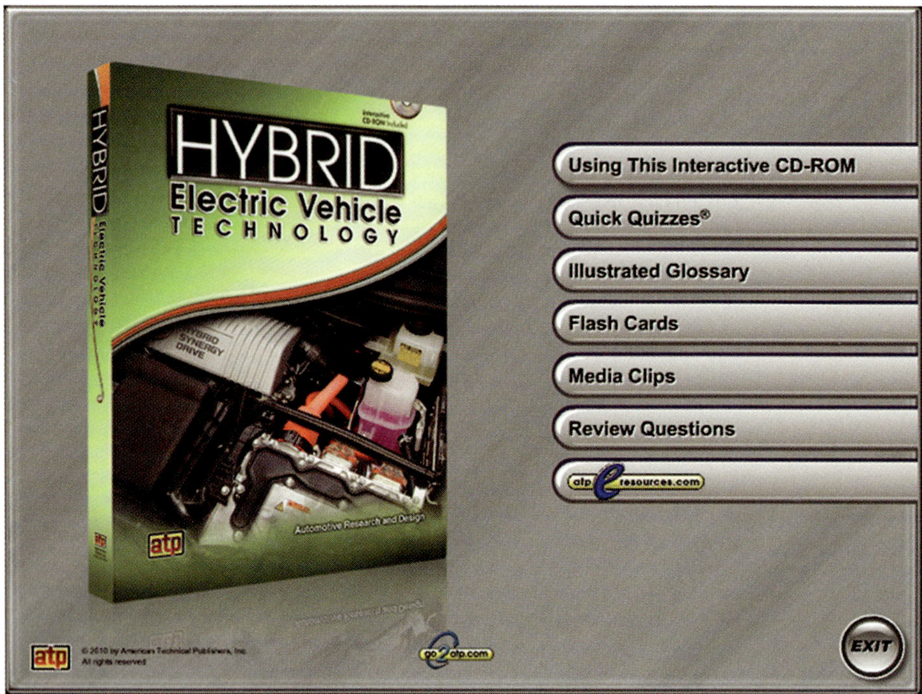

Media Clips *that illustrate fundamental hybrid vehicle concepts using animated graphics*

Review Questions *from each chapter in Microsoft® Word format for easy printing and review*

A link to ***ATPeResources.com****, which provides a comprehensive array of instructional resources*

HYBRID
Electric Vehicle
TECHNOLOGY

Introduction to HEV Technology

1

The idea of a hybrid system for automotive applications is not new. Automotive designers have been thinking of combining an internal combustion engine and an electric machine for a long time. Recent improvements in power electronics and advanced battery technologies allowed designers to efficiently integrate these two technologies into current hybrid production vehicles.

CHAPTER OBJECTIVES

- Identify the main types of hybrid vehicle technologies.
- Define series, parallel, and series/parallel power flow.
- Identify the common hybrid powertrain subsystems.

KEY TERMS

hybrid electric vehicle (HEV): A vehicle that uses more than one propulsion technology to provide power to the vehicle drive system.

mild hybrid system: A hybrid electric vehicle system that is typically characterized by system voltages at 36 V or less, may provide electric propulsion assist to the engine, and may have a limited electric-only drive mode.

strong hybrid system: A hybrid electric vehicle system that is typically characterized by system voltages between 144 V and 600 V, an electric-only drive mode, and/or electric propulsion assist to the engine.

series power flow system: An engine system connected in series to a generator that provides electrical power to the electric drive motor and the battery pack. With this system, the engine is not directly connected to the transmission, transaxle, or driveline.

parallel power flow system: An engine system connected directly to the transmission, transaxle, or driveline with supplemental electric machines connected in parallel. With this type of system, the electric machine torque is added in parallel to the engine torque and cannot be used by itself to drive the vehicle.

series/parallel power flow system: An engine and electric transaxle system that permits the electric transaxle to provide propulsion when the engine is not operating (series mode) or permits engine torque to be added to the electric transaxle and then transmitted through the drive axles (parallel mode).

TYPES OF HYBRID VEHICLE TECHNOLOGIES

A *hybrid electric vehicle (HEV)* is a vehicle that uses more than one propulsion technology to provide power to the vehicle drive system. Hybrid drive systems include a source of power, such as an engine or a fuel cell, and a way of using that power to drive the vehicle, such as a transaxle or an electric machine. An electric machine is motor/generator technology that provides torque to drive a vehicle and/or generates electricity during regenerative braking. Hybrid vehicle systems can be designed in many different configurations to fit many different vehicle platforms. The systems are commonly classified as strong, mild, and fuel-cell systems.

Mild Hybrid Systems

A *mild hybrid system* is a hybrid electric vehicle system that is typically characterized by system voltages at 36 V or less, may provide electric propulsion assist to the engine, and may have a limited electric-only drive mode. **See Figure 1-1.** Mild hybrid systems may have a flywheel or belt-driven design and usually have the capability to stop and start the engine as needed. Mild hybrid systems often provide a fuel economy increase of 10% to 15%. Common vehicles with mild hybrid systems include the GM Silverado/Sierra and Saturn VUE vehicles.

Strong Hybrid Systems

A *strong hybrid system* is a hybrid electric vehicle system that is typically characterized by system voltages between 144 V and 600 V, an electric-only drive mode, and/or electric propulsion assist to the engine. **See Figure 1-2.** Strong hybrid systems include propulsion assist and full electric drive capability. They have electric machines integrated into the transmission or transaxle and have the capability to stop and start the engine as needed. These systems often provide a fuel economy increase of 25% or more and reduced engine emissions.

Strong hybrid systems are becoming increasingly popular with all automotive manufacturers. Many manufacturers offer hybrid propulsion systems in more than one of their models and across multiple platforms. Common vehicles with strong hybrid systems include the Ford Escape, GM Allison Bus, Honda Accord/Civic/Insight, and Toyota Prius.

Fuel-Cell Hybrid Systems

A fuel-cell vehicle can also be classified as a hybrid design when a battery pack is used to increase the power of the fuel-cell electrical system. **See Figure 1-3.** A fuel-cell hybrid system is typically characterized by battery pack voltage of approximately 300 V used to increase specific power during acceleration and a fuel-cell stack. Fuel-cell vehicles all have electric-only drive systems. They do not use internal combustion engines.

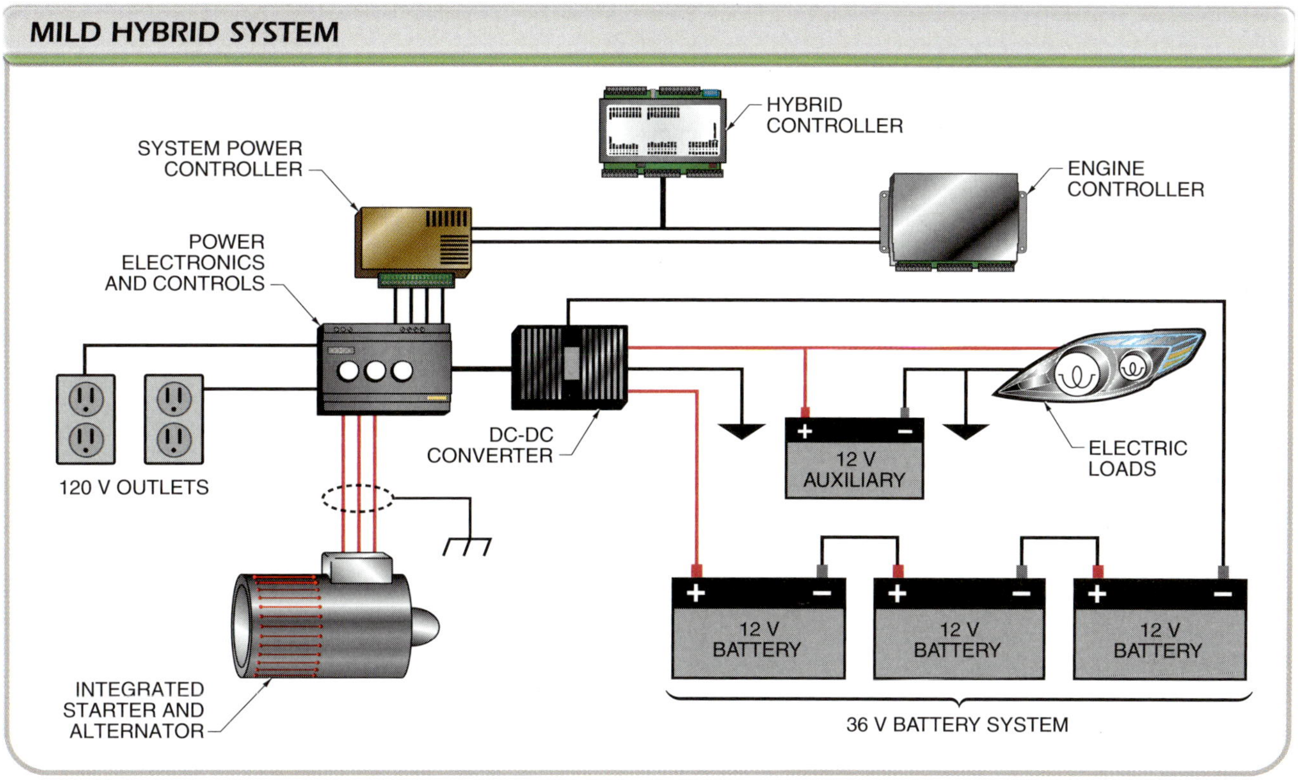

Figure 1-1. A mild hybrid system provides electric propulsion assist to the engine.

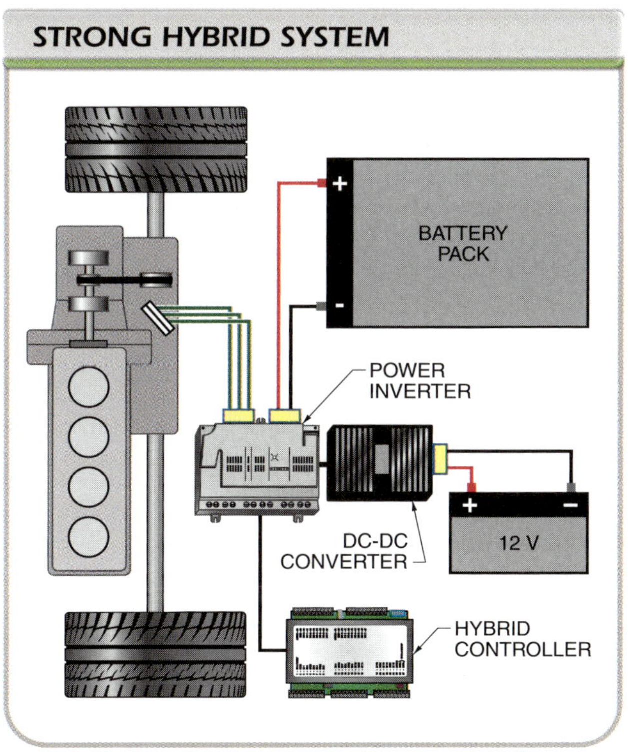

Figure 1-2. A strong hybrid system includes electric-only drive.

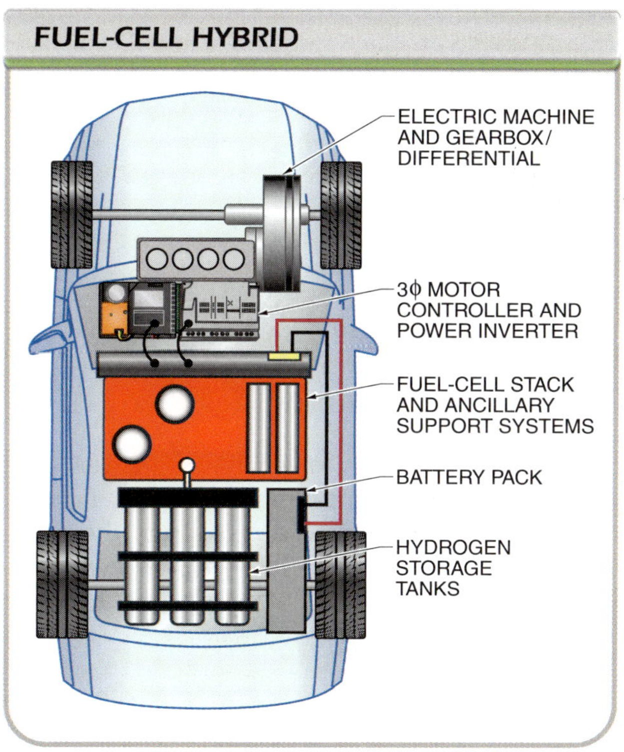

Figure 1-3. A fuel-cell hybrid includes a fuel stack and a battery pack.

TYPES OF POWERTRAINS AND POWER FLOW DESIGNS

Understanding the different types of power flow is important when maintaining a hybrid electric vehicle. The power flow designs in HEVs include series, parallel, and series/parallel systems.

> Strong hybrid systems typically have an electric machine designed as an integral part of the transmission/transaxle. This allows the vehicle to operate on battery power alone for a short time.

Series Power Flow Systems

A *series power flow system* is an engine system connected in series to a generator that provides electrical power to the electric drive motor and the battery pack. **See Figure 1-4.** With series power flow, the engine is not directly connected to the transmission, transaxle, or driveline. With sufficient battery power, the electric drive motor can drive the vehicle alone. Series power flow is not considered a mainstream design strategy because the large number of required batteries increases the vehicle weight. However, future vehicle designs are expected to use series designs as better batteries are developed.

Parallel Power Flow Systems

A *parallel power flow system* is an engine system connected directly to the transmission, transaxle, or driveline with supplemental electric machines connected in parallel. **See Figure 1-5.** With parallel power flow, the electric machine torque is added in parallel to the engine torque and cannot be used by itself to drive the vehicle.

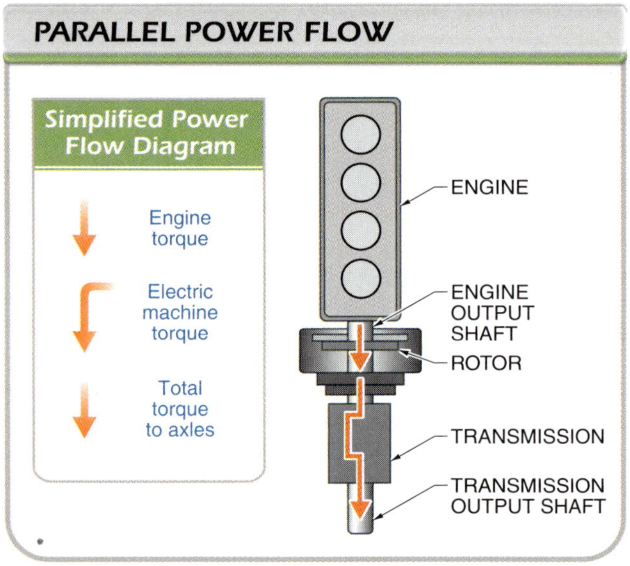

Figure 1-5. A parallel power flow system uses an engine connected to the drive train with supplemental electric machines connected in parallel.

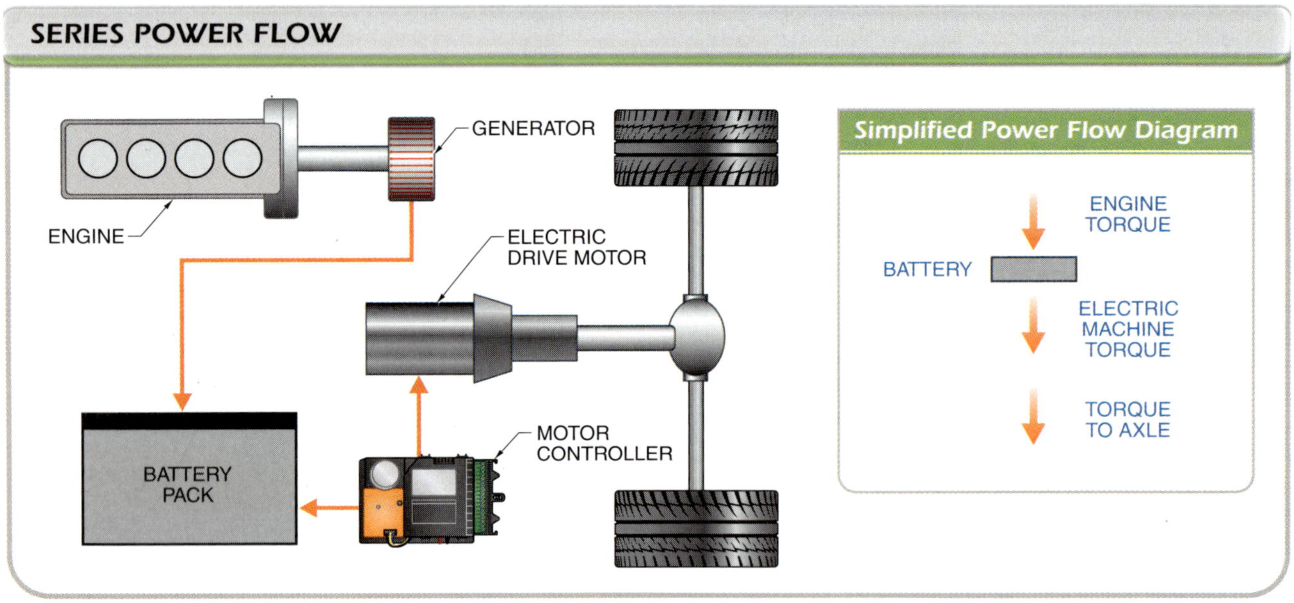

Figure 1-4. A series power flow system uses an engine to run a generator to charge a battery pack, which powers the electric motor.

Parallel power flow systems function as two separate systems that provide torque to the transmission or transaxle. The total torque is the sum of the torque from the engine and the torque from the electric machine. The electric machine is connected mechanically to the engine crankshaft output and to the input of the transmission. This provides extra torque to the transmission in parallel to the torque from the engine.

The engine torque is transferred from the crankshaft directly to the input of the transmission. The rotor of the electric machine is connected to the input of the transmission at the same point. At this point, the combined torque of the engine and electric machine provide torque to the transmission. The torque is transferred through the transmission gearing to the output shaft and then to the final drive gears and axles.

All torque from the electric machine is provided in parallel to the engine torque. In most parallel systems, the engine and electric machine are always operating in tandem. This means that the electric machine cannot operate independently of the hybrid engine. Therefore, most parallel systems do not allow electric-only drive. However, some newer systems do allow electric-only drive.

Parallel power flow is considered to be the mainstream design strategy for many automotive manufacturers. Parallel power flow also allows lower system mass. Common vehicles with parallel power flow include GM Sierra/Silverado 42 V systems, the Saturn Vue, and Honda hybrid products from years 2000 to 2005.

Series/Parallel Power Flow Systems

A *series/parallel power flow system* is an engine and electric transaxle system that permits the electric transaxle to provide propulsion when the engine is not operating (series mode) or permits engine torque to be added to the electric transaxle and then transmitted through the drive axles (parallel mode). **See Figure 1-6.** The GM Allison as well as Ford, Lexus, Toyota, and newer Honda models include series/parallel flow systems. A series/parallel power flow system is more complex than a parallel power flow system because the hybrid engine and transmission are constructed as an integrated unit.

For example, the engine torque is transferred from the crankshaft to the input of the electric transaxle into the planet carrier and through the planetary ring gear. The drive motor and the generator electric machines are connected to the same shaft. The drive motor-electric machine adds electric power. The generator that is connected to the sun gear determines the gear ratio electrically by varying its torque and speed. Torque is applied to a chain gear, the drive chain, and then to the final drive gears and axles.

As a result, the engine and electric drive motor can be operated in series or in parallel to provide torque to the planetary ring gear. Series/parallel power flow allows the overall system to be very efficient. Complex control software must be used to integrate the engine and electric transaxle systems.

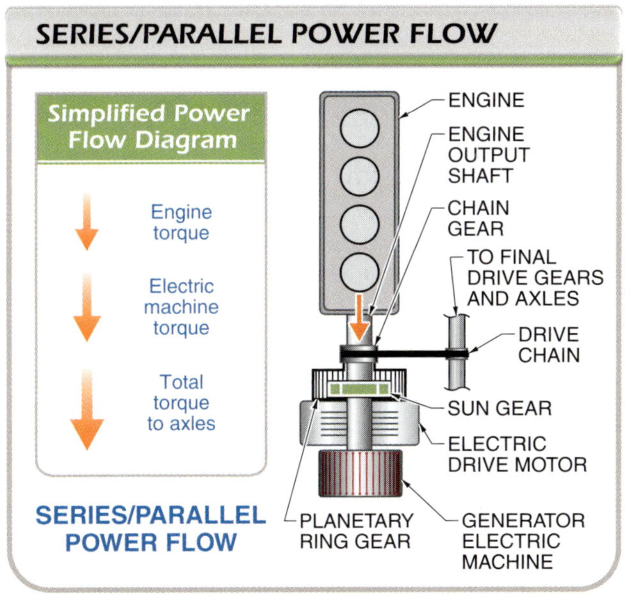

Figure 1-6. Series/parallel power flow can provide propulsion in series or in parallel.

HYBRID VEHICLE SUBSYSTEMS

Most hybrid electric gas/electric vehicles commonly consist of similar subsystems. Understanding the different types of subsystems is important when maintaining a hybrid electric vehicle. The subsystems may not always have the same size and shape but the basic functions are very similar. Common hybrid vehicle subsystems include battery pack and module systems, electric machines, power inverters, power converters, and control systems.

Ford Escape Hybrid Subsystems

The Ford Escape can be used to show the main hybrid electric vehicle subsystem. **See Figure 1-7.** The Ford Escape is a parallel-hybrid power flow system. The 2004 Ford Escape was the first production hybrid SUV on the market. It provides a 50% improvement in fuel economy over the nonhybrid model.

The battery pack is located in the rear of the vehicle under the cargo area. The battery pack is a nickel metal hydride (NiMH) system rated at 330 V DC. It has an orange service disconnect switch that is used to isolate the battery pack during service or emergency operations.

The electronically controlled, continuously variable transmission of the Ford Escape is unique. It contains two permanent magnet electric machines and a power inverter system. A planetary gear set and the final drive gearing system provide torque to the drive shafts. A controller, DC-DC power converter, and control system are integrated into the system.

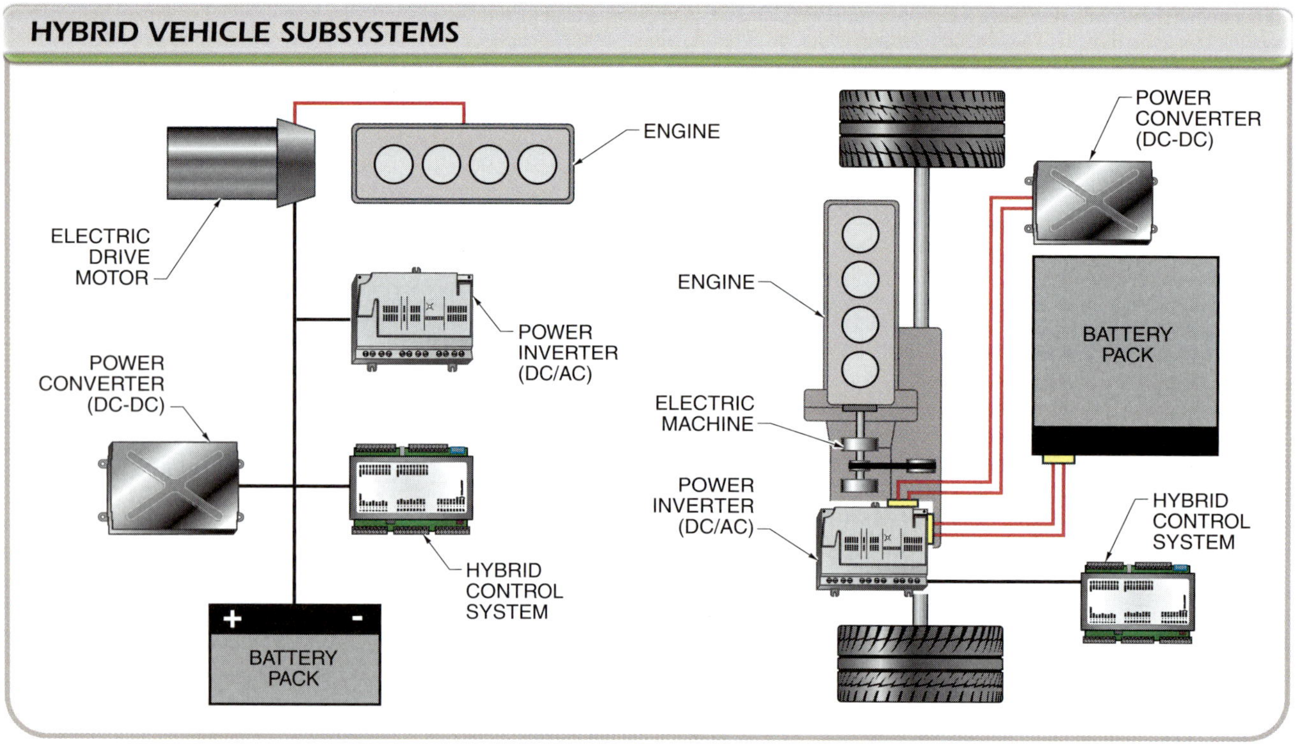

Figure 1-7. Most hybrid electric vehicles consist of similar subsystems.

Refer to the *Hybrid Electric Vehicle Technology* CD-ROM for a Quick Quiz® on the chapter contents.

POINTS TO REMEMBER

- Hybrid electric vehicles (HEVs) can be categorized as strong or mild.
- The power flow of HEVs includes series, parallel, and series/parallel designs.
- Common subsystems used in HEVs are battery pack and module systems, electric machines, power inverters, power converters, and control systems.

Refer to the *Hybrid Electric Vehicle Technology* CD-ROM for the Review Questions in Microsoft® Word format.

REVIEW QUESTIONS

1. Define the term "hybrid electric vehicle."
2. What is the difference between a mild hybrid system and a strong hybrid system?
3. What is the difference between series, parallel, and series/parallel power flow systems?
4. List at least four hybrid subsystems.

1 Introduction to HEV Technology

Trade Test

Name ______________________________ Date ______________

______ **1.** The two main types of HEV technologies are ___ systems.
A. high-torque and low-torque
B. high-speed and low-speed
C. strong and mild
D. strong and mid-torque

______ **2.** The Ford Escape hybrid electric vehicle is an example of a ___ hybrid system.
A. mild
B. strong
C. series
D. longitudinal

______ **3.** The Saturn Vue hybrid electric vehicle is an example of a ___ hybrid system.
A. mild
B. strong
C. series
D. longitudinal

______ **4.** Common types of hybrid power flow include series, parallel, and ___.
A. series/parallel
B. strong
C. mild
D. torque-leveled

______ **5.** A(n) ___ is motor/generator technology that provides torque to drive a vehicle and/or generates electricity during regenerative braking.
A. electric machine
B. parallel generator
C. series generator
D. parallel/series generator

______ **6.** A ___ hybrid system is characterized by high system voltages, an electric-only drive mode, and/or electric propulsion assist to the engine.
A. mild
B. strong
C. series
D. parallel

_______________ **7.** A(n) ___ power flow system is an engine system that is connected directly to the transmission, transaxle, or driveline with supplement electric machines connected in parallel.

A. longitudinal

B. axial

C. series

D. parallel

High-Voltage Electrical Safety

2

Hybrid vehicles often have high-voltage electrical systems. The ability to identify a high-voltage circuit, use the proper personal protection equipment (PPE), and understand the electrical characteristics of high-voltage systems will reduce the possibility of electrical hazards. Hybrid vehicles are different than traditional internal combustion engine powered vehicles. The ability to properly use an insulation tester is needed to perform safe and accurate tests on high-voltage systems.

CHAPTER OBJECTIVES

- Identify a high-voltage cable or wire by its color.
- Explain how to use the "one hand rule" for working on electrical circuits.
- Define the difference between high-voltage, intermediate-voltage, and low-voltage systems.
- Explain why there are different classes of electrical-safety gloves and why Class 0 gloves are usually appropriate for working on hybrid vehicles.
- Explain how a grounded oscilloscope creates a hazard for a technician working on a high-voltage system.
- Describe how different levels of electrical currents affect the human body.
- Describe the different category (CAT) ratings and explain how to select the proper multimeter for use on a hybrid vehicle electrical system.
- Describe how to use an insulation tester to test high-voltage circuits or components.

KEY TERMS

isolation fault: A condition that results when the resistance between a high-voltage system and the vehicle chassis has degraded to a point that may no longer be safe. The vehicle diagnostic system senses isolation faults and commands the malfunction indicator lamp to turn on to notify the operator and service technician.

high voltage: A level of system voltage that is greater than 60 V DC or 30 V AC rms.

intermediate voltage: A level of system voltage that is greater than 30 V DC and less than or equal to 60 V DC, or that is greater than 15 V AC rms and less than or equal to 30 V AC rms.

low voltage: A level of system voltage that is less than or equal to 30 V DC, or less than or equal to 15 V AC rms.

Class 0 glove: An electrical-safety glove designed to protect the wearer from electricity up to a maximum of 1000 V. Class 0 is the electrical-safety glove rating recommended by most automotive manufacturers.

differential scope probe: An instrument that is used to isolate an oscilloscope from a circuit. It ensures that the high-voltage electrical bus and earth/chassis ground are isolated from each other.

category (CAT) rating: A standard rating provided by the International Electrotechnical Commission (IEC) that outlines safety categories for electrical meters.

insulation tester: A diagnostic instrument that is used to detect a reduction in resistance between high-voltage circuits, components, or systems and the vehicle chassis.

device under test (DUT): A device that is being tested or analyzed by an instrument or test equipment.

HIGH-VOLTAGE ELECTRICAL SAFETY

There are safety precautions that should always be followed when working with any electrical circuit. However, extra precautions may be needed with high-voltage systems. The voltage used with hybrid vehicles can be up to about 600 V. It is never safe to assume that any high-voltage hybrid circuit or component is discharged.

First, whenever practical, the "one hand rule" should be used when working on any electrical circuit, especially circuits over 12 V. The rule is to work with only one hand and to keep the other hand away from the circuit. Second, probes with insulated alligator clips should be used. Also, all jewelry should be removed because these items could accidentally come in contact with a circuit. Last, proper personal protective equipment (PPE) should be worn.

There are some safety precautions that are specific to hybrid electric vehicles. For example, orange wire or cable coverings are used for high-voltage components. In addition, high-voltage systems are mounted on the vehicle chassis. The components are electrically isolated from the vehicle chassis and operate as a contained system. In the event of failure, voltage can leak to the chassis and cause an isolation fault.

An *isolation fault* is a condition that results when the resistance between a high-voltage system and the vehicle chassis has degraded to a point that may no longer be safe. It can occur for reasons such as component failure, collision damage, or improper repairs. The vehicle diagnostic system senses isolation faults and commands the malfunction indicator lamp to turn on to notify the operator and service technician.

Voltage Levels

Although there are no universal definitions for the actual voltages that correspond to voltage classifications, voltages are commonly classified as high, intermediate, or low. **See Figure 2-1.** Various safety studies have provided evidence that permit definitions to be applied to automotive safety. With the application of these definitions, automotive manufacturers can assess when and where to engineer high-voltage safety systems on a vehicle. For this book, voltage classifications provided by the General Motors Corporation are used.

High voltage is a level of system voltage that is greater than 60 V DC or 30 V AC rms. High voltage is common in hybrid powertrains and may be as high as about 600 V. *Intermediate voltage* is a level of system voltage that is greater than 30 V DC and less than or equal to 60 V DC, or that is greater than 15 V AC rms and less than or equal to 30 V AC rms. Intermediate voltage is common in mild hybrid systems, such as in 36/42 V systems. *Low voltage* is a level of system voltage that is less than or equal to 30 V DC, or less than or equal to 15 V AC rms. Low voltage is common in standard 12 V electrical systems.

VOLTAGE LEVELS

Voltage Level	Definition	
	DC	AC
High	>60 V	>30 V rms (>42.4 Peak)
Intermediate	>30 V and ≤60 V	>15 V rms (>21.2 Peak) ≤30 V rms (>42.4 Peak)
Low	≤30 V	≤15 V rms (>21.2 Peak)

Note: At this time there is insufficient data to draw any conclusions on the relative severity between switched DC and pure DC or AC waveforms. Until such data is available, it is recommended that a switched DC waveform frequency of 10 Hz to 200 Hz be treated as a DC waveform. Nominal peak voltage (excluding) ripple and transients) may be used for determining voltage classification as low, intermediate, or high voltage.

SYMBOLS:

> GREATER THAN
< LESS THAN
≤ LESS THAN OR EQUAL

Figure 2-1. Voltages are commonly classified as high, intermediate, and low.

PERSONAL PROTECTIVE EQUIPMENT

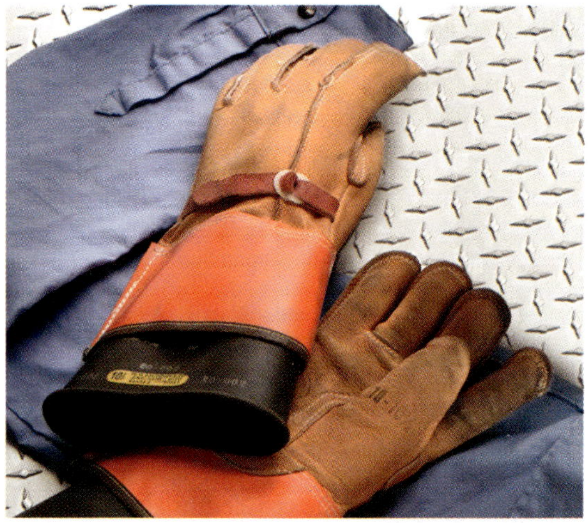

Fluke Corporation

Class	Maximum Voltage
00	500
0	1000
1	7500
2	17,000
3	26,500
4	36,000

Figure 2-2. Manufacturers of hybrid vehicles recommend the use of personal protective equipment (PPE) as part of many service procedures.

Personal Protective Equipment (PPE)

Common personal protective equipment includes appropriate nonsynthetic clothing, safety glasses, and electrical-safety gloves. Manufacturers of hybrid vehicles recommend the use of personal protective equipment (PPE) as part of many service procedures. **See Figure 2-2.** PPE is necessary because a technician may not know the condition of the vehicle safety systems and certain conditions, such as isolation faults, could cause injuries. In addition, the technician may not know if the high-voltage bus has been discharged to a safe level in order to perform any necessary repairs.

Electrical-safety gloves used for high-voltage protection are typically rated Class 0. A *Class 0 glove* is an electrical-safety glove designed to protect the wearer from electricity up to a maximum of 1000 V. Class 0 is the glove rating recommended by most automotive manufacturers.

The glove itself is made of a rubber compound that is highly resistant to electrical current. Leather protectors are worn over the gloves to protect them from being damaged by sharp edges or pointed objects. Gloves must be tested regularly to ensure that there are no holes in the material. Instructions on how to test gloves are provided by the manufacturer.

Electrical Shock

Electrical shock occurs when current is conducted through the human body. The amount of current and duration of electrical exposure through a body determine the severity of electrical shock and injuries. It should be noted that it is the current that causes injury. Voltage only provides the electrical pressure that forces current through the human body.

Although a given warning about the presence of high voltage is important, it is equally important for anyone who engages a high-voltage vehicle to understand the effects of electrical current on the human body. **See Figure 2-3.** Electrical shock occurs when a human body is placed in a circuit. An electrical circuit developed from hand-to-hand or finger-to-finger can cause injury by permitting electrical current to flow through the body (hand-to-hand) or in the hand (finger-to-finger) and back to the circuit. Either condition can cause serious electrical shock.

ELECTRICAL SHOCK

Current*	Effect†
0.5 to 1.5	Perceptible shock
1 to 5	Muscle contraction
3 to 10	Pain
10 to 40	No-let-go threshold
30 to 70	Respiratory paralysis
75 to 100	Ventricular fibrillation
250 to 300	Heart paralysis
5000 to 6000	Organ burns

* in mA AC
† current conducted through 150 lb body; effect varies with frequency

Figure 2-3. The severity of electrical shock is dependent on the level of current and the body's length of exposure to the current.

CURRENT FLOW

Resistance	Current*	Description
3000 Ω	$\frac{125\text{ V}}{3000\ \Omega} = 0.042\text{ A} = 42\text{ mA}$	Large person, 95th percentile
1750 Ω	$\frac{125\text{ V}}{1750\ \Omega} = 0.072\text{ A} = 72\text{ mA}$	Average person, 50th percentile
1230 Ω	$\frac{125\text{ V}}{1230\ \Omega} = 0.102\text{ A} = 102\text{ mA}$	Petite person, 5th percentile

* at 125 V DC

Figure 2-4. Higher voltage causes higher current. For a given voltage, current is limited by the electrical resistance.

One of the characteristics of electrical shock is the no-let-go threshold level. This is the point at which muscles begin to contract and lock onto the circuit. The point at which this occurs depends on the current and frequency of the current. In addition, the heart can be damaged if the electrical pathway is from hand-to-hand or hand-to-foot and there is enough current to cause ventricular fibrillation. Ventricular fibrillation is a condition in which heart convulsions stop the flow of blood through the body. It occurs at a current of about 75 mA to 100 mA.

According to the NEC®, resistance to electrical flow in a large person could be about 3000 Ω. Since current is the voltage divided by electrical resistance, for a given voltage, current is limited by the resistance. **See Figure 2-4.** When voltage is 125 V DC, current is 42 mA. For a petite person, the resistance could be about 1230 Ω and the current would be 102 mA. High-voltage electrical systems actually reach up to about 600 V. Since this voltage is four times as high as in the example, the current for these systems would also be four times as high.

Methods used to increase resistance and prevent electrical shock include the use of rubber gloves, rubber-soled shoes, and insulated tools. Current can be reduced by increasing the resistance in the circuit through the body and to ground. The benefit of using these methods is to minimize conduction paths across the human hands and body and to reduce the potential electrical current through the body.

> Although it is considered an intermediate voltage level, a 42 V mild hybrid system can cause serious discomfort or pain.

Ground Circuits

Ground, or return, circuits are used to connect the negative ends of electrical circuits through a ground to a common conductor, such as to a vehicle chassis or to the earth. Safety is the primary reason for establishing ground circuits. Establishing a ground circuit is necessary for an electrical circuit to function properly. In addition, they are necessary to reduce ground potential differences that can cause electrical currents to flow in other areas of the vehicle (ground loops).

There are two types of grounding that must be considered by a technician: earth ground and chassis ground. Earth ground is used for AC power systems where the negative side of the circuit is physically connected to the earth. A chassis, or common, ground is used for vehicle electrical systems where the negative side of the circuit is physically connected to the vehicle chassis.

Electrical circuits can develop across the human body when it is placed between a voltage source and the earth ground. For example, if a person were to place one hand on earth ground and touch the high-voltage positive connection of a circuit with the other hand, their body would be placed electrically in the high-voltage circuit. A similar situation occurs when a person places a hand on a vehicle chassis. If high voltage is leaking to the chassis of the vehicle and a person were to contact a high-voltage positive connection, their body would be electrically placed in the high-voltage circuit. In either case, the result could be a devastating electrical shock. **See Figure 2-5.** Risk of shock can be minimized by using only one hand when touching any part of the vehicle.

If a technician uses a 120 V oscilloscope to test a circuit, the negative side of the oscilloscope circuit and the ground can be operating at the same potential. **See Figure 2-6.** If the technician touches the circuit on the negative side of the battery pack, the technician is connected to the circuit. However, the negative side of the vehicle high-voltage system and the oscilloscope ground are at the same potential. Therefore, no current flows through the body and the technician is safe.

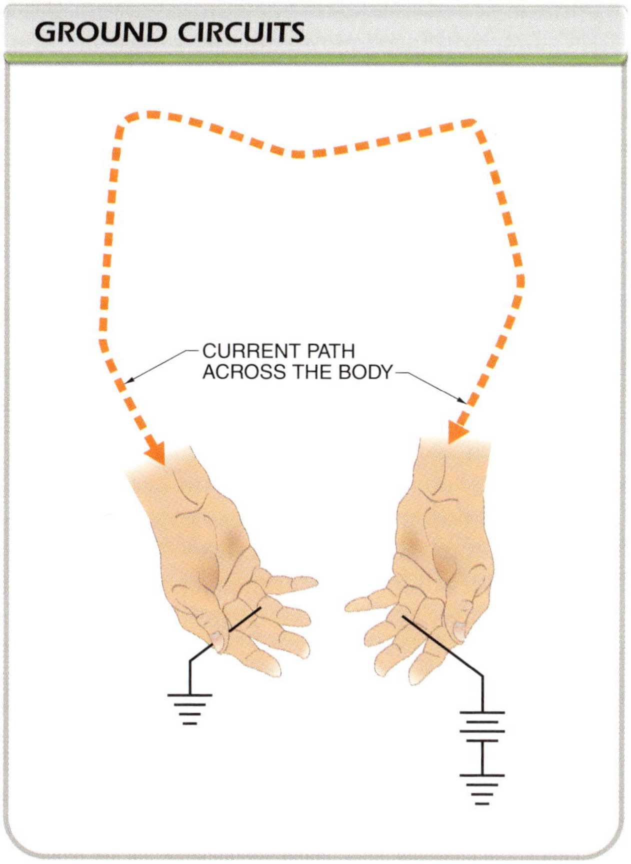

Figure 2-5. A current path across the body is created as it is placed in the high-voltage circuit.

If the technician touches the circuit on the positive side of the battery pack, the circuit and the ground are not at the same potential. The oscilloscope provides a ground path through the technician to the negative side of the high-voltage bus through earth ground. When the technician touches the positive side of the high-voltage bus and some part of the body is on the same ground potential as the negative side of the bus, the technician is in a direct current path with the high-voltage battery. The current conducts through the high-voltage components and the technician's body. As a result, the technician receives an electrical shock and the possibility of fatal injuries or severe burns.

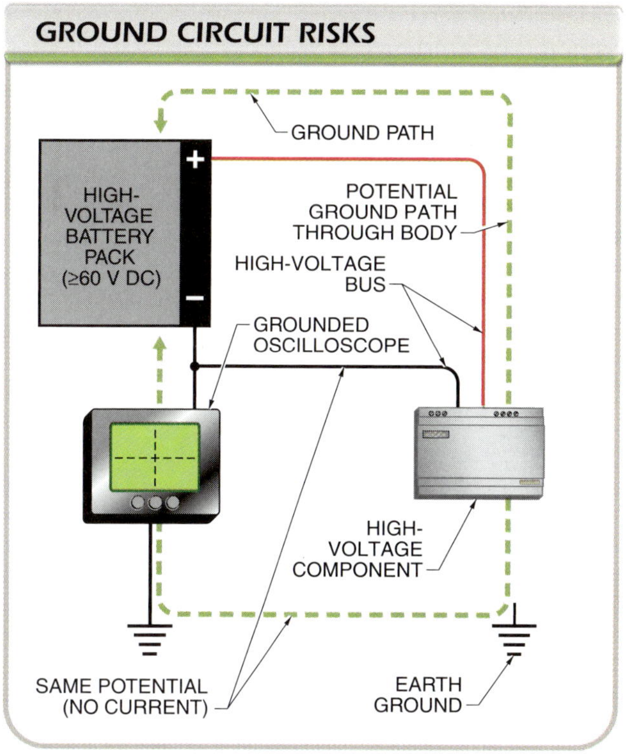

Figure 2-6. A grounded oscilloscope can create a ground path through a body in a circuit.

TOOLS AND EQUIPMENT

Unlike traditional automotive 12 V electrical systems, high-voltage electrical systems require additional safety precautions and tools in order to safely perform electrical measurements. Analysis tools, such as oscilloscopes, multimeters, and insulation testers, are valuable for diagnosing circuit failures.

Oscilloscopes

Automotive training typically does not cover the subject of grounding and the differences between earth ground and chassis ground. However, this information is important for knowing when it is appropriate to use certain instruments such as an oscilloscope. This instrument poses a hidden danger to automotive technicians and instructors because it can provide an electrical path if it is not isolated from the circuit that is being tested. Therefore, a differential scope probe should be connected to an oscilloscope if one is not built into the oscilloscope.

A *differential scope probe* is an instrument that is used to isolate an oscilloscope from a circuit. It ensures that the high-voltage electrical bus and earth/chassis ground are isolated from each other. The probe minimizes the creation of a ground path through the oscilloscope circuits in a high-voltage system. **See Figure 2-7.**

DIFFERENTIAL SCOPE PROBE

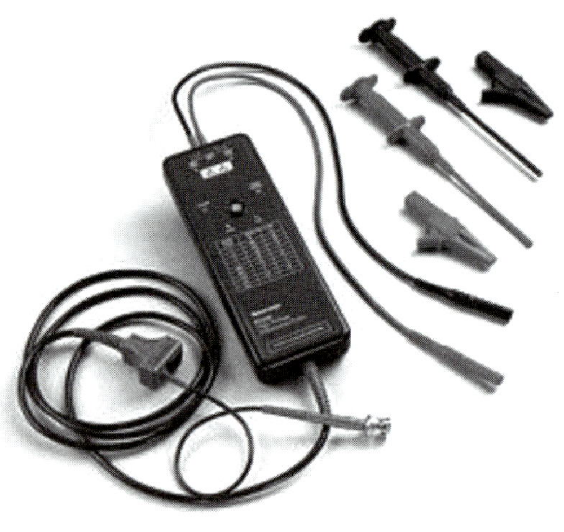

Tektronix, Inc.

Figure 2-7. A differential scope probe should be used to minimize the creation of a ground path when connecting an oscilloscope to a high-voltage system.

If the oscilloscope or its probe leads are not designed to float, differential probes should be used with any intermediate- or high-voltage circuit. This allows the oscillosope to float with no reference to chassis or earth ground. The probes isolate the oscilloscope from the circuit, ensuring that the high-voltage electrical bus, earth ground, and chassis ground are isolated from the high-voltage system.

An oscilloscope may be self powered or powered by the vehicle 12 V battery. The oscillosope battery ground is connected to the vehicle chassis and, therefore, can use the chassis as a medium to transfer high-voltage electrical leakage currents. Differential scope probes should be used regardless of whether the oscilloscope is powered by a 120 V AC or DC battery with internal or external scope DC power. Furthermore, Class 0 gloves should be worn when operating an oscilloscope.

Multimeters

As with oscilloscopes, the role of digital multimeters is unique when it is applied to high-voltage electrical circuits. Typically, technicians and instructors in the automotive repair industry are not concerned about the ratings of a multimeter. However, when applying a multimeter to a high-voltage electrical circuit, the rating of the meter becomes extremely important.

Meter Category Levels. Contemporary digital multimeters are labeled with a category (CAT) rating. A *category (CAT) rating* is a standard rating provided by the International Electrotechnical Commission (IEC) that outlines safety categories for electrical meters. **See Figure 2-8.** The main concerns with multimeter usage are the maximum steady-state voltage range and the operation of the meter in both steady-state and transient overvoltage conditions.

Transient protection is particularly important because transients can cause extremely high electrical currents and arc-over conditions. An arc-over occurs when voltage is high enough to permit an arc from a high potential to low potential. Arc-over can cause an explosion that may lead to serious injuries.

Multimeters include CAT ratings and an associated voltage rating. Although there are a number of technical aspects to consider when studying the categories of electrical power, it is important to remember standards when purchasing a multimeter. A higher CAT number reflects higher energy transients. Therefore, a multimeter with a CAT III rating is resistant to a much higher energy transient than a multimeter designed for a CAT II standard. A CAT III–600 V multimeter can withstand transients better than a CAT II–1000 V multimeter even though the voltage rating is lower. In addition, meter leads should have ratings that match or exceed those of the meter.

Within a category, a higher voltage rating implies that the multimeter can withstand higher transients. For example, a CAT III–1000 V multimeter has better protection than a CAT III–600 V multimeter. A CAT III–1000 V multimeter will function with excellent performance in analyzing and diagnosing high-voltage circuits while providing superior safety to the user. This applies to automotive use on hybrid, electric, and fuel-cell vehicles with system working voltages in the range of 300 V to 400 V.

The Fluke Corporation pioneered the CAT electrical safety equipment specification for testing electrical systems.

IEC METER CATEGORIES

Category	In Brief	Examples
CAT I	Electronic	• Protected electronic equipment • Equipment connected to source circuits in which measures are taken to limit transient overvoltage to an appropriately low level • Any high-voltage, low-energy source derived from a high-winding-resistance transformer, such as the high-voltage section of a copier
CAT II	1ϕ receptacle-connected loads	• Appliances, portable tools, and other household and similar loads • Outlets and long branch circuits • Outlets at more than 30′ (10 m) from CAT III source • Outlets at more than 60′ (20 m) from CAT IV source
CAT III	3ϕ distribution, including 1ϕ commercial lighting	• Equipment in fixed installations, such as switchgear and polyphase motors • Bus and feeder in industrial plants • Feeders and short branch circuits and distribution panel devices • Lighting systems in larger buildings • Appliance outlets with short connections to service entrance
CAT IV	3ϕ at utility connection, any outdoor conductors	• Refers to the origin of installation, where low-voltage connection is made to utility power • Electric meters, primary overcurrent protection equipment • Outside and service entrance, service drop from pole to building, run between meter and panel • Overhead line to detached building

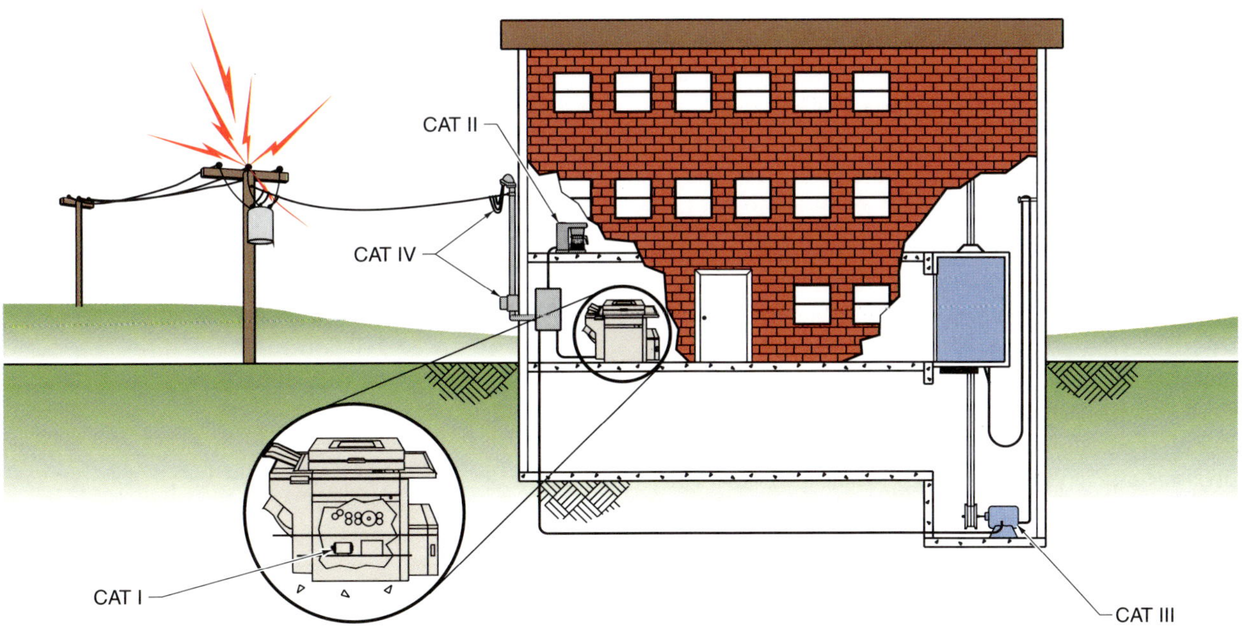

Figure 2-8. A category (CAT) rating is a standard provided by the International Electrotechnical Commission (IEC) that outlines safety-level categories for electrical meters.

Insulation Testers

A common tool used by technicians in testing and analyzing high-voltage systems is an insulation tester. An *insulation tester* is a diagnostic instrument used to detect a reduction in resistance between high-voltage circuits, components, or systems and the vehicle chassis. **See Figure 2-9.** They are recommended or required by most vehicle manufacturers to assist a technician in determining the location of an isolation fault.

INSULATION TESTERS

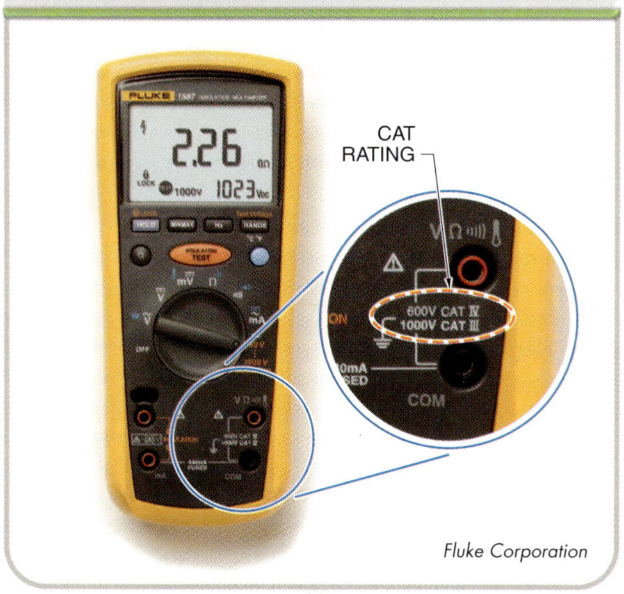

Figure 2-9. Insulation testers are recommended by most vehicle manufacturers to assist a technician in determining the location of an isolation fault.

A high-quality insulation tester has a CAT III–1000 V rating and measures insulation deterioration between conductors and ground. It applies high voltage to a circuit and measures the resistance between two test points. Manufacturers may also combine a digital multimeter (volt-ohm meter) with the insulation tester so it can be used for more than one application.

The tester is capable of providing several different voltage levels for testing. The voltage test level will be nearly equivalent to the maximum dynamic voltage at which the system will operate. This is typically 500 V for most vehicles. However, a technician should always refer to the service information provided by the vehicle manufacturer to confirm the correct voltage test level.

Electric-vehicle control and diagnostic systems can be designed to detect when there is a decrease of resistance between the high-voltage circuit and the vehicle chassis to determine the location of the fault. Insulation testers are used after the vehicle diagnostic system has detected voltage leakage to the chassis. A technician can use the insulation tester, coupled with an off-board diagnostic (scan) tool, to determine which high-voltage component caused the low-circuit resistance and associated diagnostic trouble codes.

An insulation tester has two leads. One lead should be connected to the device under test (DUT) and the other lead should be connected to chassis ground. A *device under test (DUT)* is a device that is being tested or analyzed by an instrument or test equipment. **See Figure 2-10.** The tester has a button to initiate the test. When the test is initiated, the tester injects electrical current at the selected voltage into the DUT or circuit. It then measures the amount of voltage drop to determine the resistance of the device or circuit. Care must be taken when using the tester because it can provide an electrical shock if not handled properly.

All high-voltage components can be tested with an insulation tester. Many high-voltage components use 3ϕ power. All three phases should be tested on the component and connector. Service information provides each of the test points for all high-voltage components. However, insulation testing on each component of an electric, hybrid, or fuel-cell circuit can be a lengthy process when the vehicle's diagnostic software is not designed to test and report the electrical health of individual high-voltage components.

DEVICE UNDER TEST (DUT)

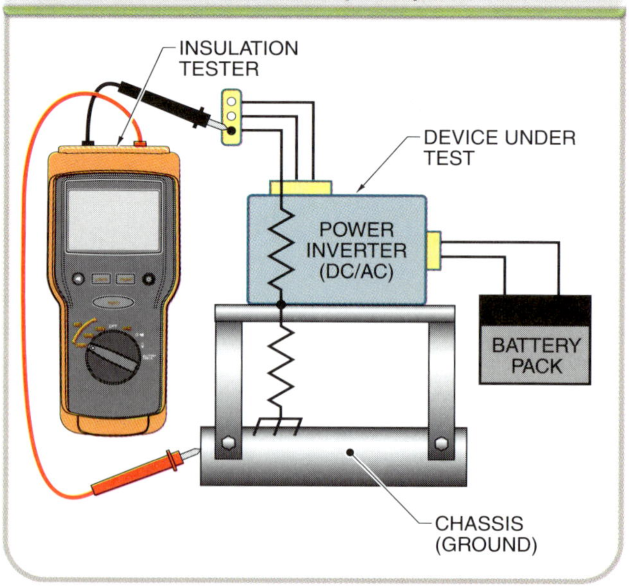

Figure 2-10. One lead of an insulation tester should be connected to the device under test (DUT) and the other lead connected to chassis ground.

Refer to the *Hybrid Electric Vehicle Technology* CD-ROM for a Quick Quiz® on the chapter contents.

POINTS TO REMEMBER

- An isolation fault is a condition that results when the resistance between the high-voltage system and the vehicle chassis has degraded to a point that may no longer be safe.
- High voltage is a level of system voltage that is greater than 60 V DC or 30 V AC rms.
- Intermediate voltage is a level of system voltage that is greater than 30 V DC and less than or equal to 60 V DC, or that is greater than 15 V AC rms and less than or equal to 30 V AC rms.
- Low voltage is a level of system voltage that is less than or equal to 30 V DC, or less than or equal to 15 V AC rms.
- Class 0 is the electrical-safety glove rating recommended by most automotive manufacturers.
- Cable or wire/cable with an orange covering contains voltage levels of greater than 60 V and the use of electrical-safety gloves is required.
- Personal protection equipment should be used when working with high-voltage circuits, components, and systems.
- Electrical current causes bodily injury. Current increases with increasing voltage.
- There are three categories of electrical testing and analyzing tools: oscilloscopes, multimeters, and insulation testers.
- A CAT III–1000 V multimeter will function with excellent performance in analyzing and diagnosing high-voltage circuits while providing superior safety to the user.
- A technician working on a vehicle should never assume that any of the vehicle high-voltage safety systems are operating correctly or that high voltage has not leaked from the high-voltage bus to the vehicle chassis.
- Insulation testers are recommended by most vehicle manufacturers to assist a technician in determining the location of an isolation fault.
- An insulation tester injects electrical current into a circuit or component and measures the resulting resistance.
- The typical setting level for a voltage test provided by most hybrid and electric vehicle manufacturers is 500 V.

Refer to the *Hybrid Electric Vehicle Technology* CD-ROM for the Review Questions in Microsoft® Word format.

REVIEW QUESTIONS

1. Explain how to use the "one hand rule" for working on electrical circuits.
2. What is the origin and dangers of an isolation fault condition?

3. List and describe the different category (CAT) ratings of high-voltage equipment.
4. Why are differential scope probes used?
5. How is an insulation tester used to identify damaged high-voltage components?

2 High-Voltage Electrical Safety

HYBRID Electric Vehicle TECHNOLOGY

Trade Test

Name ______________________ Date ______________

______ **1.** High-voltage electrical systems and components are always ___.
A. electrically isolated from the vehicle chassis ground
B. electrically connected to the vehicle chassis ground
C. rated at CAT III
D. rated at CAT IV

______ **2.** Multimeters should be rated at ___ when used to analyze a high-voltage hybrid system.
A. CAT I–600 V
B. CAT II–6000 V
C. CAT III–1000 V
D. CAT IV–2000 V

______ **3.** A technician should assume that a circuit contains high-voltage when a wire, cable, or cable covering is colored ___.
A. blue
B. red
C. orange
D. gray

______ **4.** ___ is the component of electricity that causes injury.
A. Current
B. Voltage
C. Resistance
D. Humidity

______ **5.** When using an oscilloscope to measure high voltage, it is recommended that a technician connect a(n) ___ probe to the oscilloscope.
A. isolation amplifier
B. logic scope
C. differential amplifier
D. differential scope

______ **6.** A Class 0 glove is rated to provide protection up to ___ V.
A. 500
B. 1000
C. 10,000
D. 15,000

_______________ **7.** An insulation tester measures the insulation ___ to check for a damaged component.
A. voltage
B. current
C. resistance
D. power

_______________ **8.** High-voltage electrical-safety gloves are used to safeguard a technician if a(n) ___.
A. isolation fault is present
B. repair is needed on a positive high-voltage circuit
C. repair is needed on a high-voltage component
D. all of the above

_______________ **9.** A(n) ___ applies high voltage to a circuit and measures the resistance between two test points.
A. multimeter
B. insulation tester
C. oscilloscope
D. differential scope probe

_______________ **10.** The typical voltage test level specified by most manufacturers for an insulation meter is ___ V.
A. 500
B. 1000
C. 1500
D. 2000

High-Voltage Vehicle Safety Systems

3

High-voltage vehicle safety systems are designed to protect anyone who intentionally or unintentionally comes into contact with the high-voltage electrical system of a vehicle. They are also designed to protect occupants and first responders in the event of a collision or the technicians that must repair the vehicle after the collision. High-voltage vehicle safety systems include interlock systems, service disconnect switches, and isolation-fault detection systems. The information contained in this chapter, along with the manufacturer service manual information, provides a technician the basic knowledge to safely begin working with high-voltage electrical systems.

CHAPTER OBJECTIVES

- Describe high-voltage safety systems.
- Explain the purpose of an interlock system.
- List and describe the operation of the common types of interlock systems.
- Explain the purpose of a bus discharge circuit.
- Describe the operation of active and passive bus-discharge systems.
- Explain the purpose of a comparators.
- Describe the operation of voltage and fuel cutoffs.
- Identify the types of manual/service disconnect switch systems.
- Explain the purpose of an isolation-fault detection system.
- List and describe the operation of the four stages of an isolation fault detection system.

KEY TERMS

high-voltage safety system: A system designed to protect the vehicle operator, technicians, and first responder (fire/rescue) personnel from injury caused by the high voltage present in electric propulsion vehicles.

active safety system: An electronically controlled system used to provide protection by isolating the battery system and discharging the bus capacitors.

passive safety system: A nonelectronically controlled system permanently connected to the high-voltage system that is used to provide protection by discharging the bus capacitors.

interlock system: A system used to detect when any high-voltage component or circuit has been breached. The controllers can signal for the bus capacitors to be discharged to ensure that an individual will be safe when working on a high-voltage circuit.

serial interlock loop (SIL) system: A series-circuit, or daisy-chain, system that connects each high-voltage device in a hybrid-vehicle power system. The controller senses if any of the high-voltage connectors or the cover of a high-voltage component are disconnected. If any devices are disconnected, the controller opens the high-voltage battery-pack contactors or relays.

local interlock loop (LIP) system: A high-voltage safety system that uses discrete switches with the high-voltage bus to signal a controller that a system has been breached. After a breach has been identified, the controller opens the high-voltage contactors to remove power from the high-voltage bus and/or components. The system can also be designed with a discrete switch that is routed directly to a controller input and the controller monitors the input for changes in voltage states.

grounded switch interlock system: A modified LIP system that uses switches located within its components to provide a ground for the hybrid-controller sense circuit. When any of the switches are opened, the battery-pack controller opens the high-voltage contactors or relays.

active bus-discharge circuit: A bus-discharge circuit that uses electronic circuits to quickly discharge the bus capacitors in a high-voltage system, usually in less than 5 seconds.

passive bus-discharge circuit: A bus-discharge circuit that uses resistors to slowly discharge the high-voltage bus capacitors in a high-voltage system, usually in about 5 minutes.

service disconnect switch: A mechanical lever or rotating switch device that is used to remove power from a high-voltage circuit.

HIGH-VOLTAGE SAFETY SYSTEMS

A very important part of the hybrid vehicle propulsion system is the high-voltage safety system. A *high-voltage safety system* is a system designed to protect the vehicle operator, technicians, and first responder (fire/rescue) personnel from injury caused by high voltage present in electric propulsion vehicles.

A safety system monitors and controls the high-voltage system by isolating the energy to the battery-pack system and controlling the discharge of the high-voltage bus capacitors located in the power inverter. Capacitors can store dangerous energy levels.

A safety system protects anyone who needs to work on the high-voltage system. A technician must know how to use a safety system to make a vehicle safe to work on. An *active safety system* is an electronically controlled system used to provide protection by isolating the battery system and discharging the bus capacitors. Not all manufacturers use active safety systems. A *passive safety system* is a nonelectronically controlled system permanently connected to the high-voltage system that is used to provide protection by discharging the bus capacitors.

All manufacturers have adopted one universal standard for high-voltage systems. The Society of Automotive Engineers (SAE) standard, J1673, *High Voltage Automotive Wiring Assembly Design,* guides manufacturers on how high-voltage wire construction should be implemented. The standard states "… wire/cable or harnesses containing high voltage shall be visually identified with a permanent orange harness covering material. Individual cables routed separately and not protected by auxiliary covering shall be primarily orange …" Therefore, the J1673 standard provides technicians, first responders, and vehicle operators with a visual means of identifying high-voltage wires and cables. **See Figure 3-1.**

HIGH-VOLTAGE ORANGE CABLE COVERING

Figure 3-1. The J1673 standard provides a visual means of identifying high-voltage wires and cables.

Interlock Systems

A hybrid vehicle can be dangerous to work on while the circuits are energized. An *interlock system* is a system used to detect when any high-voltage component or circuit has been breached. The controllers can signal for the high-voltage relays to open and the bus capacitors to be discharged to ensure that an individual will be safe when working on a high-voltage circuit. High-voltage safety systems include interlock circuits and discharge systems that remove the electrical energy from the circuit. Common high-voltage safety system designs are the serial interlock loop (SIL), local interlock loop (LIP), and grounded interlock switch systems.

Serial Interlock Loop (SIL) Systems. A *serial interlock loop (SIL) system* is a series-circuit, or daisy-chain, system that connects each high-voltage device in a hybrid-vehicle power system. **See Figure 3-2.** The controller senses if any of the high-voltage connectors, or the cover of a high-voltage component, are disconnected. If any devices are disconnected, the controller opens the high-voltage battery-pack contactors or relays. This isolates the energy from the electric propulsion system, cables and wiring, and all high-voltage components. As a result, a safe working voltage level is established for the technician.

The manual disconnect switch may, or may not be, mechanized as a component in the SIL circuit. This type of circuit is typically designed to operate at very low-voltage levels (about 5 V to 12 V DC).

SERIAL INTERLOCK LOOP

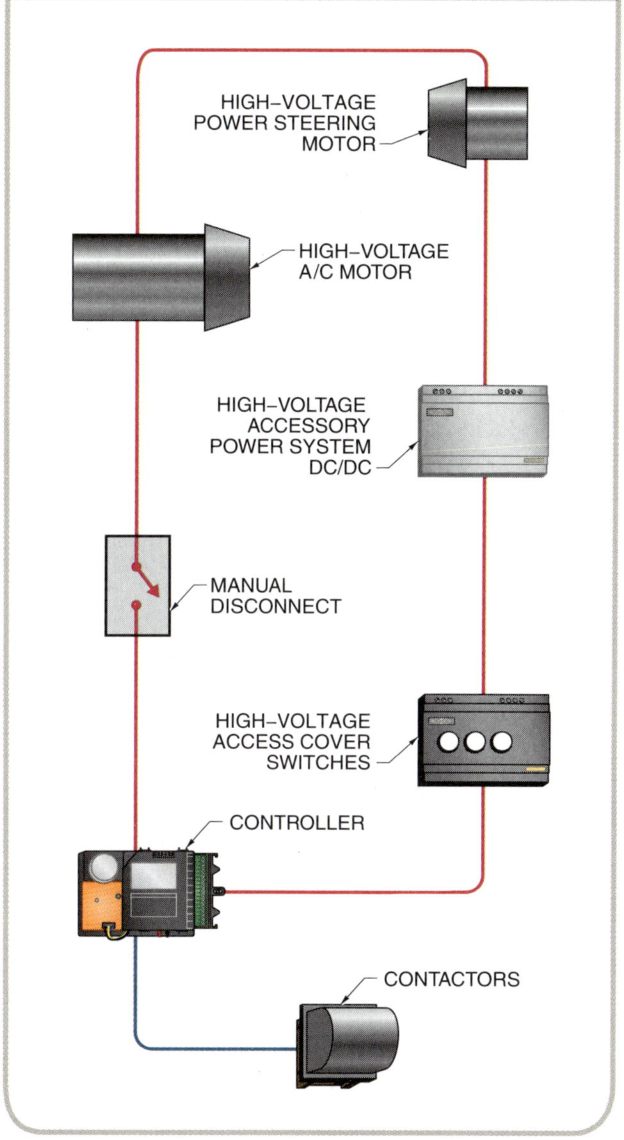

Figure 3-2. A serial interlock loop (SIL) connects each high-voltage device in series.

The high-voltage connector system includes the two 300 V positive and negative terminals. It also includes the two SIL terminals that monitor the connector status on the high-voltage device. In most designs, when the connector is disconnected, the interlock terminals will be the first to disengage. If an active bus discharge is used, the interlock system causes the high-voltage relays to open and the high-voltage bus capacitors to discharge. This mechanization ensures an extremely safe condition for the service technician or first responder.

For example, the Ford Escape Hybrid uses a serial interlock system. In this system, each high-voltage cable and connector has an SIL circuit that reports the status to the battery-pack controller via two discrete, dedicated wires. **See Figure 3-3.** In the event of failure, the battery-pack controller senses the circuit status and commands the high-voltage contactors or relays open so that the high-voltage bus capacitors can be discharged to remove the electrical energy.

The circuit is a series circuit that enters and exits each high-voltage connector or component. Each circuit has two terminals and wires to monitor the status of the connector. This circuit communicates regardless of whether the connector has been disconnected to expose high voltage.

Local Interlock Loop (LIP) Systems. A local interlock loop system can perform the same functions as a serial interlock loop system. However, a local interlock loop (LIP) system is not a series system and its components are connected to a local area network (LAN) communication bus. A *local interlock loop (LIP) system* is a high-voltage safety system that uses discrete switches with the high-voltage bus to signal a controller that a system has been breached. **See Figure 3-4.**

After a breach has been identified, the controller opens the high-voltage contactors to remove power from the high-voltage bus and/or components. The system can also be designed with a discrete switch that is routed directly to a controller input. The controller monitors the input for changes in voltage states.

The status of high-voltage connections or covers of the components is sent to the controller via the LAN bus. If the controller receives a message that the high-voltage system has been breached or the manual disconnect has been opened, the controller commands the battery-pack controller to open the battery contactors and isolate battery energy. An LIP system also provides diagnostic codes, information, and messages to alert an operator or technician of high-voltage safety system status.

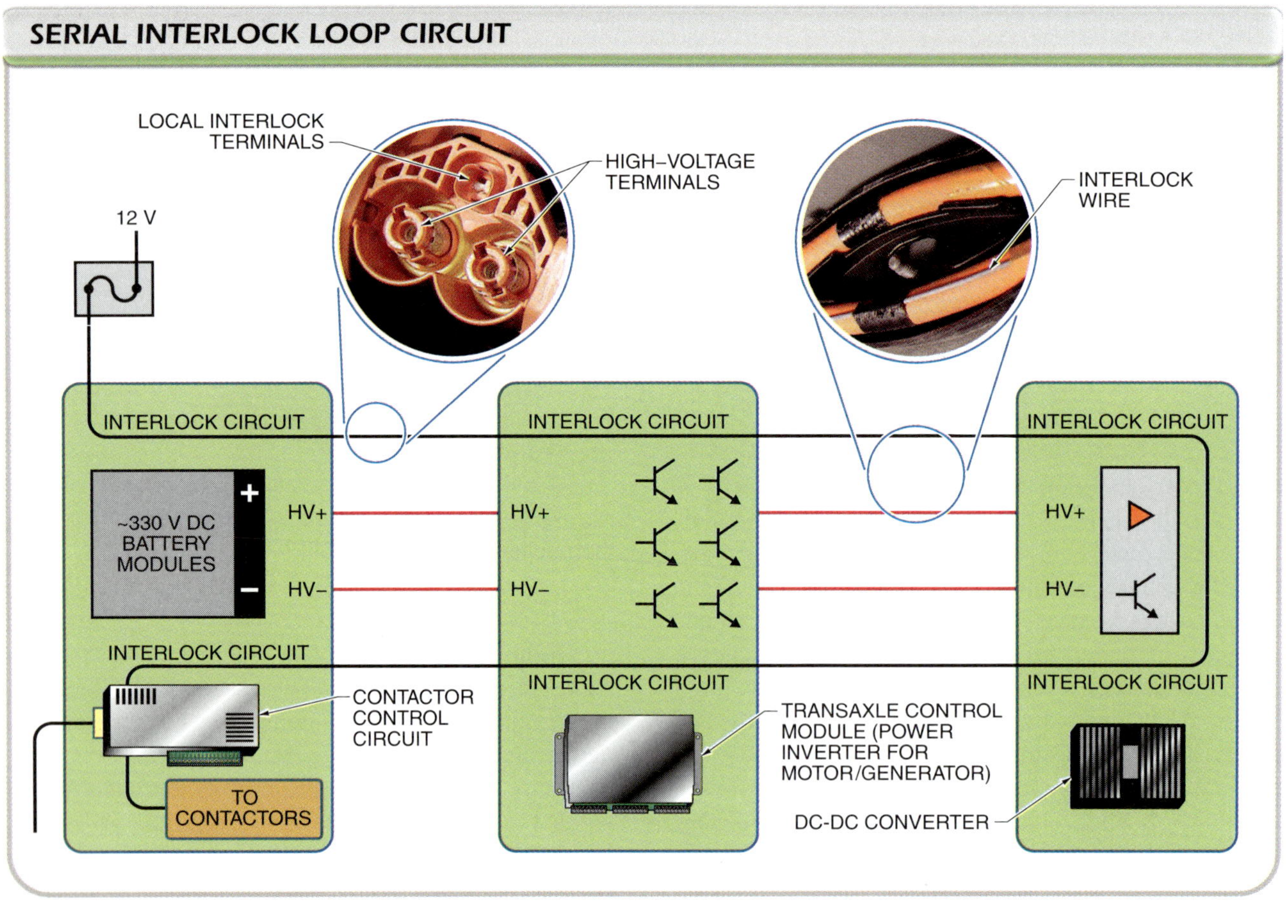

Figure 3-3. Each wire that carries high voltage has a serial interlock loop circuit that reports the status to the battery-pack controller.

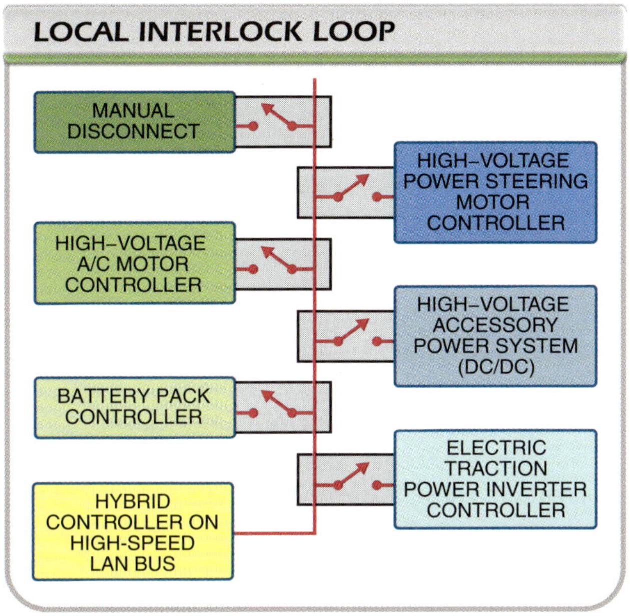

Figure 3-4. A local interlock loop (LIP) connects each high-voltage device on a LAN bus.

Grounded Switch Interlock Systems. A *grounded switch interlock system* is a modified LIP system that uses switches located within its components to provide a ground for the hybrid controller sense circuit. **See Figure 3-5.** When any of the switches are opened, the battery-pack controller opens the high-voltage contactors or relays and discharges the bus capacitors. A grounded switch interlock circuit is mechanized differently than SIL and LIP systems but the result is identical.

This type of interlock circuit is used on some of the newer Toyota Prius, Highlander, and Lexus 400h high-voltage components. The Toyota Prius hybrid controller provides power for the circuit. This series circuit contains male and female connectors located inside the battery-pack service disconnect-plug system. The female plug receptacle is wired in series with the male and female connectors in the power inverter, which provides the ground connection for the circuit. When the cover or service disconnect is removed, the circuit opens, the battery-pack controller opens the high voltage contactor, and the bus capacitors are discharged.

The Lexus RX400h and Toyota Highlander systems use a microswitch located in the service disconnect that opens when the service disconnect is removed. The microswitch is wired in series with a ring terminal located on the power inverter cover that provides ground for the circuit. When any part of a grounded switch interlock system is opened, a signal is sent to the controller that the circuit has been breached. As with all other circuits on an EV, HEV, or fuel-cell vehicle, manufacturer service information should be consulted prior to working on the system.

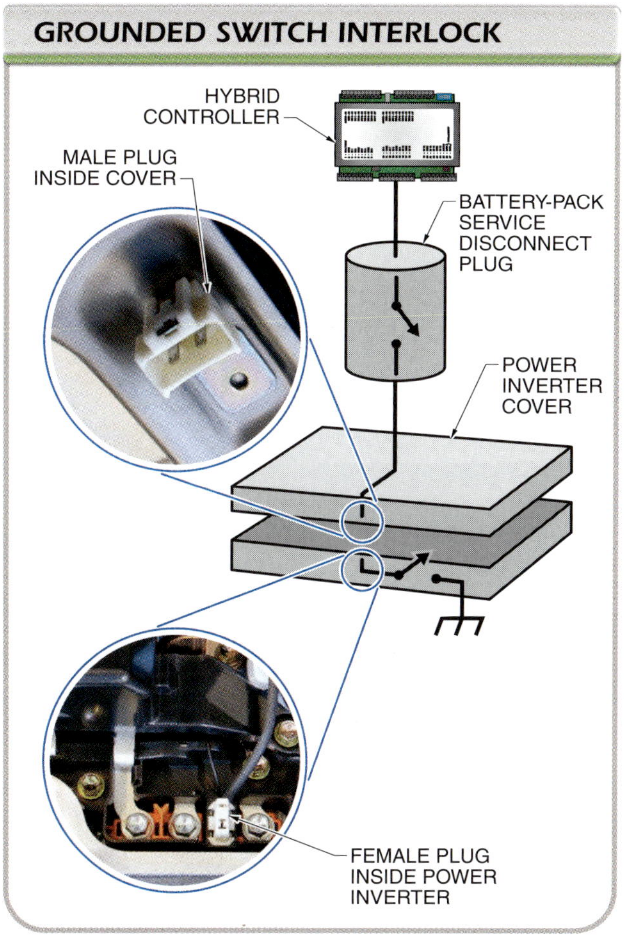

Figure 3-5. A grounded switch interlock is a modified LIP system with switches located within its components.

Bus-Discharge Circuits. The high-voltage energy from bus capacitors is not discharged when the high-voltage capacitors are opened. The elements of a high-voltage system remain at a dangerous energy level until the capacitors are discharged. There are two types of bus discharge circuits: active bus-discharge circuits and passive bus-discharge circuits.

An *active bus-discharge circuit* is a bus discharge circuit that uses electronic circuits to quickly discharge the bus capacitors in a high-voltage system, usually in less than 5 seconds. The active bus-discharge circuit is typically located within the power inverter. **See Figure 3-6.**

BUS DISCHARGE CIRCUITS

GM EV1 BUS CAPACITOR DISCHARGE CARD

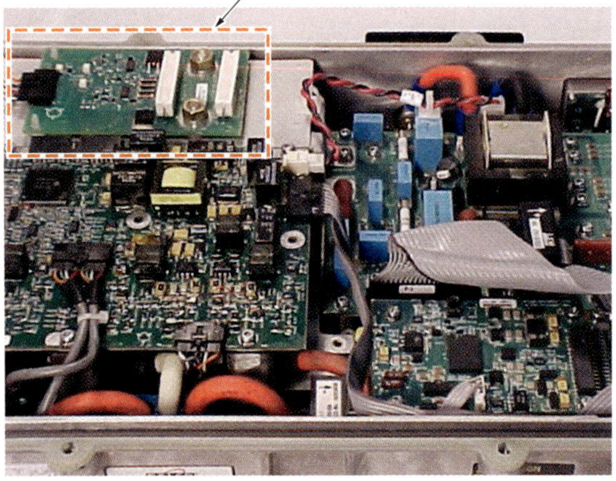

ACTIVE BUS DISCHARGE CARD

RESISTOR FOR BUS DISCHARGING (PASSIVE PARALLEL CONNECTED CIRCUIT)

PASSIVE DISCHARGE RESISTOR

Figure 3-6. An active bus-discharge circuit is typically located within the power inverter.

An active bus-discharge system must wait to discharge the high-voltage bus capacitors until after the contactors or relays are opened. This is because the capacitors continue to hold high-voltage energy levels supplied to it by the battery pack with the contactors closed. It is only after the contactors or relays are opened that the capacitors can be safely discharged and permit the active system to safely discharge the capacitor energy. Damage to the bus-discharge control circuit could result if the active circuit is turned ON when the battery contactors or relays are closed. This is because the bus discharge circuit is designed as an intermittent, not continuous, duty system.

In a system with an active bus-discharge circuit, the capacitors are typically discharged within about 2 to 5 seconds. Vehicle systems that use an active bus-discharge system typically include diagnostics to report the status of the active discharge system. The control and diagnostic system could also route the status visually to an indicator lamp on the instrument panel. This alerts the operator/technician if or when high voltage has decreased to a safe level of less than about 60 V DC.

Not all systems use an active bus-discharge and voltage-monitoring system. A *passive bus-discharge circuit* is a bus discharge circuit that uses resistors to slowly discharge the high-voltage bus capacitors in a high-voltage system, usually in about 5 minutes. For example, the 2001 to 2003 Toyota Prius uses a passive bus-discharge system in which the energy from three bus capacitors is routed through a resistor connected in parallel to the bus capacitors.

However, a passive bus-discharge system may also be used in conjunction with the active bus-discharge system to discharge high-voltage energy if the active bus-discharge circuit fails to operate. Manufacturers typically advise waiting for a specified period of time to allow the bus energy to discharge. For example, the 2001 Toyota service manual recommends waiting 15 minutes for the Prius circuit to discharge prior to performing high-voltage system service or maintenance.

Comparators. An equivalent circuit can be used to provide a more detailed look at an interlock circuit. The interlock circuit contains operational amplifiers (op-amps) used as a voltage comparator. A comparator is used to determine whether a circuit is operating correctly.

Each op-amp contains two inputs and one output. **See Figure 3-7.** The two inputs are termed inverting and noninverting. When used as a comparator, the two output voltages are termed high and low. The op-amp output voltage in the high state is equal to the supply positive voltage. This is termed saturated. The op-amp output voltage in the low state is approximately equal to the voltage on the op-amp ground, typically about 0 V.

The inverting and noninverting input voltage level states determine the output voltage. When the noninverting input voltage is higher than the inverting input, the output voltage will be saturated. When the noninverting input is lower than the inverting input, the output voltage will be 0 V.

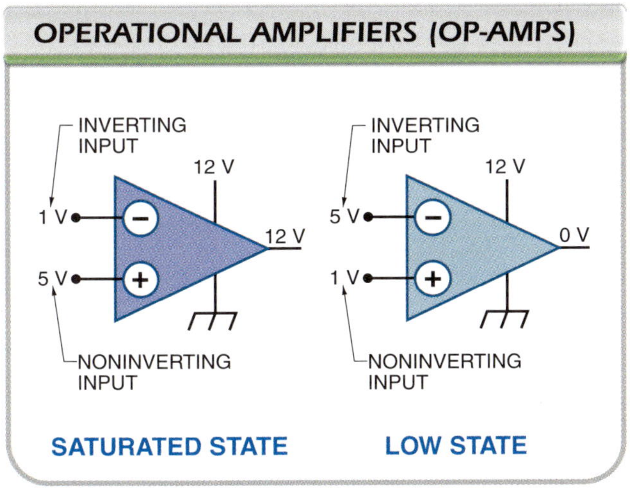

Figure 3-7. The op-amp output voltage in the saturated state is equal to the supply positive voltage. The op-amp output voltage in the low state is approximately equal to the voltage on the op-amp ground.

A practical example of a comparator uses two op-amps. **See Figure 3-8.** This type of circuit is called a window comparator. A window comparator permits the op-amps to sense whether a voltage input is within a specific voltage range, or window. The resistors limit electrical current to the inputs to reduce the electrical current to a safe operating level. The diodes block any feedback from the circuit into the amplifier output stages. The diodes restrict each op-amp circuit to operate as outputs that source (send current out) and not sink. Without the diodes, the op-amp could source or sink a circuit, which could cause the circuits to operate in undesirable states.

Hybrid electric vehicles can be designed to have large capacitors that store energy during braking and release that energy for acceleration.

COMPARATOR CIRCUITS

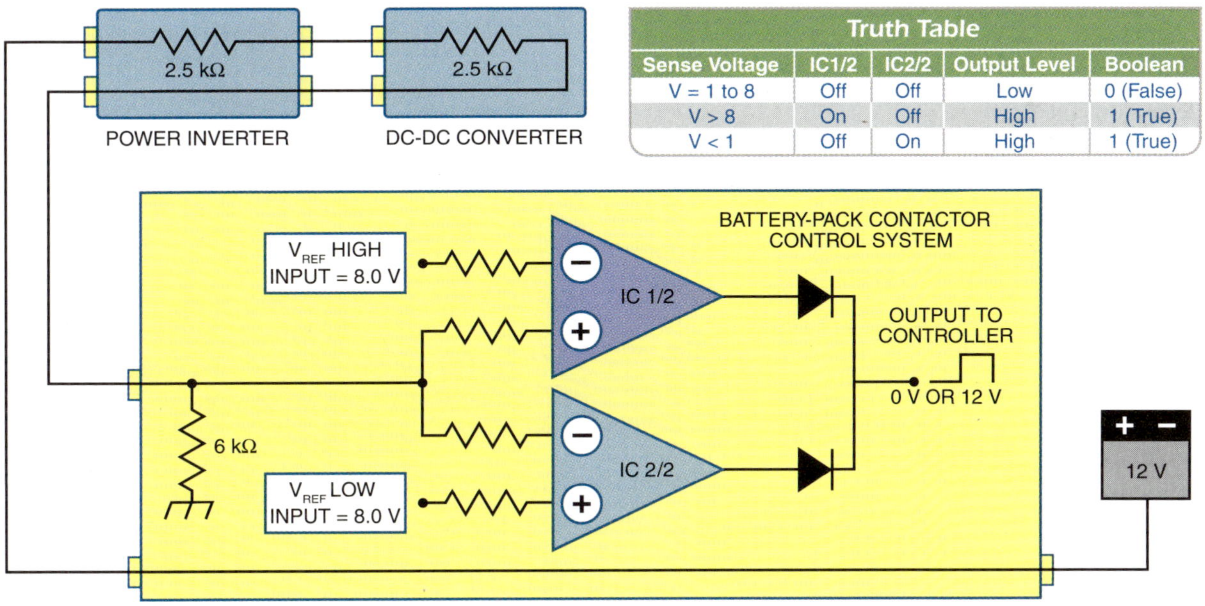

Truth Table				
Sense Voltage	IC1/2	IC2/2	Output Level	Boolean
V = 1 to 8	Off	Off	Low	0 (False)
V > 8	On	Off	High	1 (True)
V < 1	Off	On	High	1 (True)

Power Inverter	DC-DC Converter	Battery Pack Contactor Control	Total Voltage Drop
2.73 V	2.73 V	6.55 V	12 V

Normal Voltage Drop

Power Inverter	DC-DC Converter	Battery Pack Contactor Control	Total Voltage Drop
3.50 V	n/a	8.50 V	12 V

Voltage Drops with Jumpered Power Inverter Pins

Figure 3-8. A window comparator permits the op-amps to sense whether a voltage input is within a specific voltage window.

This window comparator is designed with constant reference voltages on two of the inputs. These voltages are termed voltage references (V_{ref}). In this circuit the V_{ref}High is set to an 8.0 V threshold and V_{ref}Low is set to a 1.0 V threshold. The 12-V DC supply and ground are not shown in the circuit. Both op-amps constantly monitor the voltage on the sense input wire (the V_{ref}High and the V_{ref}Low inputs). When the sense input voltage is outside the window of 1.0 V to 8.0 V, the output is driven to saturation, or 12 V. When the voltage on the sense input wire is within the window of 1.0 V to 8.0 V, the output is driven to 0 V.

A truth table is used to communicate the operation of the electronic logic circuit. A truth table provides Boolean expressions, such as ON/OFF, high/low, or true/false. It is used so that a system controller can interpret the circuit outputs.

When the op-amp sense input wire is connected to an external circuit, the comparator provides binary-logic (low or high) output to an electronic controller. A low output of 0 V indicates that the circuit is operating normally. A high output of 12 V indicates that there is a malfunction of the connected external circuit that the window comparator is sensing. For example, a window-comparator circuit can be used to describe the hybrid-SIL circuit of the Ford Escape.

The circuit includes resistors, used to represent components, located inside each component block section. This provides the varied voltage levels to the window comparator sense input line. Placing the serial interlock in a failure mode, such as an open or short-circuited connector, changes the voltage drop across the connector. This changes the voltage on the sense input line of the window comparator. In turn, this changes the output voltage level of the window comparator. The truth table is used by the controller to determine whether the high-voltage circuit is operating properly.

For example, the voltage on the sense input line of the window comparator changes if the power-inverter pins are jumpered to bypass the DC-DC converter. The total circuit voltage drop would always equal 12 V determined by calculations based on Ohm's law. The window comparator is driven to a high output because the voltage drop across the battery-pack contactor is outside the window of 1.0 V to 8.0 V. The high output tells the controller that a malfunction has occurred. Based on the controller design, the controller would take appropriate action by opening the high-voltage contactors and discharging the bus capacitors.

Voltage and Fuel Cutoffs. Some hybrid vehicle designs may include additional devices, such as high-voltage cutoff (HVCO) and inertia fuel shutoff (IFS) switches, in an interlock circuit. These devices increase the safety margin for people operating, diagnosing, or repairing a high-voltage system. HVCO and IFS switches are mechanized as closed switches that open in the event of a collision. **See Figure 3-9.**

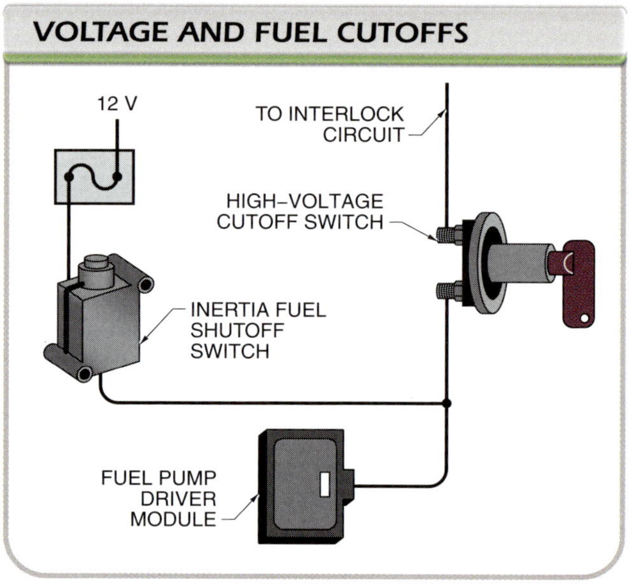

Figure 3-9. A high-voltage cutoff switch and an inertia fuel shutoff switch work with the interlock circuit and open in the event of a collision.

An HVCO is monitored by the battery system controller. The HVCO communicates the status of the vehicle to the controller, with respect to collisions, that could involve the high-voltage battery pack. A collision could result in isolation faults and expose a service technician or first responder to high voltage through the vehicle body or chassis.

If a collision is detected, the switch opens to report that a collision event has occurred. When the battery system controller receives a status change from the HVCO, it opens the battery contactors to isolate the high-voltage energy in the battery pack. The IFS also detects a collision and de-energizes the fuel pump circuit to stop engine operation. Special care should be taken when analyzing and repairing any vehicle that does not use active discharge or high-voltage interlock systems. Refer to the respective manufacturer service information for additional data on interlock and safety systems before attempting any maintenance or repair.

Service Disconnect Switches

Anyone intending to repair or perform an analysis on an electric propulsion system must properly isolate the battery pack and discharge the capacitors prior to working with the high-voltage system. Most manufacturers caution that a technician should wait at least 5 minutes before attempting to remove high-voltage components or repair the high-voltage system. This allows time for the high-voltage bus capacitors to discharge to a safe level. While wearing Class 0 electrical gloves, the technician should use a suitable voltmeter to confirm that there is no high voltage remaining on any high-voltage circuit. Service manuals provided by vehicle manufacturers explain the proper methods for removing high voltage from a particular vehicle.

Hybrid electric propulsion and related power systems can provide more electrical power than the traditional 14-volt system because they operate at higher voltage. Therefore, additional care must be exercised when servicing these systems in order to provide an extra margin of safety. One method used to reduce the possibility of electrical arcing (the grounding of an active component or conductor) during vehicle servicing is to include a service-disconnect switch system in the hybrid and electric power system.

A *service disconnect switch* is a mechanical lever or rotating switch device that is used to remove power from a high-voltage circuit. **See Figure 3-10.** The service disconnect system is one possible element of a layered vehicle safety system. The service disconnect system permits a technician to manually isolate the battery-pack system from other vehicle systems. However, the service disconnect will neither automatically discharge energy in the high-voltage bus capacitors, nor automatically isolate the battery pack from other propulsion subsystems.

SERVICE DISCONNECT SWITCHES

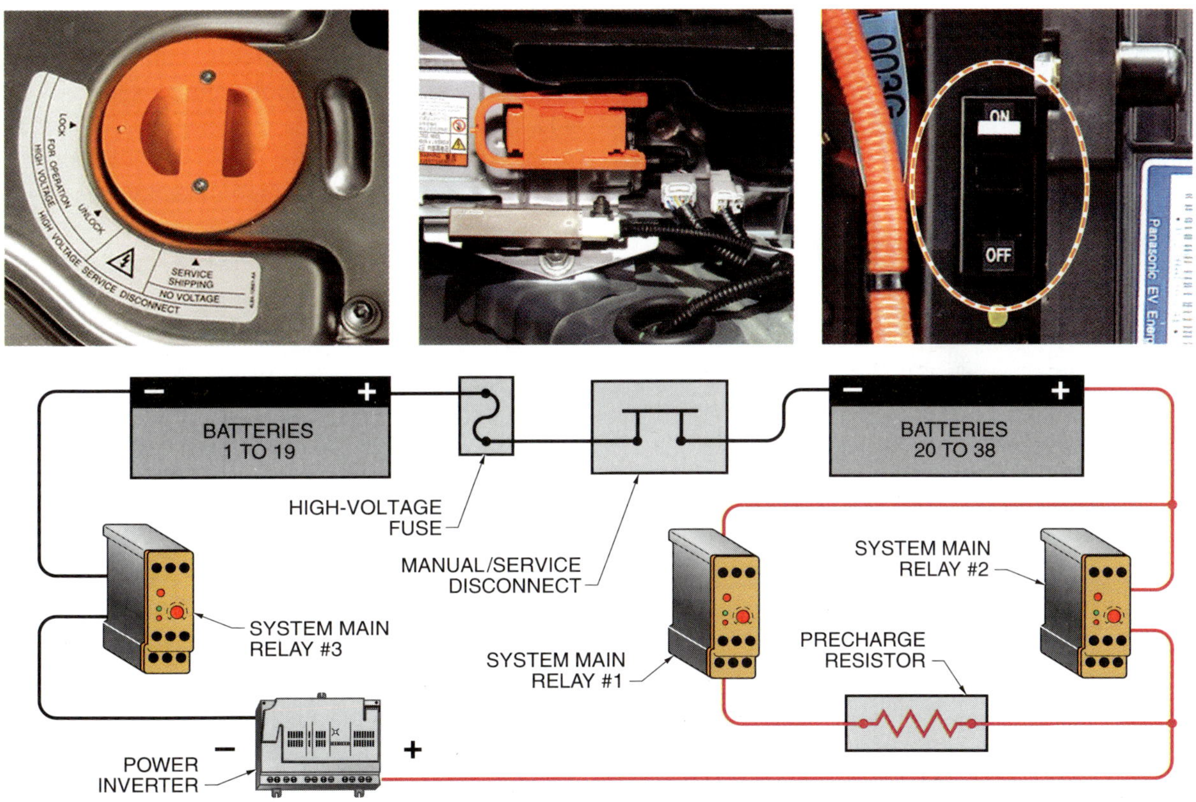

Figure 3-10. The service disconnect system permits a technician to manually isolate the battery-pack system from other vehicle systems.

A service disconnect switch is used to disconnect power to the electric propulsion and ancillary high-voltage system components. This minimizes the possibility of electric arcing due to component failure from technician negligence during a system analysis, diagnosis, or service repair.

The disconnect switch is typically located near the battery pack. Service technicians must become familiar with the location of the disconnect switch on each model to be serviced. It must be placed in the OFF position or removed when providing service to any component, wire, or cable on the hybrid and electric system.

Isolation-Fault Detection Systems

The isolation-fault detection system is an important element of the high-voltage safety system. The purpose of this system is to monitor the vehicle chassis ground to determine if high voltage is leaking to the chassis. **See Figure 3-11.** Any high-voltage component can have a fault to the vehicle chassis.

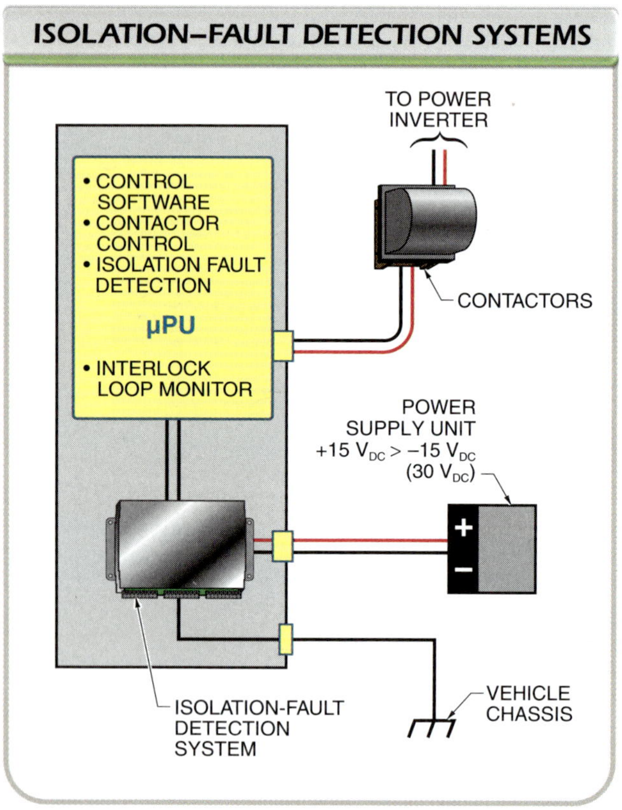

Figure 3-11. An isolation-fault detection system monitors the vehicle chassis ground to determine if high voltage is acquiring an electrical path to the chassis.

High-voltage leakage to the chassis is a potentially dangerous situation for anyone who contacts high-voltage components and the vehicle chassis at the same time. This could result in a potentially fatal electrical shock. Therefore, if an isolation fault exists and current leaks to the chassis, the isolation-fault detection system reports this status to the system controller.

Although there are many causes of high-voltage leakage, many of the failure modes are vehicle- and manufacturer-specific. However, the root cause of a fault is not necessarily important. The goal is to locate the failed component so that it can be replaced. It is also important to test each high-voltage component to confirm that the component does not have a reduction in resistance that would have caused the isolation fault.

For example, the battery-pack controller often contains an isolation-fault detection system. The power supply associated with the battery pack can generate both positive and negative voltages. This is accomplished by using a center-tapped transformer or special integrated circuit (IC). One secondary inductor (coil) can be referenced to positive voltage and the other to negative voltage. However, not all systems will use a split positive and negative voltage power supply. Many isolation-fault detection circuits can be designed to function from a single supply source, such as a 0 V to +12 V system, that does not use negative voltage levels to measure isolation faults.

Furthermore, isolation-fault detection can be implemented in more than one vehicle controller. For example, the power inverter controller could monitor DC isolation faults and the battery-pack controller could monitor AC isolation faults. AC isolation faults may be circuits that contain varying positive waveforms that do not become negative.

Control Strategies. There are many possible control system strategies for isolation fault detection. Some strategies can be very complex. Regardless of the design, there are a number of simple strategies that can be used to understand the isolation-fault detection system/circuit of any electric propulsion system. It is important to know the fundamentals of how the circuit operates and how it can sense a fault.

One strategy can occur when a vehicle is placed in park and shut down. The controller can disable propulsion by not allowing the battery-pack contactors to close on the next key or operating cycle. As another strategy, if the vehicle speed is 0 mph for more than 180 sec, the controller opens the battery-pack contactors and disables propulsion. Additionally, if the vehicle hood is open with high-voltage systems operating, the controller opens the battery-pack contactors and disables propulsion.

These are just a few examples of how an isolation-fault detection system can disable a high-voltage system vehicle in the event that dangerous levels of high voltage are sensed on the vehicle chassis. This detection will ensure that occupants, first responders, or service technicians are safe when interfacing with a vehicle that is functioning with an isolation fault.

DC Isolation Fault Detection. DC isolation fault detection measures the DC leakage from high-voltage components and cables on the high-voltage bus circuits to chassis ground. The controller monitors voltage to determine if electrical current on the vehicle chassis exceeds a specific threshold value. **See Figure 3-12.** The total resistance of all high-voltage components determines total system-leakage current. If the leakage current is greater than the threshold value, the controller sets a diagnostic trouble code (DTC) and illuminates the malfunction indicator lamp (MIL). DC isolation detection typically does not include measuring the battery-pack module string. DC isolation fault detection is very fast compared to AC isolation fault detection.

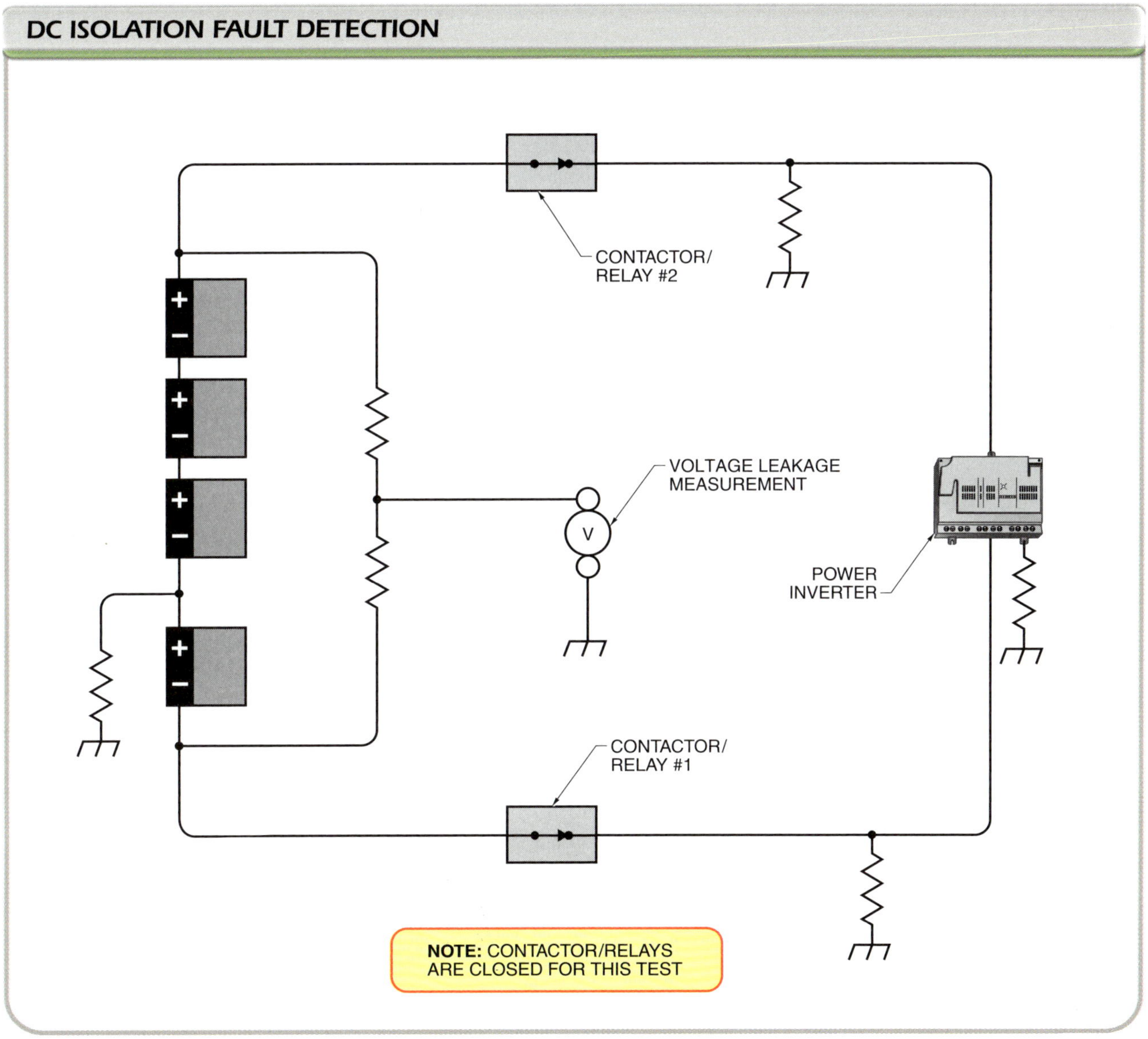

Figure 3-12. DC isolation fault detection measures DC leakage to chassis ground.

AC Isolation Fault Detection. AC isolation fault detection measures both resistance and capacitance of the high-voltage bus circuits by injecting an AC signal onto the bus. **See Figure 3-13.** This is accomplished by using a resistor-capacitor (RC) circuit. AC isolation detection typically is used to measure the isolation integrity of the battery-pack module string with reference to the vehicle chassis. The time required to execute the AC method is much longer than the time required for the DC method because the AC system senses amplitudes of an AC signal rather than electrical currents.

As high-voltage bus resistance increases, voltage drop increases. The maximum V_{out} is considered a normal condition. The high-voltage system resistance decreases when an isolation fault is present. The value of V_{out} decreases and the controller senses an isolation-fault condition when V_{out} decreases below the threshold. Isolation detection monitoring can be continuous, whenever the vehicle is powered, or initiated by testing during vehicle power-up or power-down.

Isolation-Fault Detection Circuits. An isolation-fault detection circuit is a collection of low-power electronic components that monitor the chassis of a vehicle in order to sense high-voltage leakage. There are many possible circuit designs. A basic hardware-only circuit contains resistors, operational amplifiers (used as differential and voltage comparators), voltage regulators, and a digital logic device. This circuit is relatively easy to understand. However, contemporary isolation-fault detection circuits use very few electronic components, typically for input filtering to a controller, and depend on software to sense the voltage condition of a vehicle chassis.

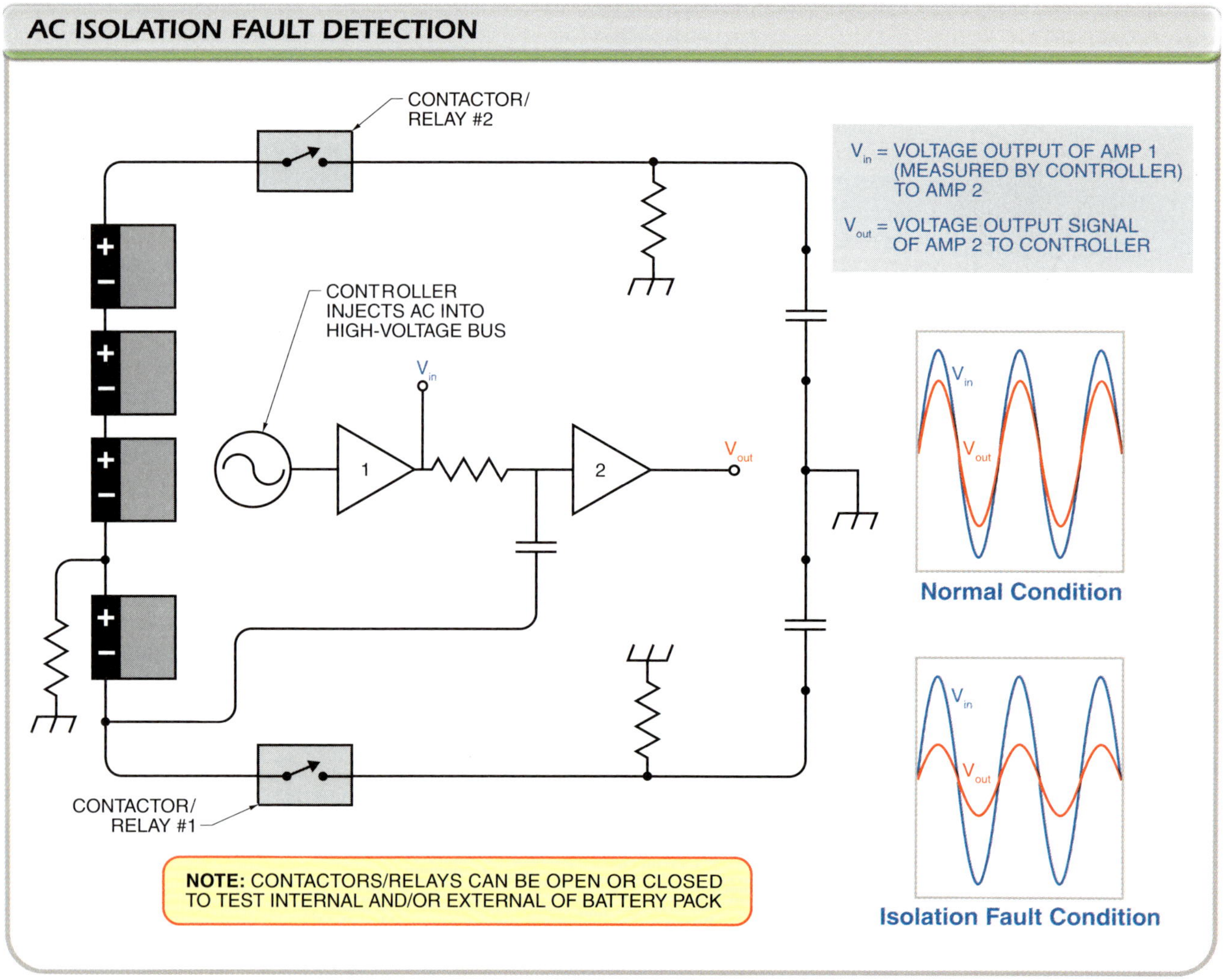

Figure 3-13. AC isolation fault detection injects an AC signal onto the bus.

Isolation fault systems can use both DC and AC signals to sense isolation faults to the vehicle chassis. The use of both DC and AC methods are necessary if the desired outcome is to sense all components and cables/wires connected to the high-voltage system. Isolation detection circuits can be mechanized to use both DC and AC circuits to detect isolation faults.

Voltage-Sensing Stage. The voltage-sensing stage of the isolation-fault detection circuit monitors the voltage on both the analog and chassis grounds. The voltage-sensing stage is made up of four equally valued resistors set up as a voltage divider circuit. **See Figure 3-14.** The voltage-sensing stage provides the necessary voltage divider circuit to monitor voltages.

The four 1 MΩ resistors are in a network that results in equal voltage drops across each resistor branch. The voltage input, or V_{in}, of about 20 V DC is provided by a small power supply or IC that receives its electrical power from the DC-DC converter. The DC-DC converter receives its electrical power from the battery pack. The analog-ground circuit is the ground circuit for the isolation-fault circuit board. The ground plane of the circuit board is always about 0 V DC. The voltage level of the chassis ground will vary depending on the voltage level and current flow leakage from high-voltage components. The difference in the analog- and chassis-ground voltage levels is monitored by the differential-amplifier stage.

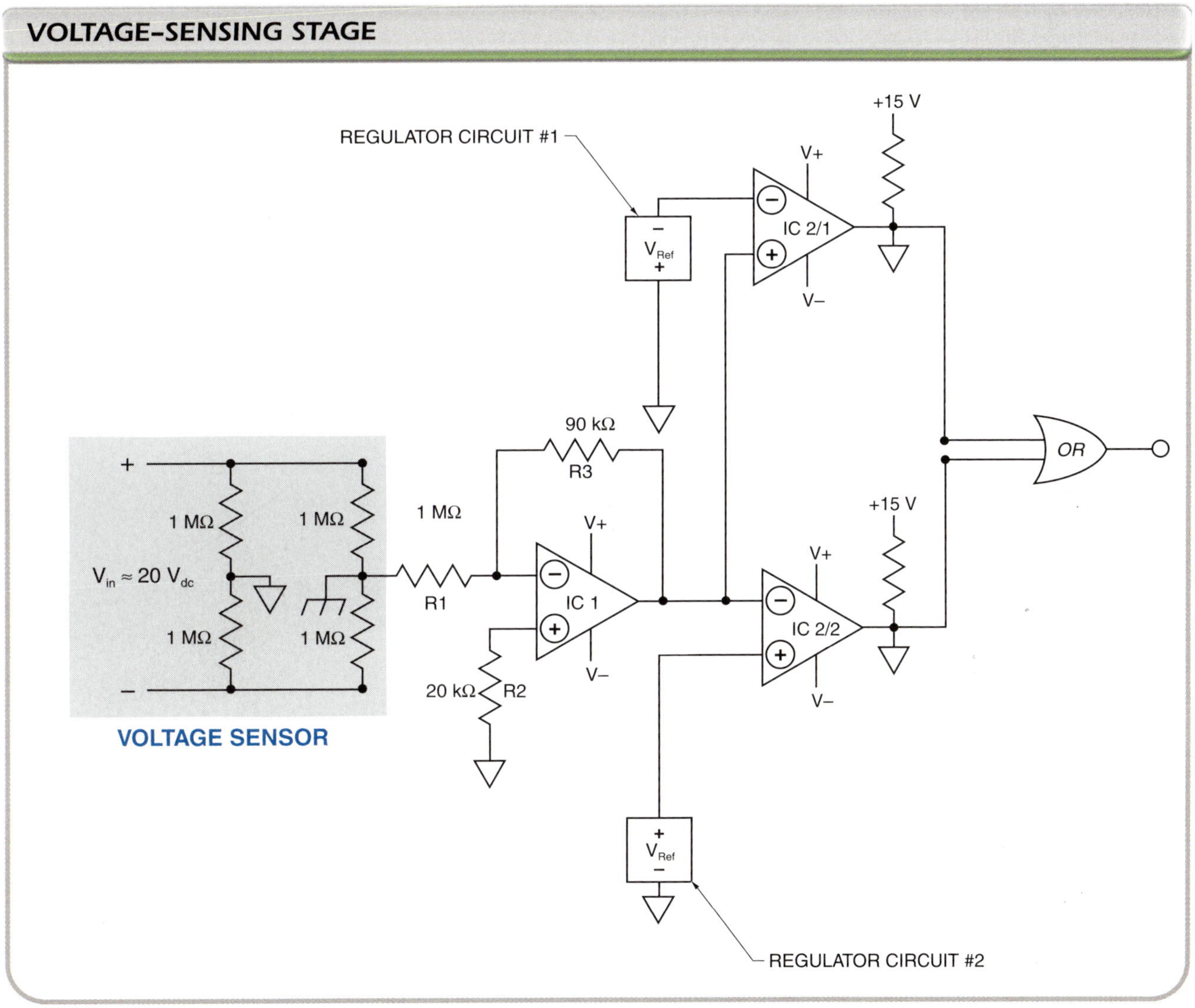

Figure 3-14. The voltage-sensing stage monitors the voltage on both the analog and chassis grounds.

Differential-Amplifier (DA) Stage. The differential-amplifier (DA) stage contains an integrated circuit, IC 1, that amplifies the difference between its inverting and noninverting inputs. **See Figure 3-15.** The maximum amplification voltage is limited to the supply voltage levels listed on the schematic.

The amount of amplification can be calibrated by varying the values of resistors R1 and R3. Resistor R2 is used for calibrating the voltage offset to eliminate voltage imbalances between the inverting and noninverting inputs. Resistor R1 is the inverting input resistor. Resistor R2 is the noninverting input resistor. Resistor R3 is the feedback resistor. When the amplifier senses voltage imbalances between the inverting and noninverting inputs by using the analog and chassis ground connections, it amplifies this difference to its output terminal. This output-voltage signal is transferred to the voltage comparator stage.

High-voltage components are protected by covers and are marked with high-voltage warning labels.

DIFFERENTIAL–AMPLIFIER STAGE

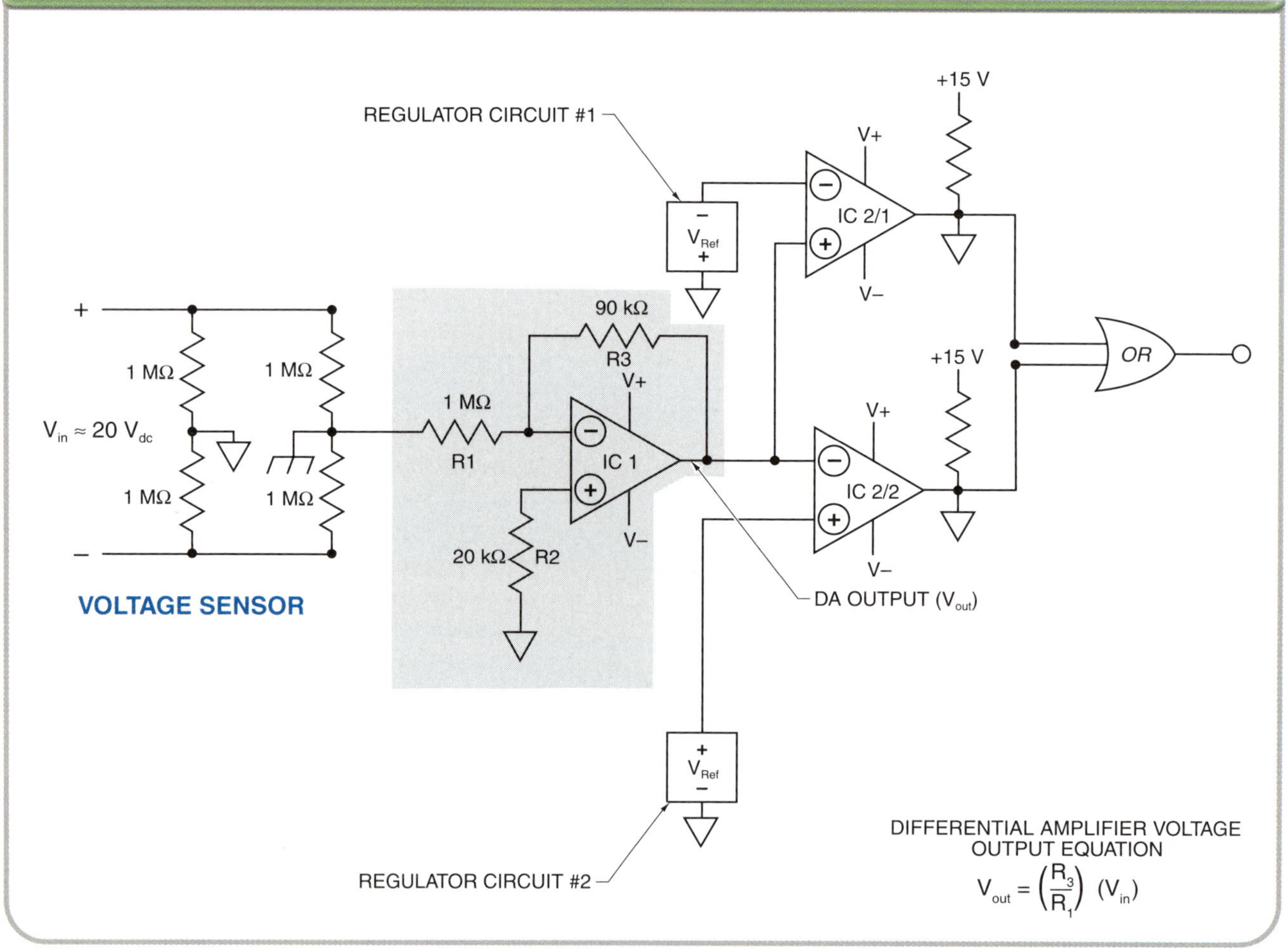

Figure 3-15. The differential-amplifier stage amplifies the difference between its inverting and noninverting inputs.

Voltage-Comparator Stage. The voltage-comparator stage senses voltages and compares it to a voltage reference value. **See Figure 3-16.** Voltage from the DA is transferred to the noninverting input of comparator IC 2/1 and the inverting input of comparator IC 2/2. Both of these comparators are within the same IC. Both IC 2/1 and IC 2/2 have voltage regulators on the remaining input pins.

The regulators provide a voltage reference, or threshold voltage level, to each comparator. The comparators compare the positive and negative reference voltages to the voltages transmitted by the DA on the inverting pin of IC 2/1 and the noninverting pin of IC 2/2. The comparators will either turn the output ON to provide a ground, or sink, for the +15 V pull-up resistor, or remain OFF and permit the output to be pulled up to +15 V.

VOLTAGE–COMPARATOR STAGE

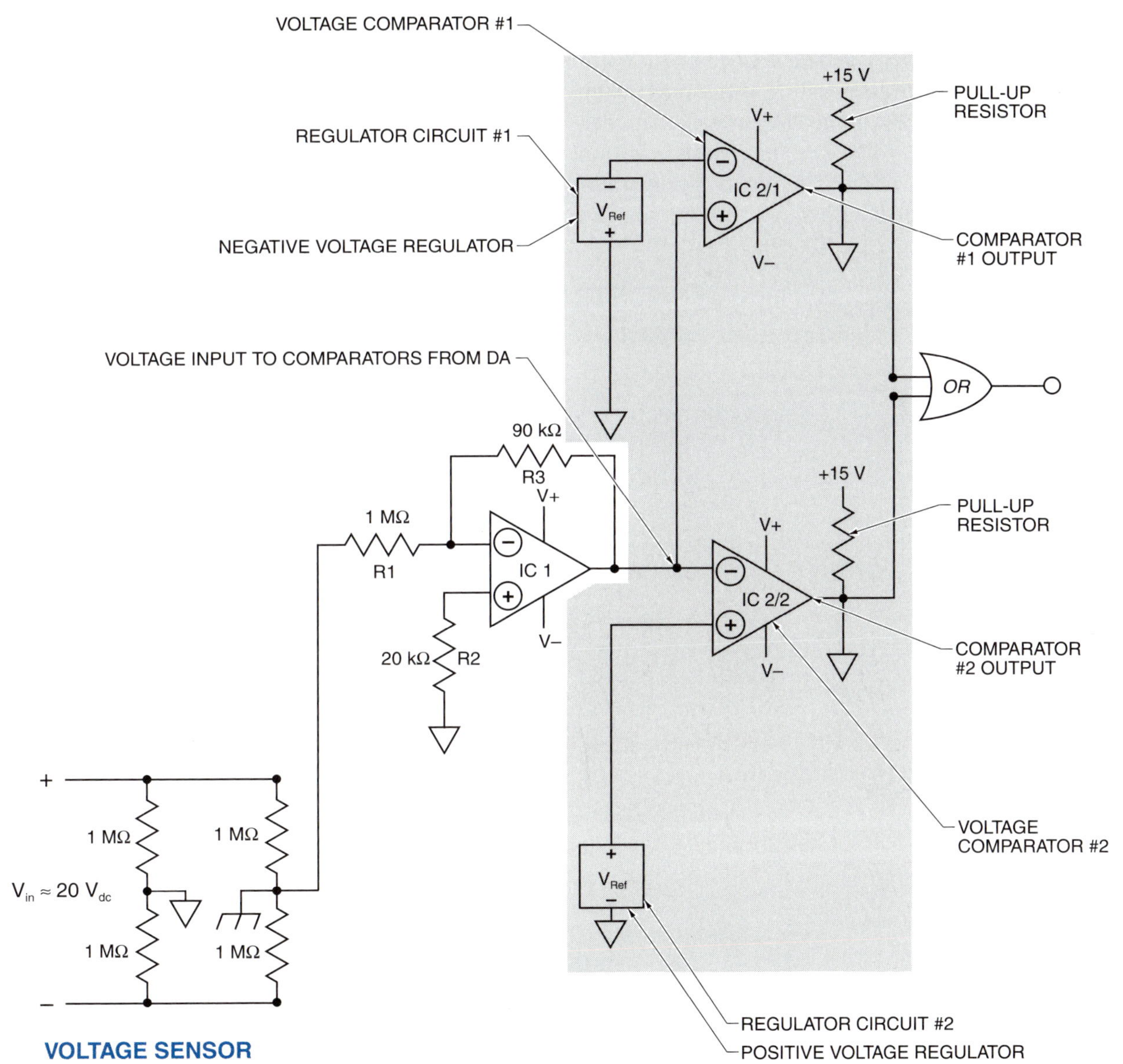

Figure 3-16. The voltage-comparator stage senses voltages and compares those voltages to a voltage reference value.

Logic-Output Stage. The final stage of the isolation-fault detection circuit is the logic-output stage. The purpose of this stage is to sense the outputs of the two comparators and provide a binary, or digital, output to the controller hardware. By using the voltage-comparator stage to monitor the output of the DA, the isolation fault detection circuit can determine either a positive or negative voltage level on the chassis. The controller uses logic to determine whether this level is within a safe voltage range. **See Figure 3-17.** In any situation where the controller senses a fault, the controller turns on the MIL and sets DTC.

For example, voltage regulator #1 reference voltage can be set to –5.1 V and voltage regulator #2 reference voltage can be set to +5.1 V. The voltage-comparator stage can provide three possible outcomes when monitoring the DA. First, if the DA output is between –5.1 V and +5.1 V, the voltage on the chassis is within normal limits and no action is required. Comparator IC 2/1 and comparator IC 2/2 outputs are turned ON and both outputs sink the pull-up resistor circuits. The voltage output of both comparators is 0 V.

Second, if the DA output is greater than +5.1 V, positive high-voltage leakage is detected on the chassis. As a result, IC 2/1 output is turned ON and IC 2/2 output is turned OFF. Only IC 2/1 output is turned ON and sinks the pull-up resistor. The voltage output of IC 2/1 is 0 V. The voltage output of IC 2/2 is 15 V.

The third possible outcome is when the DA output is less than –5.1 V. In this case, negative high-voltage leakage is detected on the chassis. As a result, IC 2/1 output is turned OFF and IC 2/2 is turned ON. Only IC 2/2 output is turned ON and sinks the pull-up resistor. The voltage output of IC 2/1 is 15 V. The voltage output of IC 2/2 is 0 V.

The OR gate function is very simple. If both of the inputs to the OR gate are 0 V, then the output is 0 V. If either of the inputs to the OR gate is 15 V, the output is 15 V. A truth table can be used to determine how the OR gate will sense each input and how this will influence the output.

LOGIC–OUTPUT STAGE

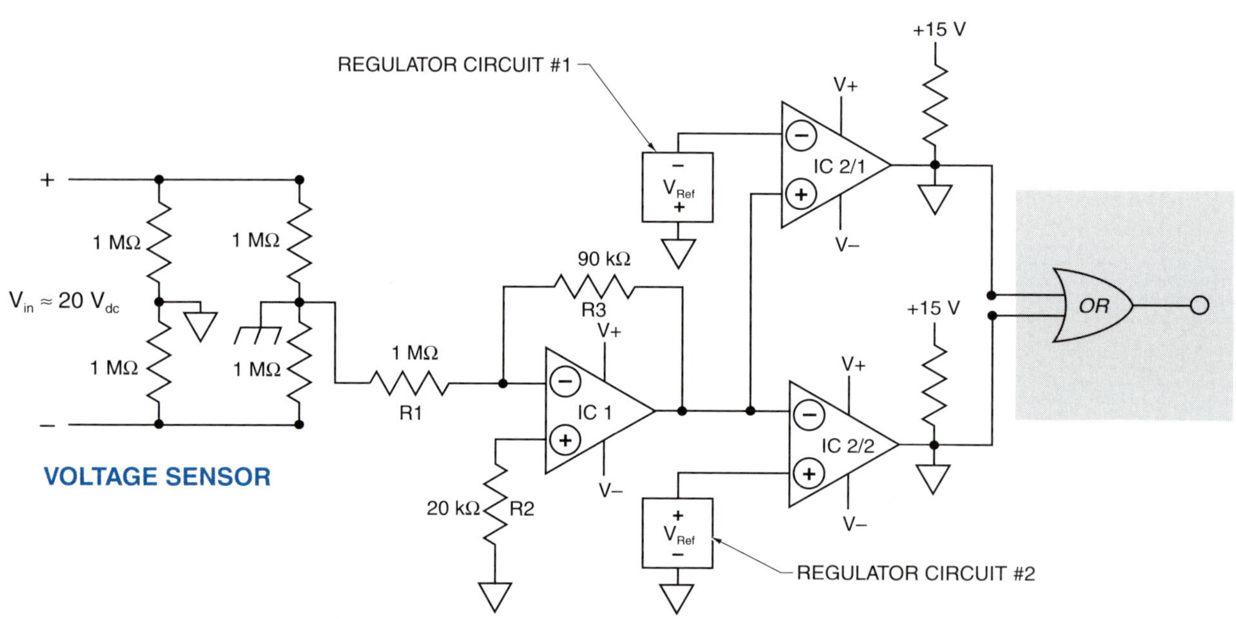

OR Gate Truth Table			
Comparator #1 Output	**Comparator #2 Output**	**OR Gate Output**	**Description**
0 V	0 V	0 V (Low)	Normal Operation
0 V	15 V	15 V (High)	Fault
15 V	0 V	15 V (High)	Fault

Figure 3-17. The controller uses logic to determine whether a voltage level is within a safe range.

Voltage-Regulator Circuits. Comparator circuits use voltage regulators to set a threshold voltage. For example, the comparators may use +5.1 V and –5.1 V as reference voltages. A simple voltage-regulator circuit can be constructed with zener diodes, resistors, and capacitors in parallel and one resistor in series. **See Figure 3-18.** The zener diodes provide the circuit with a constant voltage drop of 5.1 V. The parallel resistor is added to stabilize the zener diode. The capacitor filters the circuit. The series resistor limits the current to the zener diode.

When the circuit is connected to a power supply, the zener diodes on each regulator operate with a 5.1 V drop. The same voltage drop is also across R1 and C1 and the regulator output terminal. The result is a +5.1 V and –5.1 V voltage drop across the respective regulators. The regulator provides this voltage to the reference inputs of the two comparators.

Testing for Isolation Faults. Although controllers have the ability to test for isolation faults, a common test instrument, such as a megohmmeter, may also be used to analyze the system. A megohmmeter, or insulation tester, supplies about 250 V to 1000 V at low-current levels. It tests for isolation resistance, and the integrity of that resistance, between the high-voltage system components and the vehicle chassis.

A technician connects one lead of the megohmmeter to the high-voltage component and the other end to the chassis of the vehicle. The technician should not touch the component when it is being tested. If the system has adequate isolation resistance between the high-voltage component and vehicle chassis, the megohmmeter reading will be very high, or even infinite (∞). If the system does not have adequate isolation resistance, the megohmmeter reading will be low. The vehicle manufacturer's service manual provides acceptable limits for test results.

Isolation-Fault Detection Circuit Software. There are two general categories of software algorithms used to control the isolation-fault detection system. First, the software can be designed to monitor for an isolation fault on system startup or shutdown. Second, the software can be designed to continuously monitor for an isolation fault.

The system startup and/or shutdown software uses fewer controller system resources to execute the functions than software that continuously monitors the system. With continuous monitoring, the controller is required to execute a check of the isolation-fault detection system at regular intervals, such as every 100 msec or every 500 msec. When a system is continuously executing a diagnostic, it provides more of an opportunity for intermittent diagnostic trouble codes to be triggered by the controller diagnostic software.

VOLTAGE-REGULATOR CIRCUITS

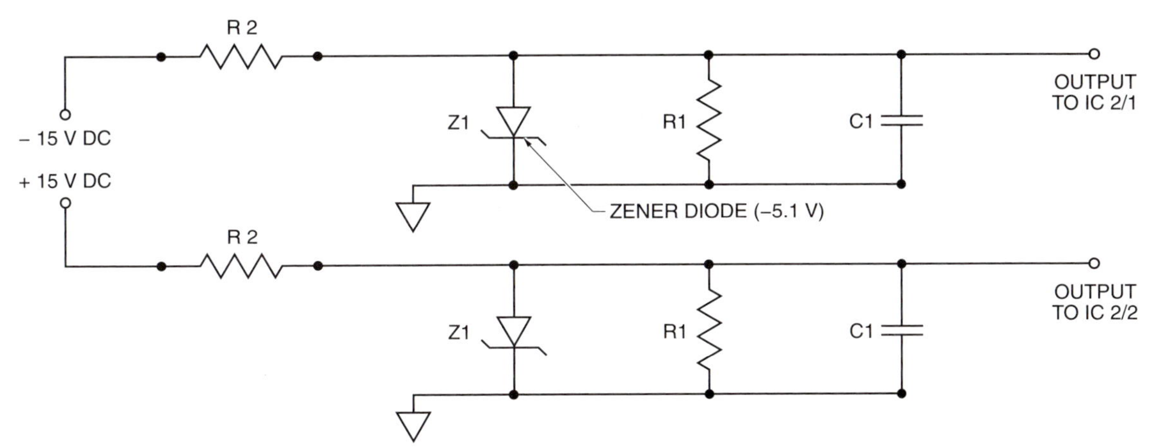

Figure 3-18. A simple voltage-regulator circuit can be constructed with zener diodes, resistors, and capacitors in parallel and one resistor in series.

Refer to the ***Hybrid Electric Vehicle Technology*** **CD-ROM for a Quick Quiz® on the chapter contents.**

POINTS TO REMEMBER

- SAE J1673 standard states that "... wire/cable or harnesses containing high voltage shall be visually identified with a permanent orange harness covering material."
- An interlock system is used to detect when any high-voltage component or circuit has been breached.
- Most electric, hybrid, and fuel-cell vehicles use a serial or local interlock loop system.
- A serial interlock loop (SIL) is a series circuit, or daisy chain, that connects each high-voltage device. The controller senses if any of the high-voltage connectors or the cover of a high-voltage component are disconnected. If any devices are disconnected, the controller opens the high-voltage battery-pack contactors or relays.
- A local interlock loop uses discrete switches with the high-voltage bus to signal a controller that a system has been breached. After a breach has been identified, the controller opens the high-voltage contactors to remove power from the high-voltage bus and/or components. The system can also be designed with a discrete switch that is routed directly to a controller input and the controller monitors the input for changes in voltage states.
- A grounded switch interlock uses connectors and/or switches that provide a ground for the hybrid-controller sense circuit.
- An active bus-discharge circuit uses electronics to quick discharge the bus capacitors in a high-voltage system, usually in less than 5 seconds.
- A passive bus-discharge circuit uses resistors to slowly discharge the high-voltage bus capacitors in a high-voltage system, usually in about 5 minutes.
- High-voltage contactors are opened before discharging the high-voltage bus capacitors to protect the discharge circuits.
- High-voltage electric, hybrid, and fuel-cell systems contain a service disconnect switch that is typically placed in the center of the battery-pack module string.
- Service disconnect switches are used to mechanically open the high-voltage battery-pack circuit string.
- Isolation-fault detection systems are connected to the vehicle chassis system to monitor any high-voltage component or circuit that may leak (acquire a path) to the chassis.
- The four stages of an isolation-fault detection system are the voltage-sensing, differential-amplifier, voltage-comparator, and logic-output stages.
- Both DC and AC sensing systems can be used to monitor and test high-voltage components and circuits in order to detect electrical paths to the vehicle chassis.

Refer to the *Hybrid Electric Vehicle Technology* CD-ROM for the Review Questions in Microsoft® Word format.

REVIEW QUESTIONS

1. What is the difference between an active safety system and a passive safety system in a high-voltage safety system?
2. Why is a waiting period recommended before removing high-voltage components?
3. List at least five high-voltage components that could provide a leak path to the vehicle chassis?
4. What is the purpose of an isolation-fault detection circuit?
5. How can a megohmmeter be used to test for isolation faults in an electric propulsion system?
6. What is the purpose of an interlock system?
7. What is the difference between a serial interlock loop (SIL) system and a local interlock loop (LIP) system?

3 High-Voltage Vehicle Safety Systems

Trade Test

Name ______________________________ Date ______________

______________ **1.** The two categories of systems that can discharge system high voltage are ___ systems.
A. primary and secondary
B. high-voltage and low-voltage
C. active and passive
D. Level 1 and Level 2

______________ **2.** SAE J1673 standard states that "all high-voltage covering and/or cabling shall be visually identified with ___ covering."
A. red
B. orange
C. blue
D. black

______________ **3.** The two types of interlock loop designs that can be used to monitor a high-voltage system are ___ systems.
A. serial and local
B. primary and secondary
C. high-loop and low-loop
D. Interlock A and Interlock B

______________ **4.** A serial interlock loop system is designed to ___ if its circuit is broken or opened.
A. immediately stop propulsion
B. reduce propulsion power by 50%
C. open battery-pack relays/contactors
D. light the MIL

______________ **5.** A ___ is used to assist in identifying the location of an isolation fault.
A. voltage tester
B. Hall effect transducer
C. wattage tester
D. megohmmeter

______________ **6.** ___ can result when isolation-fault detection, diagnosis, and repair are not performed.
A. Damage to the hybrid controller
B. Electrical shock
C. Power inverter damage
D. Delay of vehicle power-up

_______________ **7.** A passive discharge circuit uses a(n) ___ to discharge energy in the high-voltage bus capacitors.
A. resistor
B. diode
C. comparator
D. op-amp

_______________ **8.** A local interlock loop uses ___ to open the contactors when the high-voltage system has been breached.
A. LED signaling
B. LAN communications
C. selective discharging
D. passive safety systems

_______________ **9.** The purpose of the isolation-fault detection circuit is to monitor the ground of the vehicle chassis for ___.
A. hybrid-controller resistance
B. battery-pack voltage fade
C. power-inverter damage
D. voltage leakage

_______________ **10.** The Lexus RX400h and Toyota Highlander hybrids use ___ interlock systems.
A. serial
B. local
C. grounded-switch
D. open-collector

Hybrid Engines

4

Hybrid electric vehicles use a combination of internal combustion engines and electric machines based on the Atkinson cycle. The Atkinson cycle is different than the traditional Otto cycle, which has been the primary combustion cycle for over 100 years. High efficiency can be obtained over all load ranges when an Atkinson-cycle engine is combined with an electric traction system.

CHAPTER OBJECTIVES

- Explain the difference between Otto-, Atkinson-, and combination Atkinson/Miller-cycle engines.
- Describe how an Otto-cycle engine compares to an Atkinson-cycle engine at low, medium, and high load ranges.
- Describe the advantages of combining an Atkinson-cycle engine with an electric traction system.

KEY TERMS

efficiency (η): The measure of how well a vehicle uses fuel and is calculated based on the ratio of output power to input power.

Atkinson cycle: A four-stroke engine cycle in which the intake valve is held open longer than normal in order to allow the expansion ratio to exceed the effective compression ratio.

Otto cycle: A four-stroke engine cycle used in traditional automotive engine applications in which the amount of time that the intake and exhaust valves open and close is nearly equal.

Miller cycle: A modified Atkinson cycle in which a supercharger is added to an engine in order to increase the compression ratio.

pumping loss: Energy loss due to restrictions in the engine that inhibit air or gases from flowing or being pumped through the engine intake, engine, or engine exhaust systems.

HYBRID ENGINES

Electric machines provide the major propulsion in a hybrid electric vehicle. The engine in a hybrid electric vehicle also provides motive power torque to propel the vehicle. The engine of a hybrid electric vehicle is more efficient than that of a traditional internal combustion engine (ICE). The major categories of engines used in hybrid electric vehicles are the Atkinson-cycle engine, Miller-cycle engine, and cylinder deactivation technologies.

Hybrid engines contain features that enhance fuel economy and reduce emissions without sacrificing vehicle performance. Powertrain systems are optimized by combining efficient engine designs and electric propulsion technology. *Efficiency* (η) is the measure of how well a vehicle uses fuel and is calculated based on the ratio of output power to input power. Optimizing vehicle efficiency requires unique engine operating modes and characteristics. Some of these modes, such as cylinder deactivation, which eliminates fuel flow and spark to one or more of the cylinders, can be used in both hybrid engines and in traditional engines.

Atkinson-Cycle Engines

The Atkinson-cycle engine is designed to provide efficient operation of a vehicle when overall power is decreased. An *Atkinson cycle* is a four-stroke engine cycle in which the intake valve is held open longer than normal in order to allow the expansion ratio to exceed the effective compression ratio. This ability leads to higher efficiency compared to the function of Otto-cycle engines. An *Otto cycle* is a four-stroke engine cycle used in traditional automotive engine applications in which the amount of time that the intake and exhaust valves open and close is nearly equal.

An Atkinson-cycle engine is designed so that the intake valve is held open to allow intake air to flow back into the intake manifold. This reduces the compression ratio. When the reduced compression ratio is combined with an increased stroke and/or reduced combustion chamber volume, the expansion ratio becomes greater than the compression ratio. Fuel economy is improved because the compression ratio in a spark-ignition engine is limited by the octane rating of the fuel. A high expansion ratio delivers a longer power stroke, which allows more expansion of combustion gases and reduces the amount of heat wasted in the exhaust. This results in a more efficient engine. **See Figure 4-1.**

The Atkinson cycle enables the compression stroke and expansion stroke of the engine cycle to be set independently of each other. This generates high power output and offers high levels of thermal efficiency. The cycle is typically developed through the use of variable valve timing and an offset crankshaft. The offset crankshaft allows the connecting rod to remain more upright during the combustion stroke. As a result, there is less piston force against the cylinder wall and friction is reduced.

A disadvantage of the four-stroke, Atkinson-cycle engine is reduced power density compared to the Otto-cycle engine. Because a smaller portion of intake stroke is devoted to compressing intake air, an Atkinson-cycle engine does not take in as much fuel and air per stroke as a similar Otto-cycle engine. Examples of production vehicles that use Atkinson-cycle engines include the Toyota Prius and the Ford Escape hybrid electric vehicles. In hybrid vehicles, electric machines (motors) make up for the lower power level of the Atkinson cycle. The electric machines can be used independently of, or in combination with, an Atkinson-cycle engine.

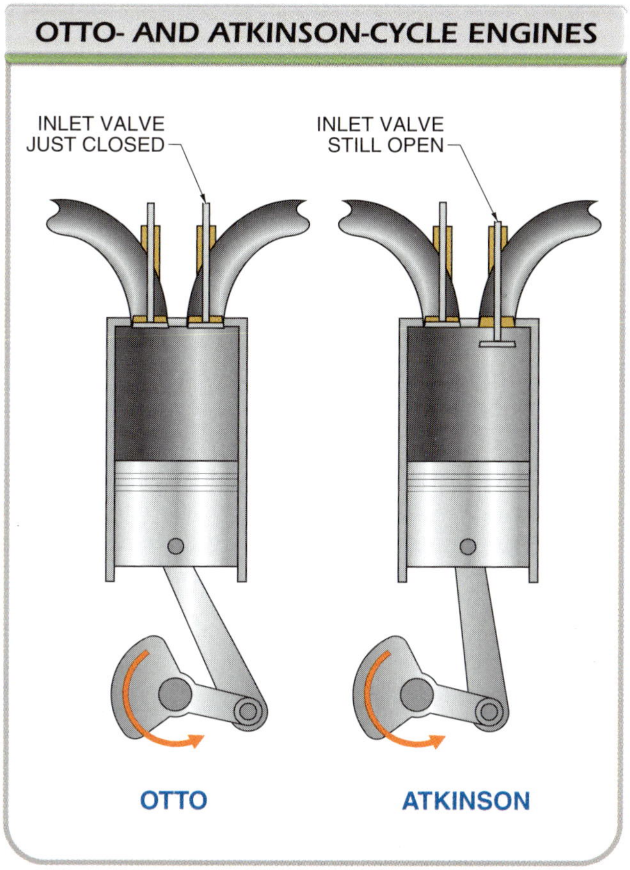

Figure 4-1. An Atkinson-cycle engine is designed so that the intake valve is held open to allow intake air to flow back into the intake manifold.

Miller-Cycle Engines

The *Miller cycle* is a modified Atkinson cycle in which a supercharger is added to an engine in order to increase the compression ratio. Efficiency increases with increasing compression ratio. A supercharger uses less energy than a piston in a naturally aspirated engine. The piston receives the same compression as it would in a standard Otto-cycle engine, but in a Miller-cycle engine it requires only about 70% of the work for compression. The supercharger generates low levels of compression where it is most efficient. The piston generates the remaining higher levels of compression in an operating range where it is more efficient than a supercharger.

In a typical spark ignition engine, the Miller cycle yields an additional benefit. In a typical gasoline engine, the compression ratio is limited because of self-ignition of the hot air-fuel mixture. In the Miller cycle, the intake air is first compressed by the supercharger and then cooled by an intercooler. This reduces the temperature of intake charge. The lower intake temperature and lower compression of the intake stroke results in a temperature of the final charge that is lower than it would be if there were an increase in the compression of the piston. This allows changes to the ignition timing that increases the overall efficiency of the engine.

The benefits of using a supercharger do not come without a cost. Approximately 15% to 20% of the power generated by a supercharged engine is usually required to do the work of driving the supercharger. A Miller-cycle engine reduces the power needed to operate the engine by about 10% to 15% compared to a standard Otto-cycle engine. Successful production engines using this cycle have typically used variable valve timing to effectively switch off the Miller cycle in regions of operation where it does not offer an advantage.

Engine Cycle Comparison

The Atkinson-cycle engine has low emissions because of its high efficiency. Modern engines use applications in order to emulate the abilities of the Atkinson cycle. These applications include late intake valve closing (LIVC), cam phasing, throttle control, and other electrical and mechanical applications.

LIVC uses precise electrical or hydraulic activation of the valves. The intake valve is commanded to remain open well into the compression stroke. This allows part of the air-fuel charge to be returned to the intake manifold. The movement of the air-fuel charge happens at atmospheric pressure and minimal pumping work is expended. Therefore, power output is controlled by LIVC.

Conventional engines control low and partial loads by throttling the incoming air with a throttle plate. Atkinson-cycle engines do not use throttle plates to regulate the air-fuel charge. Cam phasing alters valve timing by substituting the camshaft lobe positioning for hydraulics or electronic throttle control. Electronic throttle control uses a small DC motor to control throttle valve position. This reduces pumping losses, where the pistons pump against a closed throttle plate, by opening the throttle at low and intermediate loads. **See Figure 4-2.**

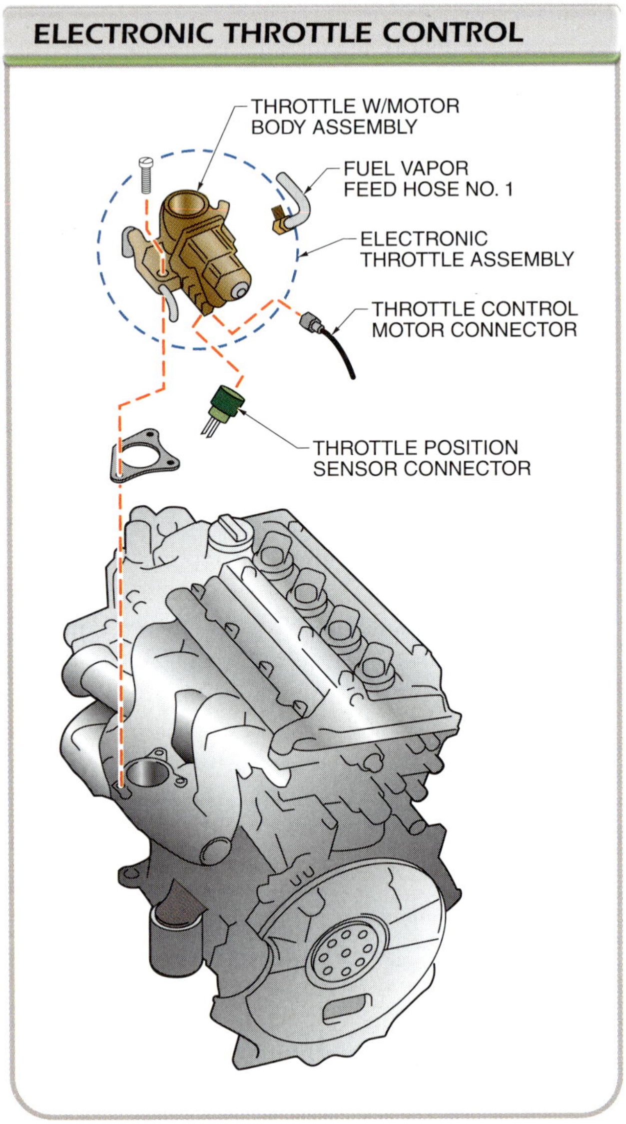

Figure 4-2. Electronic throttle control uses a small DC motor to control throttle valve position.

The air-fuel charge of an Atkinson-cycle engine reaches near maximum capability at low and intermediate loads. At low load, the efficiency at a given compression ratio is substantially higher than with the Otto cycle. This is because cylinder pressure is generated with a lower volume of air-fuel charge and cylinder power output is lower. At intermediate load, cylinder pressure is higher in the Atkinson-cycle engine than in the Otto-cycle engine because of the increase in the cylinder air-fuel charge volume. However, peak cylinder pressure decreases from the low-load cylinder pressure and power output is lower than for the Otto cycle.

Electric traction components are very efficient with power inverters, electric machines, and battery-pack systems all over 90%.

At high load, there is little difference between the Atkinson and Otto cycles. Cylinder pressure, volume of fuel in the cylinders, and power output are practically identical. Therefore, there is no advantage with the Atkinson cycle at high loads.

Engine and Electric Machine Optimization

The Atkinson cycle, Miller cycle, and combination of the two cycles are the primary configurations used to optimize engine operation, increase fuel economy, and reduce airborne emissions. A hybrid engine typically has an electronic throttle system that reduces pumping losses and increases total efficiency of the engine. *Pumping loss* is energy loss due to restrictions in the engine that inhibit air or gases from flowing or being pumped through the engine intake, engine, or engine exhaust systems.

Toyota Motor Company

A hybrid vehicle houses an engine, transaxle, and electric machines in a tightly integrated package.

Pumping losses are minimized when the electronic throttle system opens the throttle plates as needed to significantly reduce the piston pumping that would normally occur when the throttle plate is closed. Maximum engine efficiency can be achieved by targeting specific operating conditions that can be controlled through electronic control of the throttle plate, electronic intake valve control, crankshaft offset, and mechanical control of intake and exhaust valve timing. These strategies, along with aggressive deceleration fuel cutoff (DFCO) modes, may also be used to increase fuel economy and reduce airborne emissions.

The LIVC and cam phasing techniques used with an idling Atkinson-cycle engine can yield fuel consumption savings of about 5% to 9% in a naturally aspirated engine. The savings at mid-load are about 4% to 6%. There are little or no savings under full load. Supercharged or turbocharged engines can attain an efficiency well above that of a naturally aspirated engine.

The combination of a hybrid engine and electric motor results in an engine system that has much higher efficiency than a system that uses only an Otto-cycle engine. **See Figure 4-3.** Both engine and electric system efficiencies can be maximized through the control of engine speed and optimization of valve timing. Hybrid control systems adjust the power contribution of engine and electric propulsion systems to develop the optimum efficiency for the entire system. The Atkinson cycle has high efficiency at high compression ratios but low power output at low and intermediate loads and high power output at high loads.

Electric traction systems have high efficiency at low and intermediate loads but low efficiency at high loads. They can provide constant torque in all ranges. Electric traction systems for automotive use are designed as constant torque machines. These systems have an efficiency range of about 83% at peak load to 89% at low loads. Engine systems operate at efficiencies ranging from 5% to 20% depending on the load.

When combining the concepts of the Atkinson cycle with high-voltage/high-power electric traction systems, the net result is high power, high efficiency, and constant torque at low and intermediate loads. An engine of this combination has acceptable efficiency levels at high loads and high speeds. It also provides emission levels that are significantly lower than the levels of Otto-cycle systems.

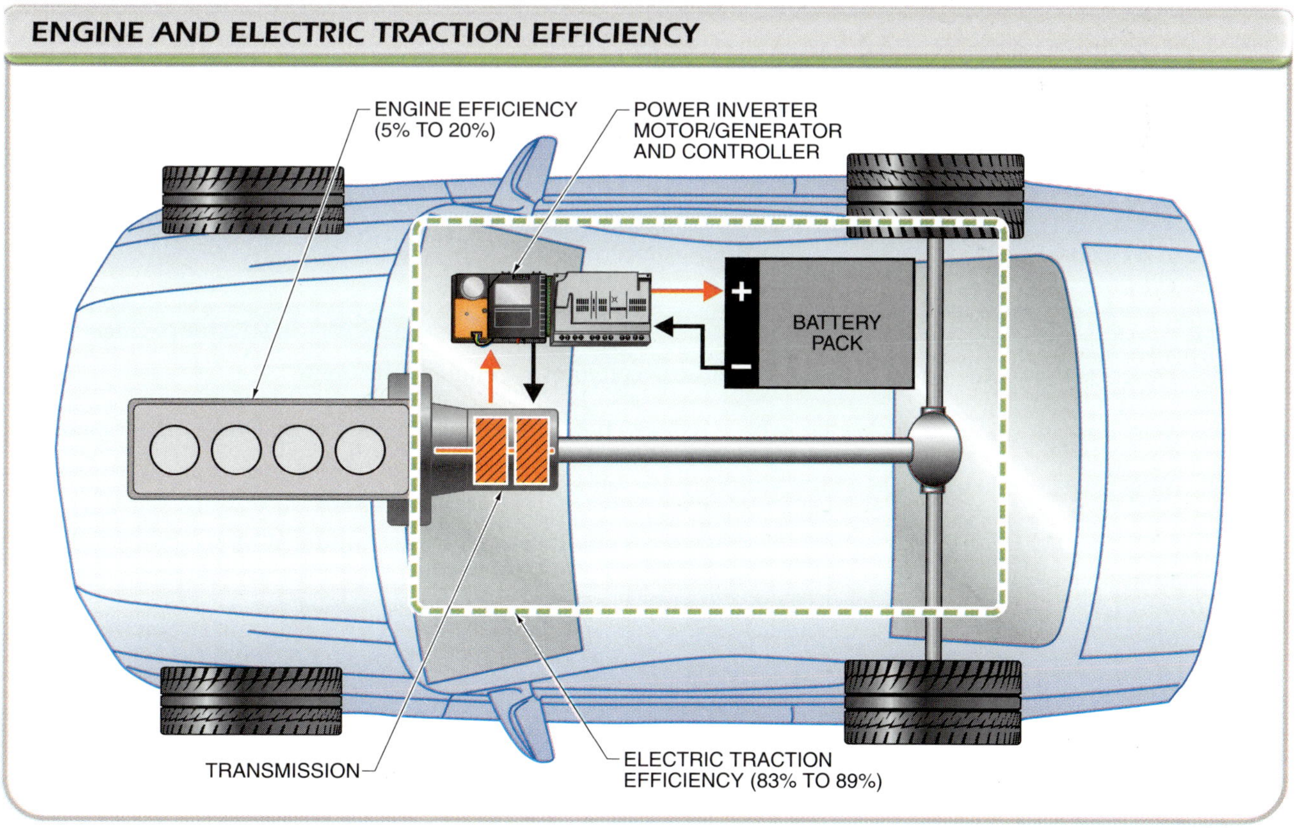

Figure 4-3. The combination of a hybrid engine and an electric motor results in efficiency that is much higher than what is achieved with an Otto-cycle engine alone.

Refer to the *Hybrid Electric Vehicle Technology* CD-ROM for a Quick Quiz® on the chapter contents.

POINTS TO REMEMBER

- Traditional vehicles are equipped with Otto-cycle engines and most hybrid electric vehicles are equipped with Atkinson-cycle engines.
- The thermal (heat) efficiency of the Atkinson cycle is substantially higher than that of the Otto cycle.
- At high load, there is almost no difference in efficiency between the Atkinson and Otto cycles.
- The Atkinson cycle refers to a four-stroke engine in which the intake valve is held open longer than normal to allow a reverse flow of intake air into the intake manifold. This reduces the compression ratio and permits the expansion ratio to exceed the compression ratio.
- The high expansion ratio of the Atkinson cycle allows an engine to deliver a long power stroke and reduces waste heat. This results in a more efficient engine.
- Electric traction systems have a high efficiency and constant torque output at low and intermediate loads.
- Atkinson-cycle engines have high efficiency and low power output at low and intermediate loads.
- High efficiency can be obtained over all load ranges when an Atkinson-cycle engine is combined with an electric traction system.

Refer to the *Hybrid Electric Vehicle Technology* CD-ROM for the Review Questions in Microsoft® Word format.

REVIEW QUESTIONS

1. Describe how delaying the intake valve closing can improve engine efficiency.
2. Why does an Atkinson-cycle engine have reduced power density compared to an Otto-cycle engine?
3. Why is a supercharger used with a Miller-cycle engine?
4. Compare the efficiency of an Atkinson-cycle engine to an Otto-cycle engine at low, intermediate, and high load.
5. How does efficiency increase when an electric machine is combined with an internal combustion engine?

4 Hybrid Engines

HYBRID Electric Vehicle TECHNOLOGY

Trade Test

Name ______________________ Date ______________

______ **1.** The LIVC and cam phasing techniques used with an idling Atkinson-cycle engine can yield fuel consumption savings of about ___.
A. 1% to 3%
B. 4% to 7%
C. 5% to 9%
D. 10% to 13%

______ **2.** The Miller cycle is a modified Atkinson cycle in which a(n) ___ is added to an engine in order to increase the compression ratio.
A. throttle plate
B. inlet valve
C. spark plug
D. supercharger

______ **3.** The ___ of an Atkinson-cycle engine at low loads is higher than with a comparable Otto-cycle engine.
A. engine efficiency
B. power output
C. air/fuel ratio
D. pumping loss

______ **4.** Electric traction systems for automotive use are designed as ___ machines.
A. constant torque
B. constant speed
C. variable-rotor angle
D. switched-pattern field

______ **5.** A high expansion ratio delivers a longer power stroke, which reduces the ___.
A. expansion stroke
B. movement of the throttle plate
C. efficiency of the engine
D. heat wasted in the exhaust

______ **6.** The ___ cycle enables the compression stroke and expansion stroke to be set independently of each other.
A. Miller
B. Otto
C. Atkinson
D. cylinder deactivation

_______________ **7.** ___ is a measure of how well a vehicle uses its fuel and is calculated based on the ratio of output power to input power.

A. Efficiency
B. Power stroke
C. Torque
D. Capacity rate

_______________ **8.** Approximately ___ of the power generated by a supercharged engine is usually required to do the work of driving the supercharger.

A. 1% to 5%
B. 5% to 10%
C. 15% to 20%
D. 50% to 60%

HYBRID Electric Vehicle TECHNOLOGY

AC Induction Electric Machines

5

AC induction machines are one of the core components of electric, hybrid, and fuel-cell propulsion technology. Induction machines are the workhorses of the industrial machine age. Hybrid vehicles can achieve good traction control and the same or better torque with fewer components than a traditional internal-combustion engine system. Although induction machines are mechanically much simpler than traditional engines, their capabilities are every bit as good. An electric machine includes a motor and components necessary to control and operate it as a motor or generator.

CHAPTER OBJECTIVES

- Describe the construction of rotors and stators.
- Explain how mutual induction is used to create a magnetic field in a conductor.
- Identify the similarities between a transformer and a motor.
- Explain the difference between wye and delta stator wiring.
- Explain how current flow through the stator coils produces rotor rotation.
- Compare positive, zero, and negative slip and explain how they relate to torque.
- Describe regenerative braking and explain how a battery pack can be recharged.
- Compare the efficiency and torque of slip motor control and vector motor control.

KEY TERMS

electric machine: A machine with induction or permanent magnet technology that provides torque for electric traction or electric energy during regenerative braking. An electric machine includes a motor and the components necessary to operate and control the motor.

rotor: The rotating part of an electric machine.

stator: The nonrotating part of an electric machine that encloses the rotor.

power inverter: An electronic component that changes direct current power to alternating current power or alternating current power to direct current power.

slip: The difference in rotation speed between the magnetic fields in the stator and the rotor.

ELECTRIC MACHINES

An *electric machine* is a machine with induction or permanent magnet technology that provides torque for electric traction or electric energy during regenerative braking. An electric machine includes a motor and components necessary to control and operate it as a motor or generator. A common type of motor is an AC induction machine.

The modern induction machine is very efficient. Its efficiency is generally about 91% to 92% at minimum load and about 85% to 86% at maximum load. Many manufacturers of electric vehicle drive systems combine the motor controller and motor efficiencies to give a total system efficiency rating.

Induction Machine Components

Induction machines have been the dominant electric machine system since Tesla's development of the motor in 1888. Induction machines are fairly simple machines compared to most other motor designs but their operation is the most abstract. The mechanical simplicity of the induction motor is what has made it the machine of choice for over a century. **See Figure 5-1.** The basic technology used for AC motors in industrial applications is used in hybrid/electric power systems.

In a DC machine, brushes are used to transfer electrical current through the armature via the commutator and field coils. This generates the interacting magnetic fields and creates the rotary motion that is used to accomplish mechanical work. An induction machine, such as that used with a cranking motor on an internal combustion engine, has no contacting components other than bearings. There is no commutator and there are no brushes. Since the induction machine has no contacting elements, it has many operational and service advantages over a DC brush-type machine. The two major components of an induction machine are the rotor and stator.

AC Induction Motors

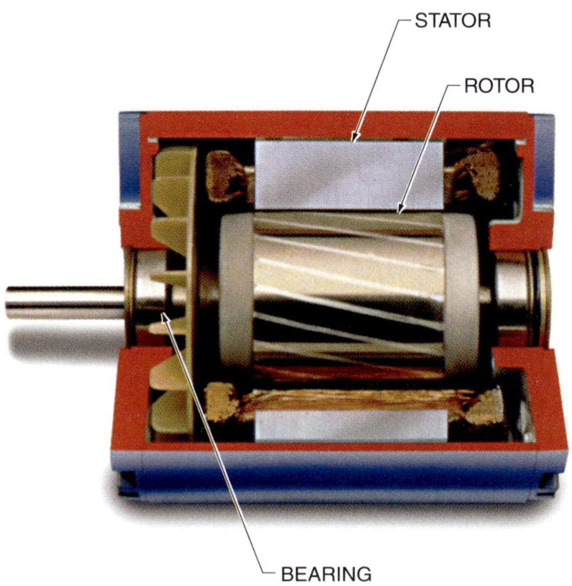

Leeson Electric Co.

Figure 5-1. The basic technology used for AC motors in industrial applications is used in hybrid/electric power systems.

ROTOR CONSTRUCTION

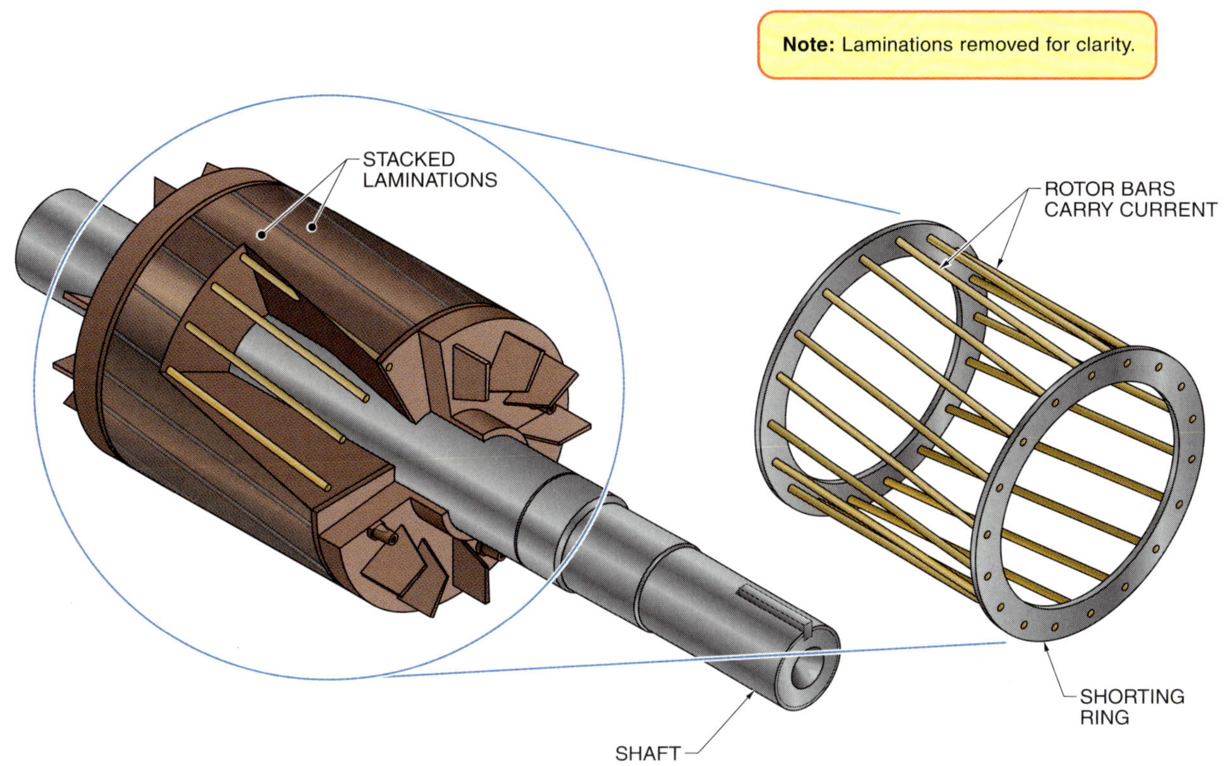

Figure 5-2. A rotor is the rotating part of an electric machine.

Rotors. A *rotor* is the rotating part of an electric machine. **See Figure 5-2.** The rotor generates rotary motion and torque in a similar manner as the armature in a DC brush-type machine. A rotor is made of hundreds of thin iron laminations sandwiched together and placed on an iron shaft. The laminations have slots into which copper bars are placed or aluminum is poured under pressure to make die cast bars. The rotor bars carry large amounts of induced electrical current. Current flows through the bars to the shorting rings that are placed at each end of the lamination stack. The shorting rings are either die cast or brazed to the rotor bars. Laminations are used instead of solid pieces of iron to reduce stray electrical eddy currents in the core of the rotor.

The rotor bars can be designed in a number of different configurations. The bar configurations must match the slot configurations that are cut into each lamination. The different bar configurations have different current flow and torque production. The shape of these different slots and bars determines rotor impedance, resistance to current flow, and torque output.

An alternate method of viewing this concept is to think of the rotor bars and shorting rings as a series circuit. If the circuit resistance is increased in a series circuit, less current flows. If the circuit resistance is decreased in a series circuit, more current flows. By changing the rotor bar shape or the number of bars, electrical current flow and magnetic fields within the rotor are changed. This alters the torque and maximum speed characteristics of the motor. For example, if the size of a 12 V cranking-motor armature increases, the resistance decreases, and total current flow increases.

In other words, simply changing the rotor bar configuration and the number of bars in the rotor can control the torque and speed curve of a motor. The electrical-current levels being supplied to the stator pole coils must also increase or decrease to control the overall torque level.

Stators. A *stator* is the nonrotating part of an electric machine that encloses the rotor. **See Figure 5-3.** The stator is similar to the field coil set in a DC motor. A cranking motor housing contains thick, copper ribbon wrapped in neat coils. The coils are held in place by pole shoes that are secured to the motor housing by large screws. An induction machine also has coils of wire placed within the slots of the stator. As with the rotor, the stator is manufactured as a stack of iron laminations to reduce eddy currents produced by the magnetic field in the stator coils. The magnetic fields around the wire coils make up the stator poles that interact magnetically with the rotor to generate rotary motion and torque.

Bearings are placed at each end of the rotor shaft for support. The air gap between the outer edge of the rotor and the inner edge of the stator is typically about 0.020″. This means that precise manufacturing of both the rotor and the stator is necessary.

STATOR CONSTRUCTION

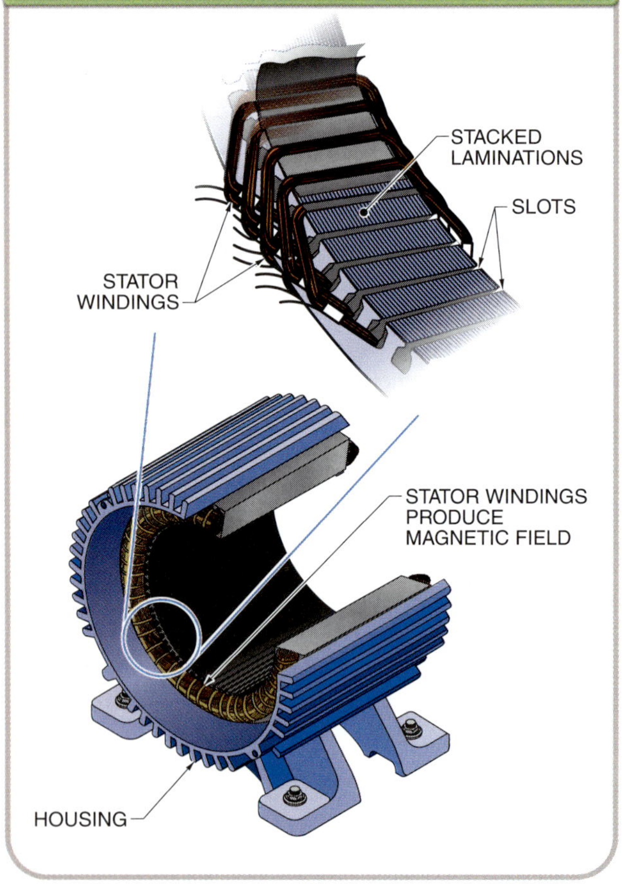

Figure 5-3. A stator is the nonrotating part of an electric machine that encloses the rotor.

ELECTROMAGNETISM AND MUTUAL INDUCTION

When magnets are placed close to each other, opposite magnetic poles attract and like magnetic poles repel. Magnetic field lines travel out from the north magnetic pole and into the south magnetic pole. Electricity flowing through a coil creates an electromagnet that results in a magnetic field. Controlling the flow of electricity controls the strength and location of the electromagnetic field.

Mutual Induction

A property of some electronic components is mutual induction. Mutual induction occurs in transformers and motors when there is relative motion between a magnetic field and a conductor. If a magnetic field is passed across, or cuts, a conductor, current flows in the conductor. Current flow causes a magnetic field around the conductor with north and south poles that are of opposite polarity to the original field. **See Figure 5-4.** The polarity of the poles depends on the winding direction and the direction of current flow. When the wire or conductor is perpendicular to the path of magnetic field lines, maximum voltage is induced in the conductor.

MUTUAL INDUCTION

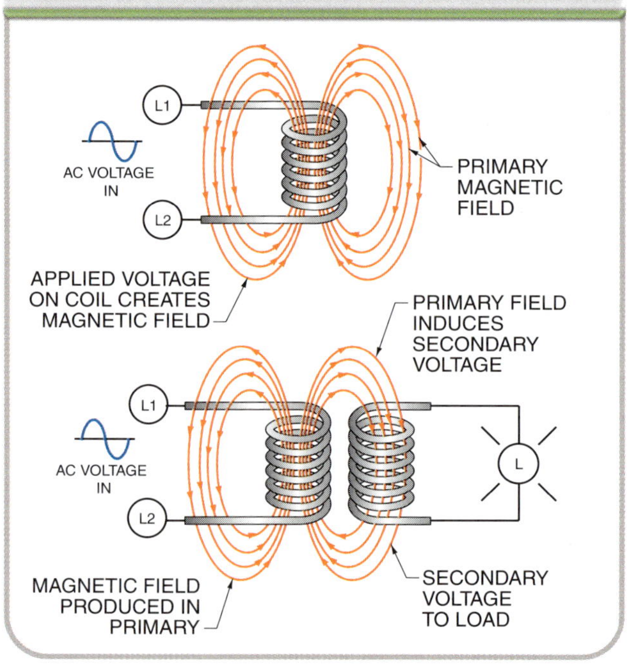

Figure 5-4. Mutual induction occurs in transformers and motors when there is relative motion between a magnetic field and a conductor.

Transformers

A transformer uses parallel primary and secondary windings (coils). The primary winding is connected to a switched source. The secondary winding is connected to a load. The load can be a spark plug, light bulb, or any other component that allows electrical current to flow.

An ignition coil is a type of transformer. The switched source is the ignition module. The current in the primary winding is pulsed ON and OFF by the ignition module. This creates a magnetic field that moves back and forth around the primary winding. The back-and-forth motion of the magnetic field around the primary cuts across the secondary winding. Mutual induction produces current flow in the secondary. The current flow generates a magnetic field around the secondary with opposite polarity to the primary.

In the case of a transformer, the magnetic field is created by an electromagnet. Electrical current flows through the primary winding to build a magnetic field. Through mutual induction, the magnetic field around the primary winding causes current to flow in the secondary winding.

A distributorless ignition system is fundamentally a transformer with the primary winding connected to battery positive and the ignition module connected to ground. **See Figure 5-5.** Secondary outputs are connected to the spark plugs and across the engine block to form a series circuit. The polarities of the primary and secondary windings are exactly opposite. This creates two magnets of opposite polarities very close to each other.

The polarity of the induced field in the secondary is opposite the polarity of the primary field. When the power is switched OFF, the magnetic field in the primary winding is no longer supplied with electrical current. The field collapses and the polarity reverses. The magnetic field around the secondary then collapses and the polarity reverses. The reversal of magnetic fields and polarities results in electrical current flowing in the opposite direction in each coil. **See Figure 5-6.**

The level of voltage and electrical current produced in the primary and secondary windings during collapse is determined by how quickly the magnetic field collapses. The collapse time is generally in the range of milliseconds or microseconds. The speed of magnetic field collapse is directly related to how quickly the switch can open and interrupt the supply of electrical current to the primary winding.

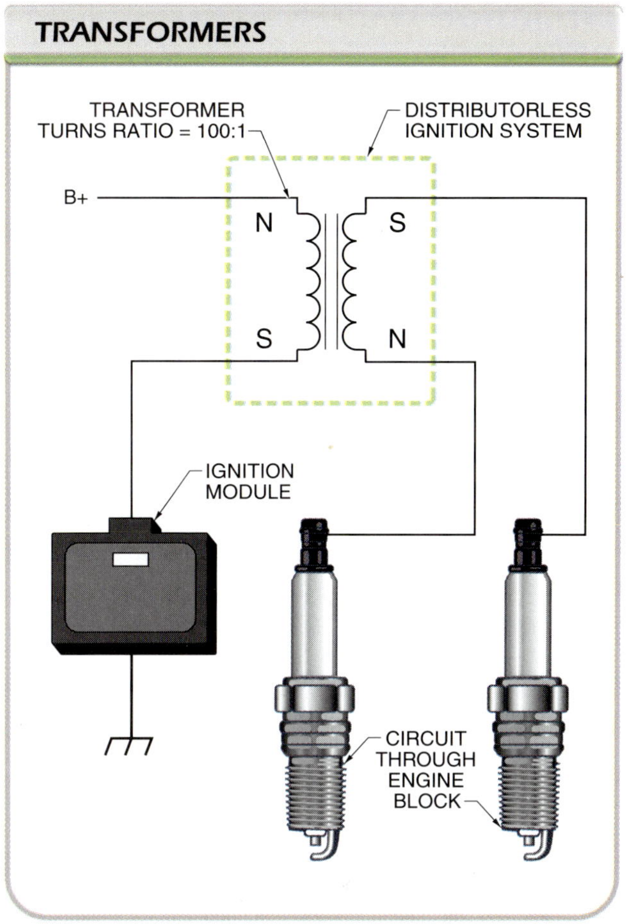

Figure 5-5. A distributorless ignition system is fundamentally a transformer with the primary winding connected to battery positive and the ignition module to ground.

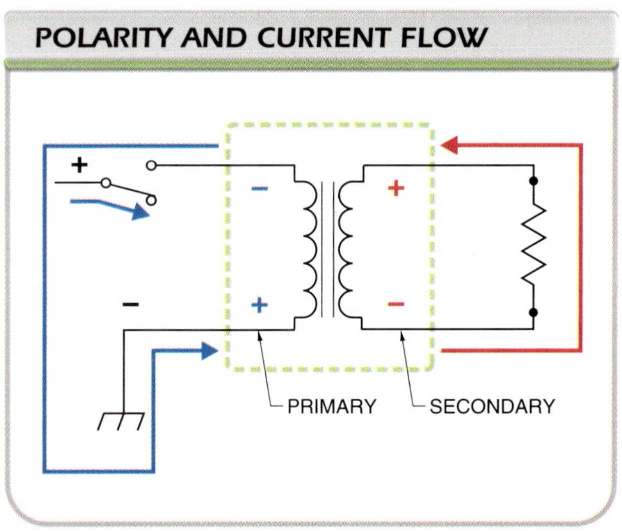

Figure 5-6. The polarity of the induced field in the secondary is opposite the polarity of the primary field.

INDUCTION MACHINE OPERATION

An induction machine is very similar to a transformer. The stator operates as a transformer primary winding and the rotor operates as a secondary winding. Although there are differences in the actual operational characteristics of the transformer and induction machine, the basic inductive principles are very similar.

Mutual Induction in Motors

The rotor must have a magnetic field so that it can interact with the magnetic field in the stator to produce motion. The rotor bars are similar to the windings of a transformer secondary. The magnetic fields and the electrical current that are generated by the stator coil magnetic fields cut across the rotor bars.

Voltage is induced in the rotor bars by mutual induction. The voltage causes electrical current to flow. This produces a magnetic field in the rotor bars. The rotor and stator are similar to a transformer where the current flow produced in the secondary winding (rotor) flows in the opposite direction of the primary winding (stator). The opposite magnetic polarities cause rotary motion. Current in the rotor bars eventually flows to the shorting rings located at each end of the rotor to complete the electrical circuit. **See Figure 5-7.**

> Induction motors are the most widely used motor technology. They are used for almost any application from hybrid vehicles, to residential ceiling fans or furnace fans, to industrial drive motors.

The current flow in the rotor bars and shorting rings makes up a series circuit. There are multiple rotor bars that produce electrical current. The current from all of the rotor bars is routed through the shorting rings. Therefore, more than one series circuit operates simultaneously during rotor operation.

The current flow in the rotor bars can be viewed as a horizontal circuit. As the magnetic field cuts across a secondary winding (the rotor), it produces an electrical current. The magnitude of the current is a function of the magnetic field strength of the primary transformer winding (the magnet). Once the magnetic field has moved past a rotor bar, the collapse of the magnetic field causes the current to continue to flow in its original direction. The movement of the stator magnetic field across the rotor causes instantaneous polarity changes in the rotor. This permits the magnetic coupling between the stator and rotor magnetic fields.

MUTUAL INDUCTION IN MOTORS

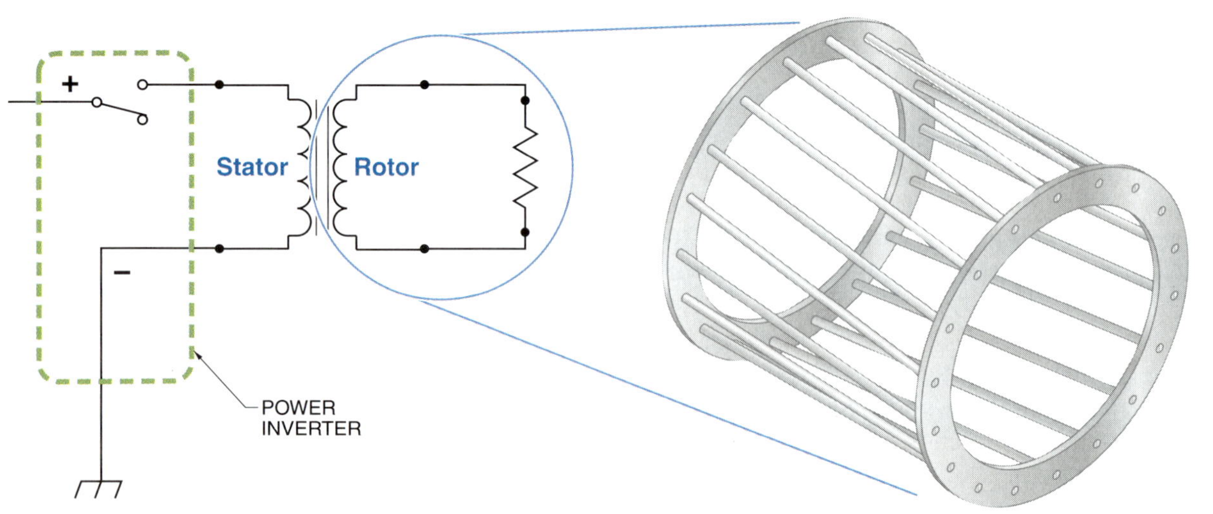

Figure 5-7. Voltage is induced in the rotor bars by mutual induction.

Poles and Phases

A *power inverter* is an electronic component that changes DC power to AC power or AC power to DC power. When the power inverter provides electrical current to each phase coil, the coils generate a magnetic field. Magnetic fields have north and south poles. The polarity of a given coil is determined by the direction of current flow and direction in which the coil is wound. **See Figure 5-8.** A coil that is wound for south pole magnetic polarity is wound in the opposite direction of a coil that is wound for north pole magnetic polarity.

POLES AND PHASES

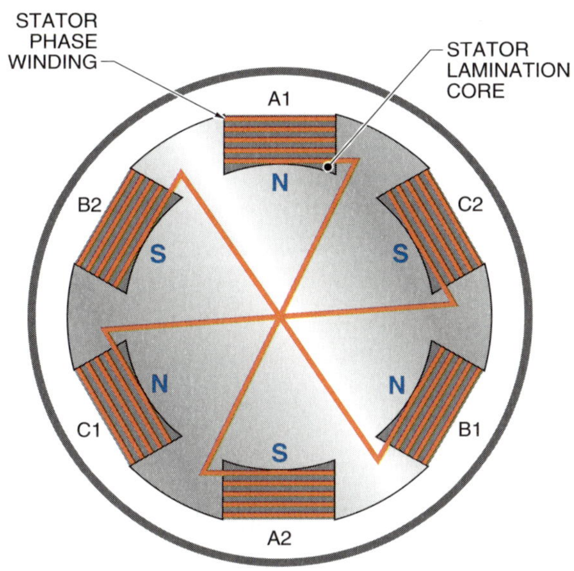

Figure 5-8. The polarity of a given coil is determined by the direction of current flow and direction in which the coil is wound.

The poles A1 and A2 make up one pole pair. The poles B1 and B2 make up another pole pair, and the poles C1 and C2 make up another pair. Actual motors may have more than three poles. Stator coil A1 is wound in a direction that gives a north pole. Stator coil A2 is wound in the opposite direction giving a south pole. Depending on torque and speed requirements, the number of stator poles, winding turns, and phases will vary greatly. However, the operation will be the same regardless of the number of poles and winding turns.

Induction machines used for hybrid/electric power systems propulsion use three-phase (3ϕ) electricity. Single-phase motors are typically used for residential and small, commercial applications, such as furnace blowers, bench tools, and small pumps. A 3ϕ motor can provide more torque than a 1ϕ motor of similar size. A typical 3ϕ motor has 12 poles and each of the 3ϕ input power lines has 4 coils wired in series. The phases are identified as A, B, and C, or as U, V, and W. This type of electric machine is categorized as a 3ϕ, 4-pole machine. Other types of induction machines may have coils wired in parallel. The wires are color coded to easily identify each of the three phases.

Wye and Delta Stator Wiring

Stators can be wired in a wye configuration or a delta configuration. The term "wye" refers to the orientation of the wires in a shape similar to the letter Y. The term "delta" refers to the orientation of the wires in a shape similar to the Greek symbol delta (Δ). **See Figure 5-9.** The physical differences between wye and delta configurations are seen in how the three separate coil windings are connected. Designers can use either configuration depending on the application of the machine and the levels of power at a rated rpm that is specified for the machine. These winding configurations are used in traditional internal combustion engine (ICE) generators and alternators.

WYE AND DELTA STATOR WIRING

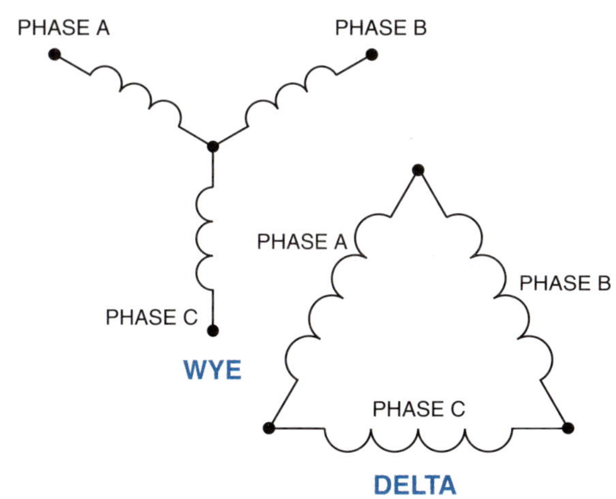

Figure 5-9. Stators can be wired in a wye configuration or a delta configuration.

Although the wye and delta winding schematic symbols are often shown as three separate coils, the actual number of windings and connections is quite different. Each phase consists of four separate coils connected in series. In a wye connection, each leg is connected to the neutral, or common, electrical connection point. In a delta connection, each leg is connected to the other legs. The four coils are then placed at even intervals around the inside of the stator housing.

The neutral point occurs where the conductors for the separate phases are brought together. In the typical hybrid electric motor, the neutral point is not used for voltage reference or as a connection for tapping power. However, the neutral connection could be used for power if the designer chooses to use it in electric vehicle power systems.

Generating Rotation

A stator operates similar to a transformer primary winding. A rotor operates similar to a transformer secondary winding. The stator magnetic field induces a voltage in the rotor bars to create electrical current flow. Induced current in the rotor creates a magnetic field of opposite polarity to the stator field. **See Figure 5-10.** When the stator magnetic field is varied, or pulsed, at a particular frequency, the opposite magnetic poles of the rotor react to the stator by following its magnetic field. As more current is provided to the stator coils, the magnetic field of each pole coil becomes stronger. A stronger stator pole induces a stronger magnetic field in the rotor bars.

The stator frequency is independent of the stator magnetic field strength. When voltage increases or decreases, so does electrical current flow of the stator. This flow controls the magnetic field strength. The controller separately increases stator frequency to increase rotor speed. It then continues to maintain constant magnetic field strength of the stator.

Controlling the rotation of the rotor is a simple process. It involves controlling the rotation speed (frequency) of the stator magnetic field. The actual switching, or energizing, operation of the poles does not allow all of the poles to be energized at the same time. The poles are pulsed, or switched, in sequential order to create a moving magnetic field. If all phase coils were allowed to energize simultaneously, there would be no rotation of the stator fields and the rotor would be magnetically locked.

GENERATING ROTATION

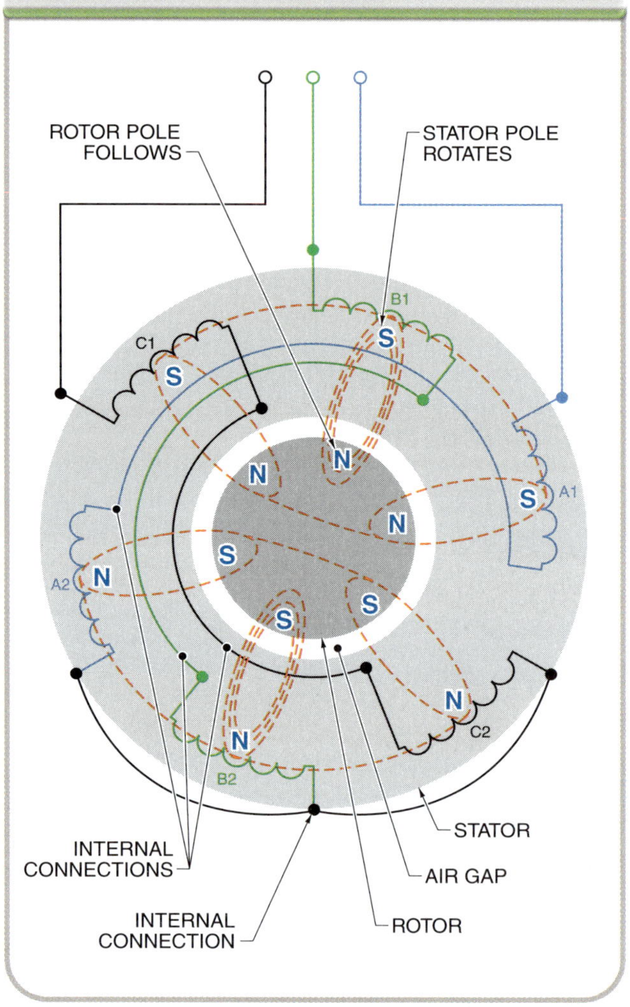

Figure 5-10. Induced current in the rotor creates a magnetic field of opposite polarity to the stator field.

Slip

Slip is the difference in rotation speed between the magnetic fields in the stator and rotor. In an induction machine, the stator magnetic field always rotates faster than the rotor magnetic field. The slip is positive and the rotor speed lags behind the stator field speed. **See Figure 5-11.**

Slip is the key element that determines how an induction machine produces torque. Torque can be produced only if there is slip. When slip increases, torque increases. When slip decreases, torque decreases. To generate torque, there must be a relative speed difference between the rotor and stator. Without slip, there is no speed difference. If the stator magnetic field does not cut across the rotor bars, no current is produced. Therefore, no torque is produced.

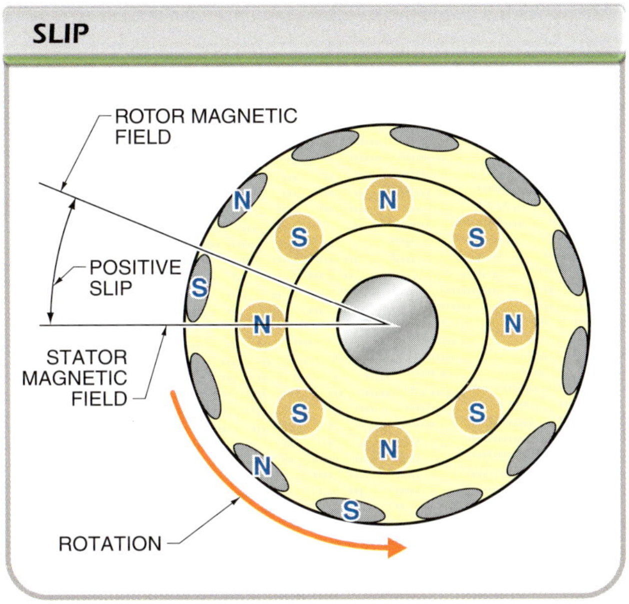

Figure 5-11. Slip is the difference in rotation speed between the magnetic fields in the stator and rotor.

An increase in slip during operation means that either the stator frequency has increased or more load has been placed on the rotor output gear. An increased load results in the rotor decreasing its speed relative to stator field speed. Regardless of whether the stator field increases its speed or the rotor slows down, more slip results. More slip means that stator fields cut the rotor bars more frequently and more torque is produced. For example, when the vehicle operator uses the accelerator pedal to request more torque for acceleration, the motor controller commands more electrical current and frequency to the stator poles. This results in a stronger coupling between the rotor and stator magnetic fields, in which delivers more torque.

Slip can be stated as the difference in speed in rpm or it can be stated as a percentage of the stator speed. For example, if the stator speed is 5000 rpm and the rotor speed is 4800 rpm, the slip is 200 rpm (5000 – 4800 = 200). The slip can also be stated as 4% (200 ÷ 5000 × 100 = 4). Slip is usually given as a percent.

Induction machines have a slip range of approximately 0% at no load and up to about 20% at full load. The typical induction machine commonly uses slip up to about 14%. Normally, any slip greater than 20% results in the motor operating in the breakdown torque (rotor stall) area.

The rotor stall area is very similar to brake torquing a vehicle equipped with an automatic transmission in a service bay. The motor is operated at a maximum rpm determined by the stall speed of the torque converter. Stall speed is the maximum speed the engine can attain when the drive wheels are stationary, transmission is in drive, and the throttle is wide open. Once stall speed is reached, no increase in engine rpm or torque can be attained. The rotor stall area of an induction machine is the equivalent of torque-converter stall rpm. If more than 20% slip is requested, torque dramatically decreases.

Zero Slip. It is possible for an induction machine to operate with zero slip. Zero slip occurs when the rotor rotates at the same speed as the stator magnetic field and magnetic field lines do not cut the rotor bars. **See Figure 5-12.** If the stator magnetic field is reduced by decreasing electrical current to the phase coils at the same instance that slip is commanded to zero, the torque achieved will be approximately zero.

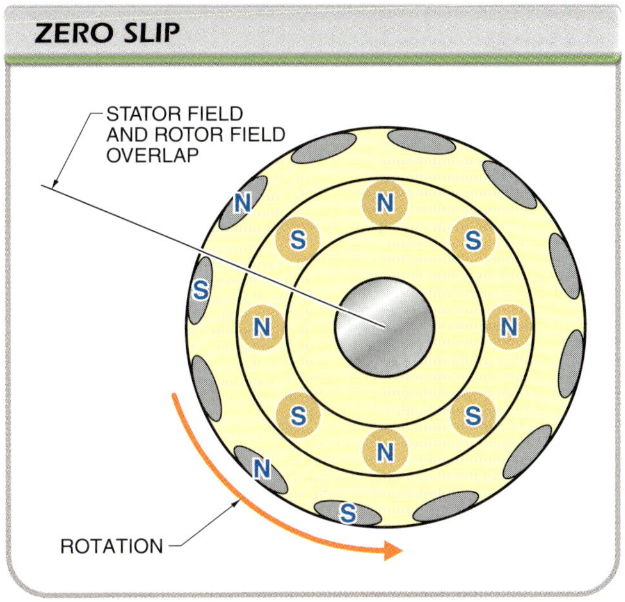

Figure 5-12. Zero slip occurs when the rotor rotates at the same speed as the stator magnetic field and there are no magnetic field lines to cut the rotor bars.

One important application of zero slip is for propulsion control. If one tire of a vehicle starts to slip on ice, the induction-machine controller can immediately be commanded to a zero slip condition. This will stop the tire from spinning and slipping on the ice.

The step-by-step processes that execute control commands are called algorithms. Since slip values are contained in the motor-controller software algorithms, commanding slip to the induction machine is virtually instantaneous. Given a particular situation, the controller can constantly alter the slip percentage as the speed of the wheel spin changes. This allows the maximum amount of torque to be transferred through the wheel, even though it is spinning on a slippery surface.

Software code is used to convert algorithms into a language that can be understood by the controller. The algorithms must be converted from a text or flowchart format to a "bits and bytes" language that the microcontroller can interpret through the use of a compiler. The programmer converts the algorithm into a binary language using the compiler so that the controller can interpret the code commands.

Negative Slip. When an induction machine operates as a generator, slip is negative and the stator field lags behind the rotor. **See Figure 5-13.** At a fundamental level, there are only minor differences between a motor and generator. When a vehicle is moving, it stores energy in the form of moving mass, or kinetic energy. This kinetic energy can be used to generate electricity and recharge the vehicle battery system.

NEGATIVE SLIP

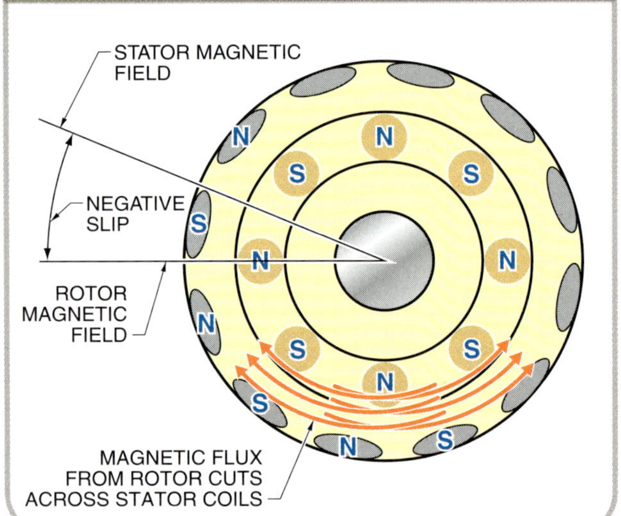

Figure 5-13. When an IM operates as a generator, slip is negative and the stator field lags behind the rotor.

Designers use the kinetic energy stored in the moving vehicle and control the slip value so that the induction machine functions as a generator. In the case of a decelerating vehicle, the induction machine can capture the kinetic energy that would have normally been lost in braking system heat. It can convert this kinetic energy to electrical energy in order to assist the vehicle braking system or to charge the battery system.

When the vehicle decelerates, the rotor will always rotate faster than the stator in order to reduce propulsion torque and reduce vehicle speed. When rotor speed is permitted to exceed the stator speed, the motor can be used as a generator. When negative slip is used to brake the vehicle, it is called regenerative braking (regen). During regen, the drive axles provide torque to the rotor. The only method of manually defeating the regenerative mode in vehicle deceleration is to disable the regenerative mode by having the controller command zero slip.

To use an induction machine as a generator, the stator frequency is decreased to a value less than rotor rpm. Alternatively, the rotor rpm is driven above the stator frequency by the vehicle driveline. In either case, the rotor rpm is greater than the stator frequency.

During negative slip, the induction machine continues to supply current to the stator phases, even though the rotor is leading the stator field. With the induction machine continuing to supply power to the stator, the stator continues to supply power into the rotor. In regen mode, the battery system is used to excite (or power) the stator. With the rotor operating faster than the stator, the magnetic fields of the rotor bar now begin to cut across the stator coils. This induces voltage back into the stator coils through the insulated-gate bipolar transistor (IGBT) flyback diodes, and eventually to the battery system.

By controlling the negative slip angle, the output power of the stator can be controlled and used to recharge the battery pack. The rotor now operates in a similar manner as the primary winding of a transformer. The stator operates in a similar manner as the secondary winding of a transformer. The stator windings now become an output for electrical power. The amount of output power generated from the stator windings depends on the amount of negative slip and the IGBT ON/OFF time that is commanded by the controller/inverter. As negative slip (torque) increases, more total current and elevated voltage levels can be generated. Negative slip frequency percentages can range from –1% to –15%. As negative slip (torque) decreases toward 0%, the total voltage and electrical current levels are reduced.

Slip can be considered analogous to the electronic spark timing on an ICE. Electronic spark timing advances and retards ignition timing to optimize power at a given engine rpm and fuel delivery rate. The same concept holds true for the slip value in induction-machine control. If the stator magnetic field is advanced up ahead of the rotor and additional electrical current is provided to the stator coils, the induction machine produces more torque. If the stator magnetic field is retarded back toward the rotor's speed, the induction machine produces less torque. If the stator magnetic field is retarded back beyond the rotor's speed, the induction machine acts as a generator and charges the battery.

The electrical power system operates sequentially. Electrical energy starts when the system excites the stator. Next, the stator supplies power to the rotor. Then, the rotor supplies power back to the battery system. If electrical power is factored into the overall system operation, then the power provided by the battery system and the power generated by the rotor and transferred to the battery system during regen can be thought of as the sum of power.

From the perspective of regen, the sum of power is the difference of potential between electrical power that the battery system provides and the electrical power that the rotor generates and is transferred back to the battery pack (regen). For example, if the battery system is providing 300 V to the stator for excitation and the rotor produces 400 V in regen mode, then there will be a difference of 100 V between the stator and rotor. The rotor has 100 V more potential than the battery system. Therefore, the sum of power results in the rotor providing electrical current to the battery system.

Slip and Current Tables. A slip (lookup) table, located in the controller software, is designed to control how much torque a motor can develop. Slip and current tables are used by the control software to determine the correct rotor and stator magnetizing currents that are needed to achieve a desired torque. The design of the slip table is based on vehicle weight, acceleration pedal position, rpm input, electrical current (ampere) level, and other factors. The table is very similar to the ignition timing and fuel delivery maps (tables) used in ICE controls. Developing the table is a tedious task and requires that the controls engineer model the slip table before actually loading it into the induction-machine controller. **See Figure 5-14.**

Both slip and current tables are designed to determine the correct rotor and stator magnetizing currents to achieve desired torque. An electric propulsion system can be designed to achieve and control a system for maximum torque or maximum efficiency. Software tables can be designed to use torque and efficiency calculations. If high torque and acceleration are commanded, the software algorithm can be designed to provide maximum torque without regard to efficiency. If efficiency is the primary goal of a control strategy, then torque and acceleration performance may be sacrificed to ensure that efficiency criteria are met.

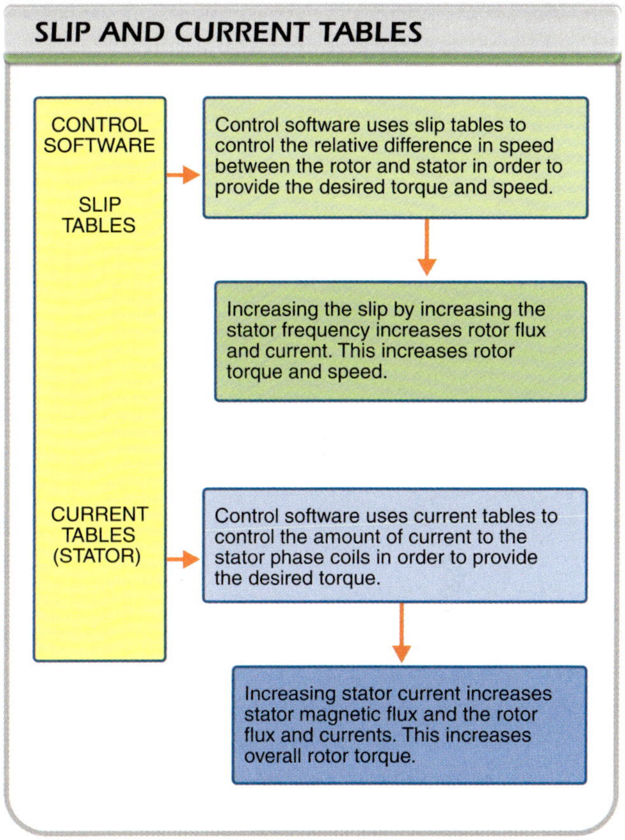

Figure 5-14. Slip and current tables are used by the control software to determine the correct rotor and stator magnetizing currents to achieve a desired torque.

Torque

An electric motor far outperforms the ICE in torque production at low and intermediate speeds. For comparison, a typical ICE may produce about 275 lb-ft of torque at 5000 rpm. For example, a typical induction machine can produce about 110 lb-ft to 150 lb-ft of torque from 0 rpm to 5000 rpm. Typically, the induction machine has a very flat torque curve from 0 rpm to its top end. This is a real advantage because high torque can be produced when it is required at the low rpm level. Additionally, the torque can be multiplied with a transmission/gearbox or differential.

The speed and torque generated by the rotor is based on the stator rotating frequency and the amount of electrical current supplied to each stator pole coil. The magnetic strength of each coil is controlled by the amount of electrical current in the coil. The strength of the induced-rotor magnetic field changes as the strength of the stator magnetic field changes. In effect, stator current is transferred to the rotor through the interaction between the magnetic fields of the rotor and stator.

Electric motors outperform the internal combustion engine in torque production at low and intermediate speeds.

As more stator current is supplied to the rotor, the electric machine produces more torque, or power. A cranking motor produces torque to generate motion and an alternator consumes engine torque to generate electricity. In an induction machine, the stator consumes torque and the rotor produces torque.

The speed in which the stator field rotates around the circumference of the stator ring is known as synchronous speed. The speed, or frequency, at which the induction-machine controller triggers stator coil poles is measured in hertz (Hz). The frequency at which the stator field cuts the rotor bars also increases the magnitude of the electrical current in the rotor bar. The frequency ultimately strengthens the magnetic coupling between the rotor and stator. An induction machine is categorized as an asynchronous electric machine because the stator coils can be triggered at different times and the speed of the stator field can be changed as needed.

Vehicle Braking with Negative Slip

An induction machine can be used for vehicle braking. When the induction machine is in generating mode, power is consumed from rotating the rotor in the stator magnetic field. Power comes from the motor as it converts the kinetic energy of the vehicle into electrical energy. This, in turn, loads the axles/tires with a reverse torque, which results in braking.

An induction machine can be used along with an antilock braking system (ABS) and traction control system (TCS) to optimize vehicle braking and torque management. This application reduces the normal wear and tear on the braking system. In addition, the battery pack can be recharged during braking or coasting. This assists in recovering lost energy in the battery pack and replenishing the energy and power levels in the battery modules.

The actual amount of energy recovery depends on the regen calibration. Blending of regen with the vehicle braking system must be carefully implemented in order to provide optimum vehicle control and efficiency. Since an induction machine is capable of generating hundreds of volts and amps, the controller must use input from the battery pack and brake system controllers. The amount of regen resides within the controller software algorithms.

Vehicle Reversing

The mechanical action of reversing the direction of the rotor is very simple. Since the rotor follows the stator magnetic field, the rotor turns in the reverse direction if the stator field is reversed. An induction machine can rotate just as fast in the negative (reverse) direction as it can in the positive (forward) direction. This is a significant advantage in hybrid vehicle design because the rotation direction can be changed by using a simple software algorithm and a line of software code. Therefore, a transmission is not necessary to reverse a hybrid vehicle.

Propelling a vehicle in the reverse direction has significant safety issues. Therefore, the stator frequency and slip must be controlled to limit the reverse speed to a safe level. For example, a reverse command could be restricted to 30 mph by limiting the allowable slip and stator current.

Maximum RPM

The induction machine in an automotive propulsion system can operate at a very high rotational speed. The maximum ratings normally range from about 9500 rpm to about 16,000 rpm. The maximum speed is based on the number of poles per phase and the maximum frequency of the stator field. The formula used to calculate the rpm of an induction machine is as follows:

$$rpm = \frac{(120)(Hz)}{P}$$

where

rpm = rotational speed, in rpm
Hz = stator field frequency, in Hz
P = number of poles per phase in the stator

The stator frequency found in a contemporary electric machine used for propulsion and power generation is about 1 Hz to 800 Hz and is designed with 4 poles per phase or more. A baseline maximum rpm can be calculated from these numbers. For example, when the maximum stator field frequency is 400 Hz with 4 poles per phase, the maximum rotational speed of 12,000 rpm is calculated as follows:

$$rpm = \frac{(120)(400)}{4}$$

$$rpm = \frac{48{,}000}{4}$$

$$rpm = \mathbf{12{,}000}$$

An advantage of a high rotational speed is that only limited gearing may be needed. In some cases, gearing may not be necessary. The driveline costs can be reduced if the gearing can be designed with a constant ratio and no gear shifting.

Vector Motor Control Strategy

Slip motor control is based on the relative speed, or angle positions, of the stator and rotor. Given system sensor inputs, a slip table is used to determine the desired magnetic field position of the stator. Although slip motor control offers excellent torque control, some designers prefer vector motor control because it offers a better mathematical solution for motor control. The efficiency of vector control is about equal to that of slip control, however, vector control has more precise torque control.

A vector has a direction and magnitude. In motor control, a vector direction is the angle at which the magnetic field is directed and the vector magnitude is the strength of the magnetic field at a particular angle. The direction, rate of change, and magnitude of the magnetic field determine the speed of the rotor and overall torque of the motor.

Slip and vector control have almost identical hardware requirements. However, it is more difficult to implement vector control than slip control. Vector control does not use slip tables and is much more calculation-intensive than slip control. It may require the addition of a digital signal processor (DSP) to operate it in conjunction with a standard microprocessor. Some advanced microprocessors have on-board DSP capability.

A control algorithm uses a mapping system as a vector to convert the 3ϕ power into a two-axis coordinate system. The control strategy for the motor of an AC induction machine is very similar to that of a DC motor when using a 3ϕ to two-axis coordinate transform system. The integrator software algorithms, along with DSP calculations, trigger power electronics in the power inverter. This generates sine waves at the optimum intervals of a 360° circle to achieve the most efficient motor speed and torque control application. **See Figure 5-15.** The illustration shows six vectors, with magnetic flux shown as an arced yellow line. The vector coordinates are coded with a binary number for use by the integrator. The DSP merely uses the table input commands from sensors to execute the proper sine coordinates.

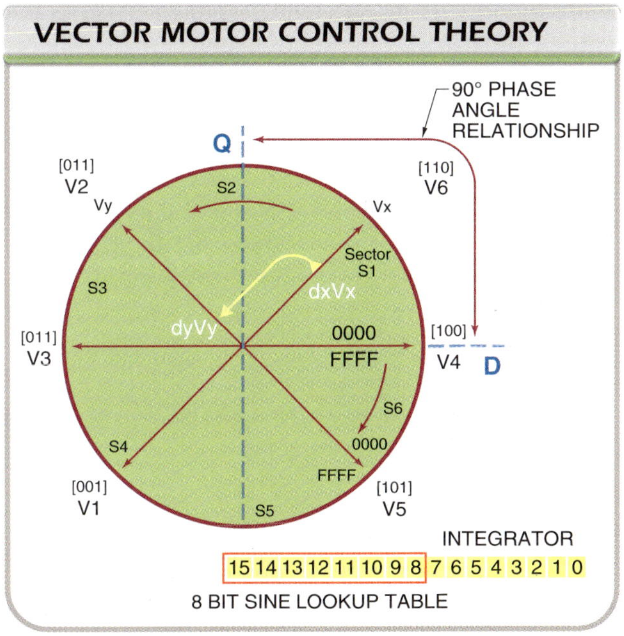

Figure 5-15. A vector algorithm uses a mapping system as a vector to convert 3ϕ power into a two-axis coordinate system.

The 3ϕ to two-axis transform calculation functions the same for the motor of an AC induction machine as it does for a DC motor. An example of a DC series motor is the traditional cranking motor for an ICE. In a series cranking motor, the two major components are the armature and the field windings. These two components are wired or connected in series.

As with any series circuit, both components receive the same electrical current. The vector control algorithm transforms the controls of the 3ϕ motor into a simple DC motor control strategy. By mapping the three phases into a two-axis coordinate system, the current of the 3ϕ induction machine can be broken down into two components. The first component is the magnetizing component (ID) that is analogous to a DC motor field winding circuit. The second component is the torque component (IQ) that is analogous to a DC motor armature circuit.

By controlling the ID and IQ, the control of the AC induction machine is treated very much like the control of a DC machine. For example, to increase the torque of an induction machine, the IQ component (armature current) is increased. In most cases, the ID current component (field winding current) is held constant to the point of field weakening.

Field weakening occurs at the point at which the rotor output voltage, or counter voltage, exceeds the voltage provided by the stator magnetic field. Therefore, increasing the electrical current strengthens the magnetic field strength of the stator. This reduces the effects of counter-voltage output of the permanent magnet rotor and thereby increases torque output.

In theory, torque of the DC motor is at its maximum when the angles of the armature and field winding magnetic field are at right angles (90°) to each other. The vector 3ϕ to two-axis coordinate control system calculates the vectors based on the 90° angle theory of a DC motor.

Therefore, the motor of an induction machine using vector control is very similar to a simple DC motor. Vector control treats the rotor and stator as a 90° angle system to generate maximum motor torque. To generate maximum torque in a DC series motor, the position of the armature is at 90° with respect to the field winding magnetic field. In the vector control system, the 90° angle is also important. Both the direction and magnitude of the 90° angle determine torque and speed of the motor.

Vector control must be tuned to the rotor of a particular motor. The accuracy of vector control depends on the rotor time constant. When a capacitor or inductor is placed into a circuit, it takes a specific amount of time for the capacitor to charge/discharge or the inductor to build/collapse its magnetic field. With an induction machine, the inductive and small capacitive time constants determine the time it takes for electrical current to flow within the rotor. This determines the overall control of the torque and speed of the rotor. A time constant is a topic that is typically discussed during a study of capacitors and inductors.

Refer to the *Hybrid Electric Vehicle Technology* CD-ROM for a Quick Quiz® on the chapter contents.

POINTS TO REMEMBER

- A 3ϕ motor has higher torque than a comparable 1ϕ motor.
- A rotor is the rotating part of an electric machine. The rotor bars are typically constructed of copper windings or aluminum placed in slots cut into the laminations.
- A stator is the nonrotating part of an electric machine that encloses the rotor. Copper conductors are wrapped in coils and placed within the slots cut into the laminations.
- Stators are typically wired in wye or delta configurations.
- Induction electric machines have no contacting components.
- Mutual induction causes current flow in the conductor if a magnetic field is passed across, or cuts, a conductor.
- The north and south poles induced in a rotor are polar opposites of the original field.
- The amount of torque in an electric machine depends on the stator current and the slip.
- Asynchronous operation is where pulsing of the stator poles is independent of the rotor speed or position.
- Stator frequency operation can generally range from 1 Hz to 800 Hz.
- Stator frequency and the number of stator poles determine the maximum speed of an induction electric machine.

- Slip is the relative speed difference between the stator and rotor.
- A slip table for controlling the speed and torque of the induction electric machine is very similar to ignition timing and fuel control tables for controlling an internal combustion engine speed and torque.
- Slip control can be categorized as positive, zero, and negative slip.
- Typical slip percentages for an induction electric machine are between 0% and 15%.
- The two major types of induction machine control strategies are slip and vector control.
- An electric machine can be reversed by simply reversing the direction of rotation of the stator magnetic field.
- An electric machine can have the same torque and speed in reverse rotational direction as it can in the forward rotational direction.

Refer to the *Hybrid Electric Vehicle Technology* CD-ROM for the Review Questions in Microsoft® Word format.

REVIEW QUESTIONS

1. What is the difference between positive and negative slip?
2. Explain how induction motor operation resembles transformer operation.
3. Explain why an electric propulsion system does not need a transmission to propel a vehicle in reverse.
4. How does an induction machine operate when the controller requests traction control because one of the drive wheels is slipping on ice?
5. How can maximum speed (in rpm) be determined?

5 AC Induction Electric Machines

Trade Test

Name ______________________________ Date ______________

______________ **1.** All electric traction systems for hybrid, fuel-cell, and electric vehicles include a(n) ___.
A. HVAC compressor
B. electric steering system
C. power inverter
D. by-wire braking system

______________ **2.** All induction machines include a ___.
A. differential and rotor
B. rotor and stator
C. transmission and stator
D. rotor and brushes

______________ **3.** The rotor of an induction machine is constructed from stacks of thin iron pieces called ___.
A. laminations
B. core bars
C. splined bars
D. washer rings

______________ **4.** The rotor bars in induction machines are typically constructed of ___.
A. copper and steel
B. copper and brass
C. aluminum and steel
D. aluminum and copper

______________ **5.** An electric machine stator can be constructed in ___ configurations.
A. wye and delta
B. wye and dual
C. delta and double-layer
D. delta and cross-wound

______________ **6.** The operation of an induction machine is similar to the operation of a ___.
A. relay
B. transformer
C. transistor
D. power inverter

_______________ **7.** A 3ϕ induction machine with 4 poles per phase has ___ total stator poles.
A. 2
B. 4
C. 12
D. 24

_______________ **8.** An induction machine operates with a(n) ___ stator that triggers a timing control strategy.
A. synchronous
B. six-step
C. asynchronous
D. axial

_______________ **9.** ___ is the electrical principle that forms the basis for induction machine stator and rotor operation.
A. Mutual induction
B. Self-induction
C. Maxwell angling
D. Oersted vectoring

_______________ **10.** The torque of an induction machine is controlled and determined primarily by waveform ___.
A. amplitude
B. frequency
C. shape
D. domain

_______________ **11.** As slip increases in an induction machine, ___ also increases.
A. speed (rpm)
B. regenerative braking efficiency
C. refractive magnitude
D. torque

_______________ **12.** To place an induction machine in a regenerative braking mode, the controller executes a ___-slip torque command.
A. positive
B. negative
C. null
D. reverse

_______________ **13.** When an induction machine is functioning in regenerative braking mode and generating electrical power, the rotor speed ___ the stator frequency speed.
A. is faster than
B. is slower than
C. is the same as
D. alternates faster or slower than

HYBRID Electric Vehicle TECHNOLOGY

Permanent Magnet Electric Machines

6

A permanent magnet (PM) electric machine is a core component of an electric, hybrid, or fuel-cell vehicle. Permanent magnet machines are similar to induction motor machines except for the use of permanent magnets. The software control of a permanent magnet stator is simpler than induction motor stators because of the synchronous conduction angles. Simpler calculations are needed to control synchronous machines. In a permanent magnet machine, the magnetic field angles are fixed. More predictable control strategies can be generated without the inefficient commutator and brushes used with series brush motor designs.

CHAPTER OBJECTIVES

- Explain how the rotor of a permanent magnet machine is different from the rotor of an induction machine.
- Describe the shape and function of a permanent magnet as it relates to generating torque.
- Explain how torque is controlled by varying the stator electrical current.
- Explain why a permanent magnet (PM) electric machine is categorized as a synchronous machine.
- Define field weakening.
- Explain how a field-weakening control strategy works to control regenerative braking power.

KEY TERMS

back electromotive force (EMF): The voltage generated by an electric machine during regenerative braking mode that results in electric current generation.

synchronous field control: A control system that causes the rotor and stator fields to interact at consistent and repeatable angles through the 360° circle in which the rotor revolves.

field weakening: The point at which the rotor output voltage (back EMF) exceeds the voltage provided by the stator-pole magnetic fields. Additional electrical current is delivered to the stator poles to increase the stator magnetic field strength and overcome the back EMF.

PERMANENT MAGNET BRUSHLESS DC MOTORS

Permanent magnet (PM) electric machines are used in electric propulsion systems. They are part of the brushless direct current (BLDC) motor family. Therefore, PM electric machines can be referred to as PM/BLDC machines. PM/BLDC machines are similar to both induction motors and DC brush motors.

With induction machines, the rotor becomes magnetized when the stator magnetic field cuts across the rotor and induces a magnetic field into the rotor bars. This creates a magnetic coupling between the stator field and the induced magnetic field around the rotor. With PM/BLDC machines, the rotor is manufactured with extremely strong permanent magnets mounted on the rotor housing. A PM/BLDC motor is designed without the brushes or commutator that are used in traditional DC motors.

Brushless DC Motor Construction

In a PM/BLDC machine, permanent magnets take the place of the rotor bars used in other motor designs. A PM/BLDC motor, as with an induction motor machine, has a stator and housing assembly and can be liquid- or air-cooled. **See Figure 6-1.** The stator is very similar to the stator of a 3ϕ induction motor. The major difference between PM and induction machines is in the rotor design.

Rotors. The magnets create the rotor magnetic field instead of mutual induction from the stator field. The magnet placement on the rotor determines the timing of interactions with the stator field.

BRUSHLESS DC MOTORS

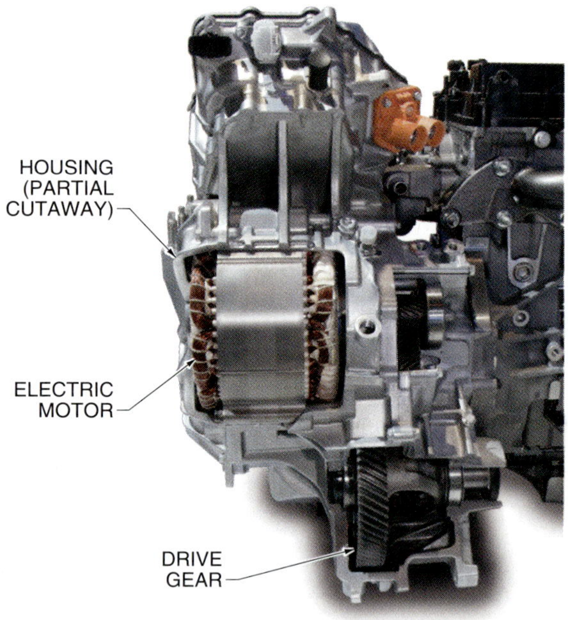

Figure 6-1. Permanent magnet motors are brushless DC motors used in electric propulsion systems.

The rotors of PM motors can have several different magnet design configurations. A common design has the magnets mounted in a V-shaped arrangement. **See Figure 6-2.** The V-shaped magnet arrangement is used in both the 2004 Toyota Prius and the Ford Escape hybrid models. This arrangement creates an air gap between the magnets to reduce the effects of the back electromotive force produced by the rotor at higher speeds. *Back electromotive force (EMF)* is the voltage generated by an electric machine during regenerative braking mode that results in electric current generation. This air gap also reduces torque ripple at low-speed/high stator-current settings.

ROTOR DESIGNS

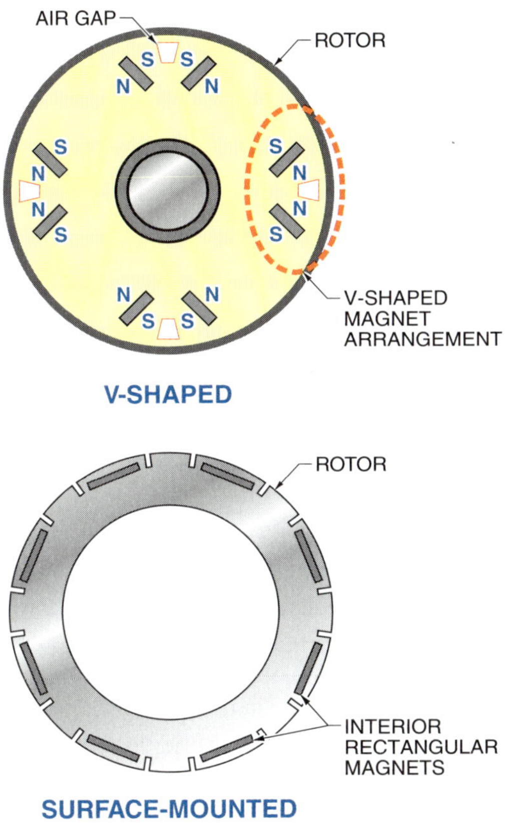

Figure 6-2. A common rotor design uses magnets mounted in a V-shaped arrangement. An alternative design uses a surface-mounted magnet per pole.

Earlier Toyota Prius models used a different rotor magnet design. This design used one interior, rectangular magnet per pole. A common design uses an 8-pole rotor and a 3ϕ, 8-pole stator for a total of 24 poles. Both the Prius drive motor and the generator are 8-pole electric machines. The surface-mounted magnets provide higher efficiency and higher torque because they are closer to the stator coils. However, this design also generates higher torque ripple and a higher back EMF.

> Chrysler Corporation is credited with the first use of the alternator in 1960. Alternators were developed because of the increased electrical loads in the vehicles of the time. DC generators could not provide enough electricity, especially at low engine speeds.

Permanent Magnets. Some of the more popular materials used to make magnets are alnico, ferrite (ceramic), rare earth samarium-cobalt, and neodymium-iron-boron (NdFeB). Neodymium magnets are more popular in high performance applications because they are less expensive than samarium-cobalt magnets. The NdFeB magnet offers the highest energy density and is the most cost effective. However, its temperature must be held well below about 300° F (about 150° C) or the magnet can be permanently damaged. Different magnets have different requirements, but it is important to maintain proper temperature control for all PM materials.

The magnets are mounted on the inside (interior mounted) or on the outer circumference (surface mounted) of the rotor housing and are evenly spaced around the circumference. The number, shape, and strength of the magnets can vary significantly from one motor design to another depending on the torque and speed requirements of the PM machine.

Brushless DC Motor Operation

Since PM machines use magnets instead of rotor bars, there is a constant magnetic field to interact with the stator magnetic field. PM machines do not depend on mutual induction to induce magnetic fields into the rotor in order to produce rotation. The poles of the magnets interact with the stator magnetic field and follow the field as it rotates. The rotor follows clockwise or counterclockwise to allow forward and reverse traction.

A PM/BLDC machine is a synchronous machine. Therefore, a PM/BLDC machine uses synchronous field control. *Synchronous field control* is a control system that causes the rotor and stator fields to interact at consistent and repeatable angles through the 360° circle in which the rotor revolves. By comparison, an induction motor machine operates in asynchronous mode. The controller can trigger the stator phases at any point in the 360° control area. In other words, the triggering of the stator field in an induction machine is continuously variable in relation to the rotor.

Rotor and Stator Interaction. A fundamental property of magnetism is that like poles repel and unlike poles attract. The stator magnetic fields in a PM/BLDC machine function much in the same manner as the stator in the induction machine. The controller switches the stator field ON at the desired time. The stator poles attract and repel the rotor magnet poles causing the rotor to turn. In the PM/BLDC machine, the direction of the rotor is determined by the direction of the stator pole rotation. The rotor magnets chase the stator field when producing positive torque in drive mode and lead the stator field when producing negative torque in regenerative mode.

A main area of concern in motor design is the air gap between the rotor and stator. **See Figure 6-3.** The magnetic lines of force must cross the gap to allow the rotor and stator to interact. As air gap increases, the magnetic field has a farther distance to travel. This weakens the interaction between the rotor and stator magnetic fields. The net effect is less power for propulsion.

As a PM/BLDC electric machine rotates, it produces electrical power. When the speed increases, the voltage increases. Eventually, the voltage reaches a point where the machine torque and speed will diminish. Field-weakening control strategies are used to overcome torque and speed loss at high speeds. *Field weakening* is the point at which the rotor output voltage (back EMF) exceeds the voltage provided by the stator-pole magnetic fields.

AIR GAP

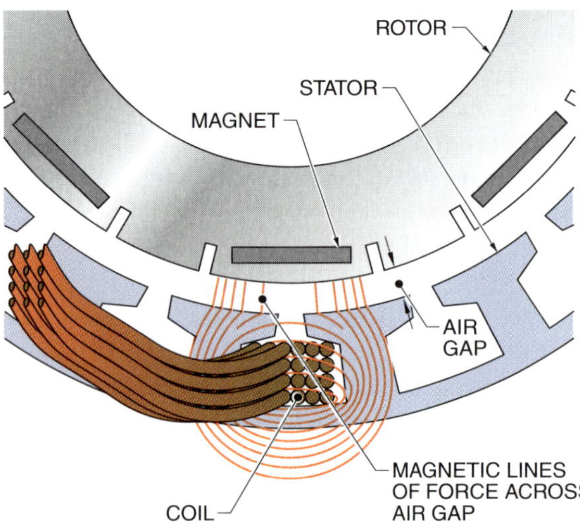

Figure 6-3. The magnetic lines of force must cross an air gap to allow the rotor and stator to interact.

The electric machine can be installed between the engine and transmission.

At this point, if the operator desires more speed or torque, the control system provides more stator current to increase the winding magnetic field strength. This counteracts the rotor back EMF and allows increased speed and torque. The stator current can be increased up to the point where the battery pack can no longer supply the energy needed to support field weakening. Therefore, magnet strength determines the design of the field weakening strategy for a PM/BLDC machine.

Torque and Speed Control. A PM/BLDC machine uses stator phase-advance or phase-retard sequential triggering angles for speed and torque control. The stator poles will lead, or advance, the rotor magnets to produce positive torque in drive mode. The stator poles will lag, or retard, the rotor to produce negative torque in regen mode.

In a PM/BLDC machine, switching occurs at the point when the stator is turned ON. Stator switching points occur at the same distance apart in an electrical cycle of 360°. **See Figure 6-4.** At the switching points, the motor controller triggers electrical power to the stator coils to generate speed and torque for propulsion. This control operation is very similar to electronic spark control systems that advance and retard spark based on changes in speed and torque requests of the operator.

Field strength of the rotor is fixed and field strength of the stator is variable. Therefore, the field strength of the stator is used to control the torque. When the electrical current flow from the power inverter to the stator increases, the magnetic field strength increases and more torque is produced. The torque limit is determined by the amount of current that can be delivered by the battery pack.

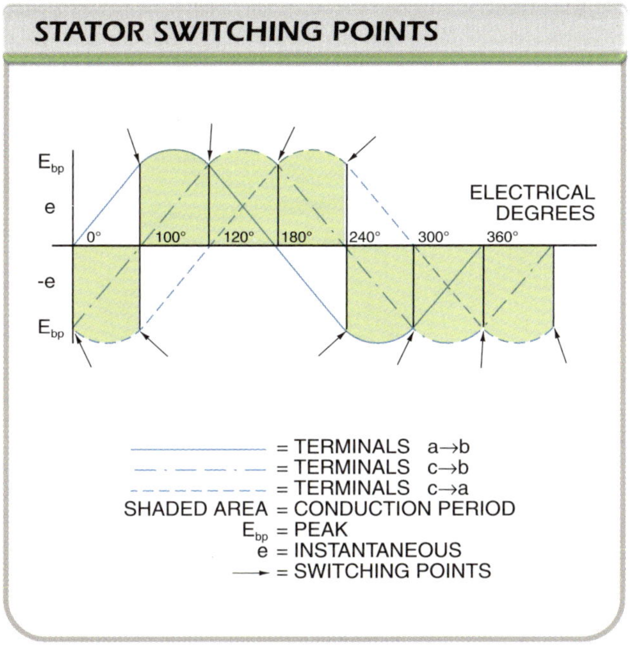

Figure 6-4. Stator switching points occur at the same distance apart in an electrical cycle of 360°.

The frequency of the stator determines the speed of the machine. When the speed increases where the three stator phases are switched, the rotational speed of the rotor increases as the rotor follows the stator. This increases motor output speed. Regardless of rotor speed, the relative locations for waveform switching remain at the same points on the circumference of the stator. As a result, the control algorithms and software configurations for a PM/BLDC machine are simpler than those of an induction machine.

With synchronous operation, the conduction angle between the stator and rotor remains the same on each cycle. **See Figure 6-5.** The stator field windings are always at the same position and the rotor magnets are at fixed points on the rotor. As the rotor spins through a 360° circle, the magnets rotate through the same angles with respect to the stator pole coils. With the stator pole coils stationary, and the location of rotor magnets determined by design and manufacturing, triggering of the stator windings can be synchronized at an optimum angle for the desired torque output.

Traction Control and Regeneration. The operation of a PM/BLDC machine is very similar to an induction machine in the areas of traction control, regen, and ABS. The stator voltage and current are controlled to generate the positive, zero, or negative torque outputs. These outputs are used to assist in controlling propulsion, regen, and blended braking. Blended braking is a braking method that combines the hydraulic brake system and the electric motor to stop a vehicle.

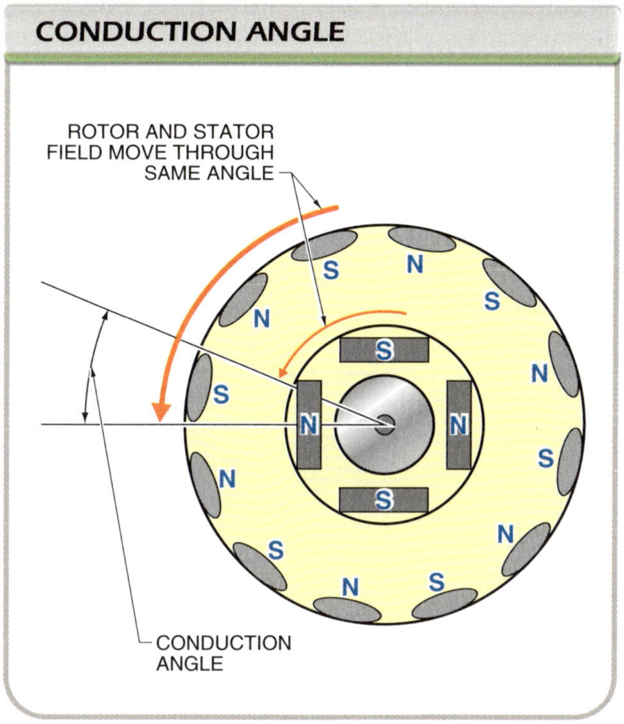

Figure 6-5. The conduction angle between the stator and rotor remains the same on each cycle.

Flux vector control for a PM/BLDC machine controls the rotational speed of the stator field. It compares a theoretical optimized slip calculation to the actual motor performance. The controller adjusts motor torque and speed based on the flux vector and slip calculation variables. **See Figure 6-6.**

The regen function of a PM/BLDC machine has different technical control issues than an induction machine. One of the dangers of regen in a PM/BLDC machine is that the stator always develops power. This is because the permanent magnets in the rotor continuously cut across the stator field windings. The power inverter will continue to provide electrical power to the PM/BLDC machine until the point of field weakening. This is the point at which the PM/BLDC motor provides more electrical power output than the input that the power inverter provides to the PM stator.

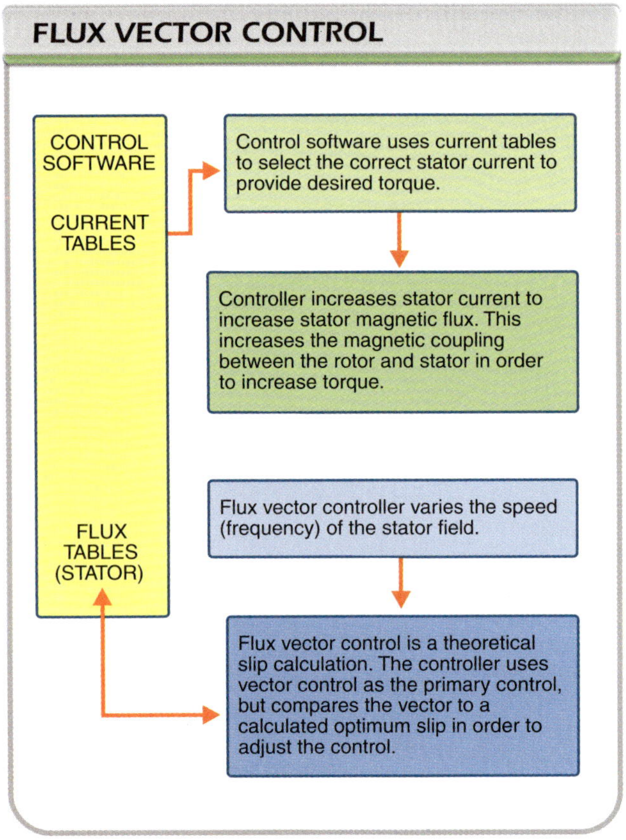

Figure 6-6. The flux vector controller adjusts motor torque and speed based on the flux vector and slip calculation variables.

If the operator requests more speed from the motor at this point, the controller must provide more electrical current from a generator or battery pack to the stator pole coils. This current is used to overcome the back EMF that is being generated by the PM rotor magnets. This is necessary so that the stator pole coils generate a magnetic field strong enough to overcome the effects of the magnetic field of the rotor magnets.

Once the rotor begins to produce more power than the stator, torque output begins to drop significantly. In addition, operating a PM electric machine past the point of field weakening is very inefficient because of the increased levels of electrical power required to operate the PM machine in the field-weakening region.

General Motors

The electric machine can be installed in parallel to the engine.

To control regen power, the power inverter transistors must be modulated, or turned ON and OFF, with different pulse width commands. This controls the power that is transferred through the transistor flywheel/flyback diodes in the IGBTs. By modulating the transistors, the battery pack supplies a counter voltage to the stator phases. This results in a lower regen output with the output current sent through the flywheel/flyback diodes to the battery system.

The controller uses a slip calculation to control regen. The controller designates the on-time of the transistors to take advantage of the linear power output of the motor when in regen. This supplies the necessary electrical power to the battery module(s) in the battery system. Field weakening may also be used with slip control during specific regen scenarios, such as prolonged downhill operation, to reduce or eliminate regen capability of the PM/BLDC electric machine. This strategy is used to ensure that the battery-pack system is not damaged from being overcharged. Finally, because the output of the PM/BLDC is linear, calculating regen is a simple matter of using voltage, current, and a time plot. Calculating regen also involves tracking the battery-system voltage level with the battery-pack controller.

Refer to the *Hybrid Electric Vehicle Technology* CD-ROM for a Quick Quiz® on the chapter contents.

POINTS TO REMEMBER

- The shape, length, width, thickness, and number of magnets determine the torque and maximum speed of a PM electric machine.
- Magnet temperatures must be held within a specified tolerance to minimize damage and maintain magnetic strength.
- Air gaps between the stator and rotor are kept to a minimum to ensure that magnetic interactions are strong.
- Magnetic field strength weakens as an air gap increases.
- Rotor magnets are placed either on the surface or in the interior of the rotor.
- V-shaped, interior-rotor magnet design can reduce back EMF.
- Stator poles lead, or advance, the rotor magnets to produce positive torque in propulsion mode.
- Stator poles lag, or retard, behind the rotor to produce negative torque in regenerative mode.
- The stator-pole electrical current is increased to increase the torque of a PM electric machine.
- The stator frequency is increased to increase the speed of a PM electric machine.
- The control system of a PM electric machine uses a field-weakening strategy to overcome torque and speed loss at high speeds.

Refer to the *Hybrid Electric Vehicle Technology* CD-ROM for the Review Questions in Microsoft® Word format.

REVIEW QUESTIONS

1. What is the difference between the rotor design of an induction machine and a PM/BLDC machine?
2. Explain why a PM/BLDC machine is categorized as a synchronous machine.
3. Define field weakening.
4. How can a field-weakening control strategy be used to assist in controlling regenerative braking?
5. How does varying the electrical current of the stator control torque?

6 Permanent Magnet Electric Machines

HYBRID Electric Vehicle TECHNOLOGY

Trade Test

Name ______________________ Date ______________

______ **1.** A PM electric machine contains all of the following except a ___.
A. rotor
B. stator
C. magnet
D. commutator

______ **2.** In a PM/BLDC machine, an increase in stator current causes a(n) ___.
A. increase in torque
B. decrease in torque
C. increase in rotor speed
D. decrease in rotor speed

______ **3.** A V-shaped magnet arrangement in the rotor decreases the ___.
A. maximum rotor speed
B. minimum rotor speed
C. rotor diameter
D. back EMF

______ **4.** A PM/BLDC operates with a(n) ___ stator to trigger timing control.
A. synchronous
B. asynchronous
C. six-step
D. axial

______ **5.** Field weakening of a PM electric machine is necessary to overcome the back EMF created by the ___.
A. stator magnetic field
B. stator speed
C. magnets
D. flux vector angle

______ **6.** One method to control the regenerative braking power generated by a PM electric machine is to ___.
A. control the lag angle
B. control the lead angle
C. increase the stator field speed
D. decrease the stator field speed

_____________________ **7.** Another method to control the regenerative braking power generated by a PM electric machine is to ___.

A. increase the rotor speed

B. increase the field-weakening current

C. increase the lead vector

D. eliminate power-inverter signaling

_____________________ **8.** The Toyota Prius PM electric machine controls ___ magnet poles on its rotor.

A. 4

B. 8

C. 16

D. 24

Power Inverter Systems

7

Power inverters control the power that a battery pack transfers to the electric machine and the energy that electric machines generate and transfer back to the battery pack. A power inverter system serves as the central control point for the high-voltage system. In many respects, a power inverter system provides the same functionality to an electric machine as the fuel and ignitions provide to an internal combustion engine. The power inverter is responsible for varying the frequency, voltage level, and type of waveform to the electric machine to control its speed and torque.

CHAPTER OBJECTIVES

- Identify common locations for power inverters.
- Identify the types of transistors used in power inverters.
- Explain why current sensors are used for power inverters.
- List the different waveforms used to control an electric machine.
- Describe how sine and six-step waveforms are created.
- Demonstrate how to calculate rotor frequency and motor speeds.
- Describe the operation of a DC brush-type motor.
- Explain why power inverter boosting is used.
- Describe the operation of a six-pack motor drive system.
- Describe the operation of the regenerative braking system.

KEY TERMS

reactive power: Electrical power that is generated during regenerative braking mode when the magnetic fields of the rotor cut across the stator pole windings to produce electrical power that is transferred through the IGBT flyback diodes and to the battery pack.

flyback diode: A diode that is connected in the reverse bias position across the emitter and collector of an IGBT in order to steer reactive current during regenerative braking mode.

electric machine control waveform: A generated, periodic voltage level used to control the voltage level to an electric machine so that the torque and speed can be controlled. The four waveforms are the sine, trapezoidal, six-step, and rectangular waveforms.

torque ripple: Rough torque delivered by an electric machine caused by the use of waveforms that have aggressive changes in voltage, such as six-step and rectangular waveforms.

inductor: An electrical device that stores energy in the form of a magnetic field. In its simplest form, an inductor consists of a wire loop or coil. In an inductor, current lags 90 electrical degrees behind the voltage.

overmodulation: A technique used to increase the available voltage to a motor by using waveforms other than sine waves.

boost converter: A special DC-DC converter that is used within the power inverter to increase the voltage from the battery pack to the electric machine.

reactor: A power inductor within a power inverter system that can boost battery-pack voltage to a higher voltage level.

capacitor: An electrical device that stores energy in the form of an electrostatic field. In its simplest form, a capacitor consists of two conducting plates separated by an insulating material called a dielectric.

six-pack motor drive: A power transistor configuration that uses six IGBTs to control the torque and speed of an electric machine.

POWER INVERTERS

The motor drives used with induction motors and permanent magnet (PM) motors need to be designed to control motor speed and torque. Contemporary electric machines and controllers use the torque method for control. Torque is the output force of an electric machine that provides the mechanical rotation to crank the engine or propel the vehicle.

In the slip control algorithm, torque control is achieved by controlling the phase electrical current in the stator coils and by controlling the slip frequency. Phase voltage is adjusted to maintain the desired electrical current level. The controller software sets the magnetic field level of the stator coil and adjusts the stator frequency to maintain desired slip frequency in an induction motor or lead/lag speed in a PM machine.

The power electronic devices and battery pack are placed in the space between the rear passenger seat and the trunk in many contemporary hybrid electric vehicles such as the Honda Civic Hybrid. **See Figure 7-1.** In the Toyota Prius, the power inverter is placed under the hood. The power inverter system of the Ford Escape Hybrid is integrated into the transaxle assembly. Integrating the power inverter system into the transaxle makes it easier to assemble the vehicle. However, it is more expensive to repair and more difficult for the technician to isolate systems for analyses and testing.

Transistors

Power inverters use transistors in the control circuits as power switches. One of the first power transistors available to designers was the bipolar transistor. **See Figure 7-2.** The design of the bipolar transistor provided the benefits of a solid-state solution for many applications. However, its performance was limited by several undesirable characteristics.

A bipolar transistor requires a high base current in order to turn on, has relatively slow shutoff characteristics, and is susceptible to thermal runaway due to its negative temperature coefficient. The lowest attainable on-state voltage or conduction loss is governed by the collector-emitter saturation voltage.

The base current of bipolar transistors must be continuously supplied with about one-tenth of the collector current to maintain saturation. Saturation occurs in a bipolar transistor when an increase in base current will not cause an increase in collector current.

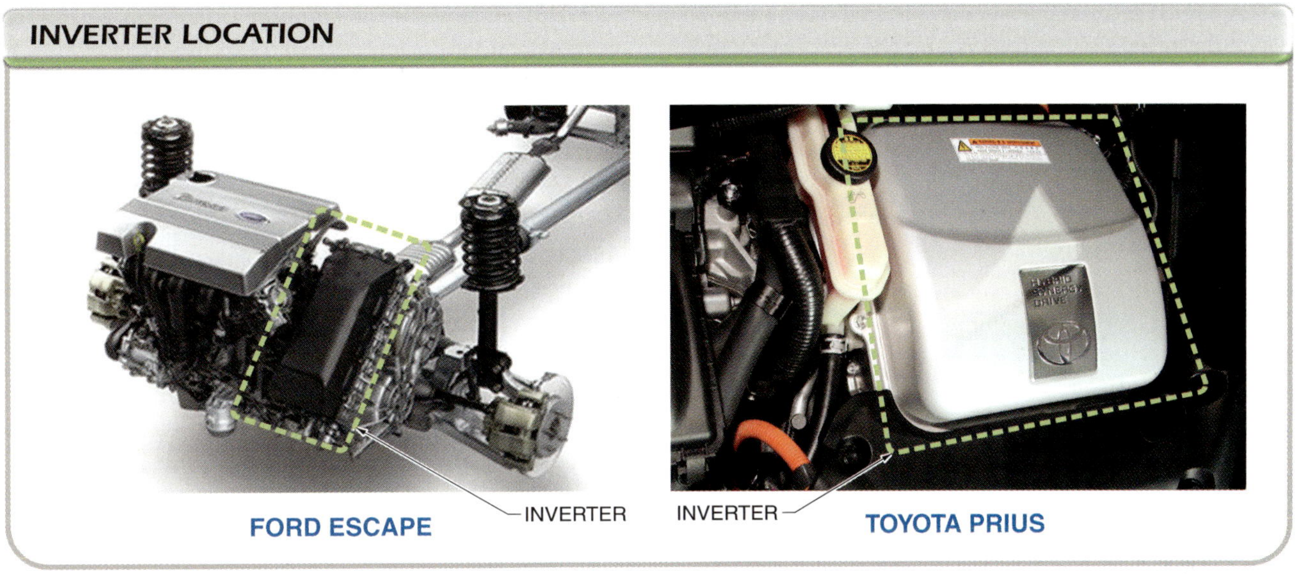

Figure 7-1. Inverters can be placed in different locations on different vehicles.

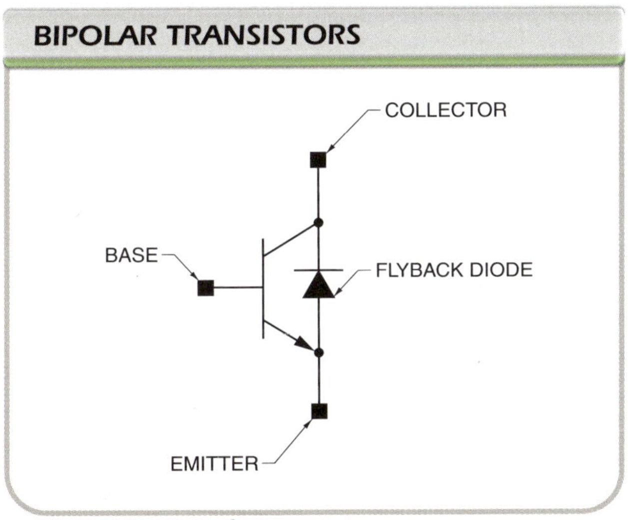

Figure 7-2. One of the first power transistors available to designers was the bipolar transistor.

Bipolar transistor drive circuits must be sensitive to variable load conditions. The base current must be kept proportional to the collector current. This prevents desaturation under high-current loads and prevents excessive base drive under low-load conditions. The high base current of a bipolar transistor causes excessive power dissipation in the drive circuit. Bipolar transistors are minority carrier devices. Charge storage effects, including recombination, slow the transistor performance when compared to the transistor performance of majority carrier devices like MOSFETs.

MOSFETs. The undesirable characteristics of bipolar transistors led to the development of the metal-oxide semiconductor field-effect transistor (MOSFET). **See Figure 7-3.** A MOSFET is controlled by voltage rather than current. It has a positive temperature coefficient that prevents thermal runaway. Furthermore, the on-state resistance of a MOSFET has no theoretical limit and its on-state losses can be much lower than those of a bipolar transistor.

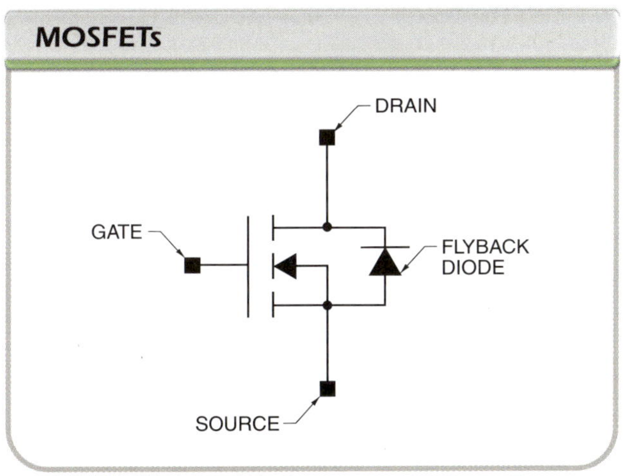

Figure 7-3. MOSFETs were designed as replacements for bipolar transistors.

A MOSFET also has a body-drain diode, which is particularly useful for redirecting the reactive power generated by collapsing magnetic fields. *Reactive power* is electrical power that is generated during regenerative braking mode when the magnetic fields of the rotor cut across the stator pole windings to produce electrical power that is transferred through the IGBT flyback diodes and to the battery pack.

IGBTs. In the 1980s, the insulated-gate bipolar transistor (IGBT) was introduced into the electronics market. An IGBT is a cross between a bipolar transistor and a MOSFET. **See Figure 7-4.** The IGBT has the output switching and conduction characteristics of a bipolar transistor but is voltage controlled like a MOSFET.

IGBTs

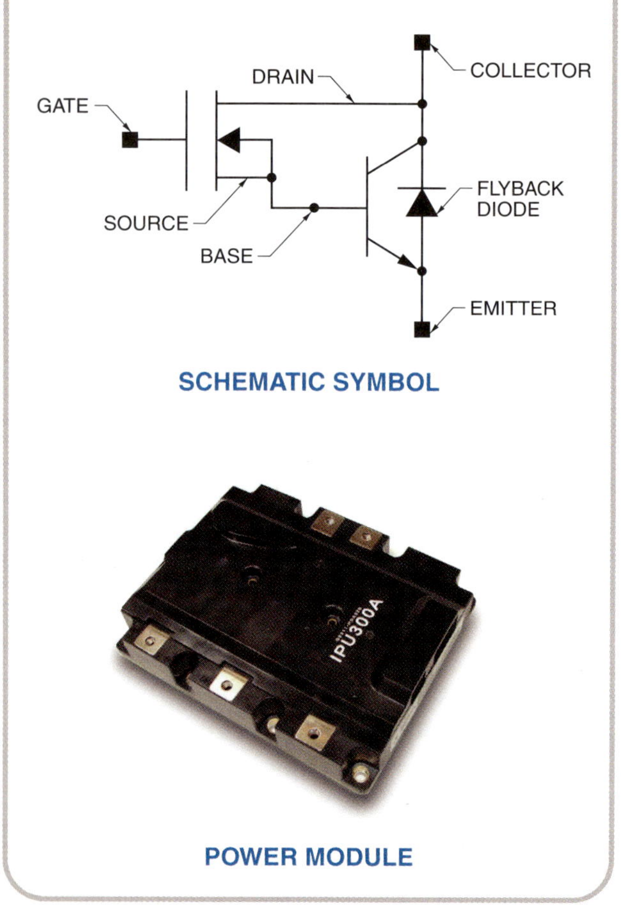

Figure 7-4. An IGBT has the output switching and conduction characteristics of a bipolar transistor but is voltage controlled like a MOSFET.

This means that it combines the high electrical current-handling capability of a bipolar transistor with the ease of control of a MOSFET. At operating frequencies between 1 kHz and 50 kHz, the IGBT offers an attractive solution to the traditional bipolar transistor and MOSFET.

An IGBT, like a MOSFET, can remain fully ON by maintaining the gate voltage above a certain threshold. It is possible to replace the MOSFET with an IGBT to improve efficiency and/or reduce costs. An IGBT uses less space than an equivalent MOSFET.

The structure of an IGBT resembles that of a power MOSFET. An IGBT has a high input impedance (resistance) and fast turn-on speed like a MOSFET. IGBTs exhibit an ON-voltage and current density comparable to a bipolar transistor while switching much faster.

IGBTs are replacing MOSFETs in high-voltage applications where conduction losses must be kept low. Since the structures of IGBTs and MOSFETs are similar, a design change to IGBTs can be accomplished without needing to redesign the gate drive circuit.

If an IGBT or MOSFET is used in a circuit, a freewheeling flyback diode must be added as a separate discrete component across the transistor. A *flyback diode* is a diode that is connected in the reverse bias position across the emitter and collector of an IGBT or MOSFET in order to steer reactive current during regenerative braking mode.

Gate Circuits. An IGBT has similar voltage and current ratings to a bipolar transistor. However, one of the more complex areas of the power inverter system, relative to controlling the IGBT, is the gate drive circuitry and fault handling. The presence of an isolated gate makes an IGBT simpler to drive than a bipolar transistor.

In an IGBT, the gate circuit drives the bipolar transistor section at 5 V to 15 V, generating extremely small currents. This is convenient because all of the amplifier sections that were required to drive a bipolar transistor are now no longer necessary with IGBT power module technology.

The control circuit, for example, consists of the microprocessor unit (MPU), an optical device to isolate the MPU from the IGBT, high voltage, electromagnetic interference, and temperature. Most IGBTs are manufactured as modules that have all of the gate drive circuitry built in. This simplifies the controller and requires fewer components on the controller circuit board.

The function of a fuel injection system can be used to describe the function of a control circuit. The injector of a fuel injection system is pulsed ON and OFF by a driver transistor located in the powertrain control module (PCM) or electronic control unit (ECU). The injector driver transistor must be turned ON and OFF by a pulse, which is delivered by the PCM/ECU. The controller in an electric propulsion system must deliver pulses to the gate of the transistors similar to that of an H-bridge circuit. The gate of an IGBT also can be considered similar to the base circuit element of a bipolar transistor.

As with the injector driver, the pulse width of the gate drive is controlled by the propulsion controller. Pulse width modulation (PWM) is 0% to 100%. The controller sets the PWM triggering of the gate circuit of an IGBT to control the amplitude and frequency of the AC signal sent to the induction/PM machine.

One of the most important aspects of the gate drive circuitry is the fault handling section. IGBT fault handling is similar to the quad driver module (QDM) circuits on a General Motors vehicle. Fundamentally, a QDM monitors the input signal coming from the PCM/ECU and compares it to the output line. The output line is usually connected to a relay, solenoid, or some other device that it must drive. If the input command does not match a specific threshold voltage on the output line, then a system fault code is set.

The gate drive circuit operates much in the same way for the IGBT. If a fault is detected in the IGBT drive circuit or the gate drive circuit, a system fault code is set. If a gate drive fault is commanded, the malfunction indicator lamp in the instrument panel is lit.

Power Inverter Design

A power inverter is the basic component of an electric propulsion system. A power inverter is controlled by a microprocessor that receives inputs from various input sensors. See Figure 7-5. Most of the input sensors are identical to the sensors that can be found in typical ICE emission and fuel control systems.

For example, the throttle position sensor system, speed (rpm) sensor, and brake sensor switch are typical sensing systems found on traditional ICE systems. Sensors unique to hybrid systems are the electrical current sensors. The power devices in a power inverter are usually IGBTs or MOSFETs. The power devices determine the power rating (in kW) of the motor device to be used. The basic design of a 3ϕ power inverter consists of the six power switches (transistors) configured as an H-bridge with gate drive circuits.

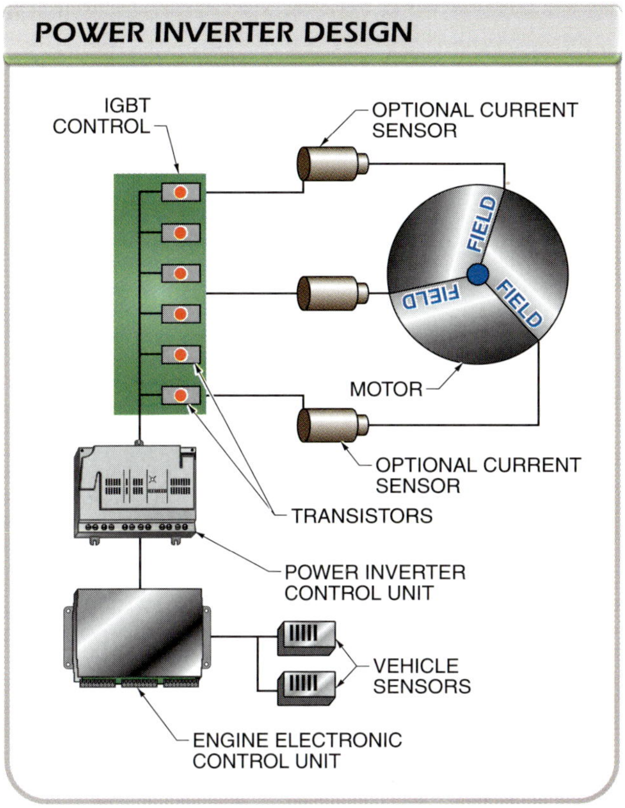

Figure 7-5. A power inverter is controlled by a microprocessor that receives inputs from various input sensors.

H-Bridge Circuits. H-bridge circuits are common in automotive applications. This type of circuit design can be found in any circuit that requires bidirectional rotation of an electric motor. This occurs whenever a motor must stop and hold a specific position.

For example, an H-bridge circuit can be found in the idle air control (IAC) circuit and automatic air conditioning circuits to control the idle speed of an ICE or the blend/defrost/air delivery doors. Both the IAC and air conditioning motors are bidirectional PM machines. Bidirectional operation is accomplished by turning ON the H-bridge transistors in tandem. See Figure 7-6.

The electrical current flow direction through the coil can be changed to allow the motor to run in two directions. By turning ON Q2 and Q3, the motor turns in one direction. By turning OFF Q2 and Q3 and turning ON Q1 and Q4, the motor turns in the opposite direction.

The induction/PM machine for a hybrid and electric vehicle is also bidirectional. The major differences between a power inverter and an IAC circuit are the amount of power being controlled in the circuit and how the power is applied.

H-BRIDGE CIRCUITS

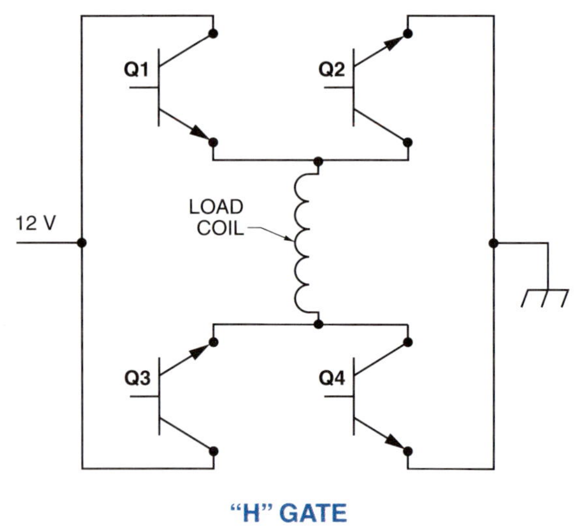

Figure 7-6. H-bridge circuits can be found in any circuit that requires bidirectional rotation of an electric motor.

MOTOR CONTROL

For a power inverter controller that uses a slip control strategy, electrical current and slip tables are calculated based on the desired torque for a given set of input signals. The input signal comes from the current, rpm, and throttle system sensors. For example, increasing electrical current increases torque when operating at a fixed slip frequency. Increasing slip frequency increases the electrical current in the rotor or increases the torque when operating at a constant electrical current.

Controllers use electric machine control waveforms to control electric machines. An *electric machine control waveform* is a generated, periodic voltage level used to control the voltage level to an electric machine so that the torque and speed can be controlled. The four electric machine control waveforms are the sine, trapezoidal, six-step, and rectangular waveforms. Waveforms are also referred to as "waves," such as sine waves. **See Figure 7-7.**

For induction/PM machines, implementing slip or phase angle control is not difficult. However, generating the slip or phase angle tables for induction machine motors is difficult to do because they are asynchronous machines. The difficulty is in determining the value of slip or phase angle for a given torque request. The relationship between the torque and slip depends on the motor design. Therefore, the slip and current tables need to be customized to the motor.

MOTOR CONTROL WAVEFORMS

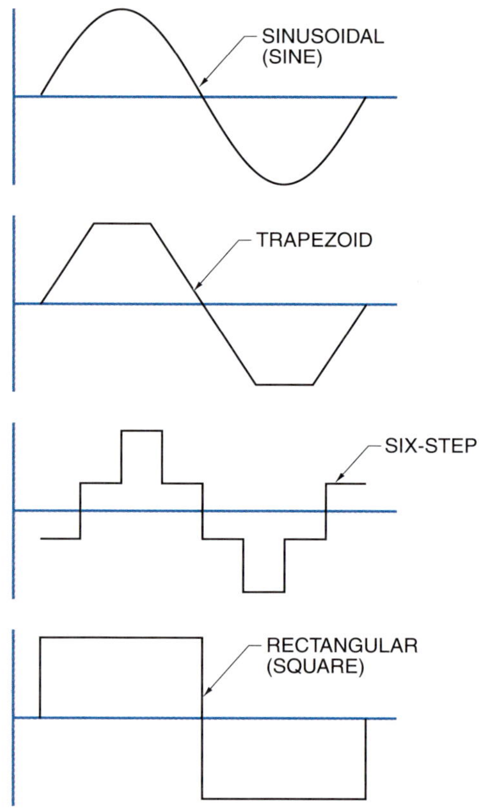

Figure 7-7. The four electric machine control waveforms are the sine, trapezoidal, six-step, and rectangular waveforms.

For PM machines, only a calculation for the amount of stator current at a given stator frequency is necessary because they are synchronous machines. The rotor in a PM machine maintains constant magnetic field strength. The controller varies the current in the stator pole coils to control the magnetic field strength. This is what controls the output torque and speed.

The induction/PM controller and inverter must be able to synthesize (create) AC power from a DC battery pack. The controller determines when control (triggering) events need to be executed and the power inverter changes the DC power into AC power. The controller must also control both the amplitude and frequency of the AC signals to manipulate motor torque and rotational speed.

Six-Step Motor Control

Six-step control is so named because of the square, six-stepped waveform that is generated when comparing the line-to-line or phase-to-phase voltages. When a pair of IGBTs is commanded to power two motor phases to rotate the motor, the controller commands succeeding pairs of IGBTs. When the controller energizes succeeding pairs of windings, the controller continues to use one of the IGBTs in the previous triggering cycle in the next switching cycle. **See Figure 7-8.**

This strategy is electrically very clean. Holding one IGBT ON for two switching events reduces the overall inductive loading (transient spiking) on each switching device. This reduces electrical transients that would invade the power electronics system and helps increase the life of the IGBT.

The power inverter system is often located underhood.

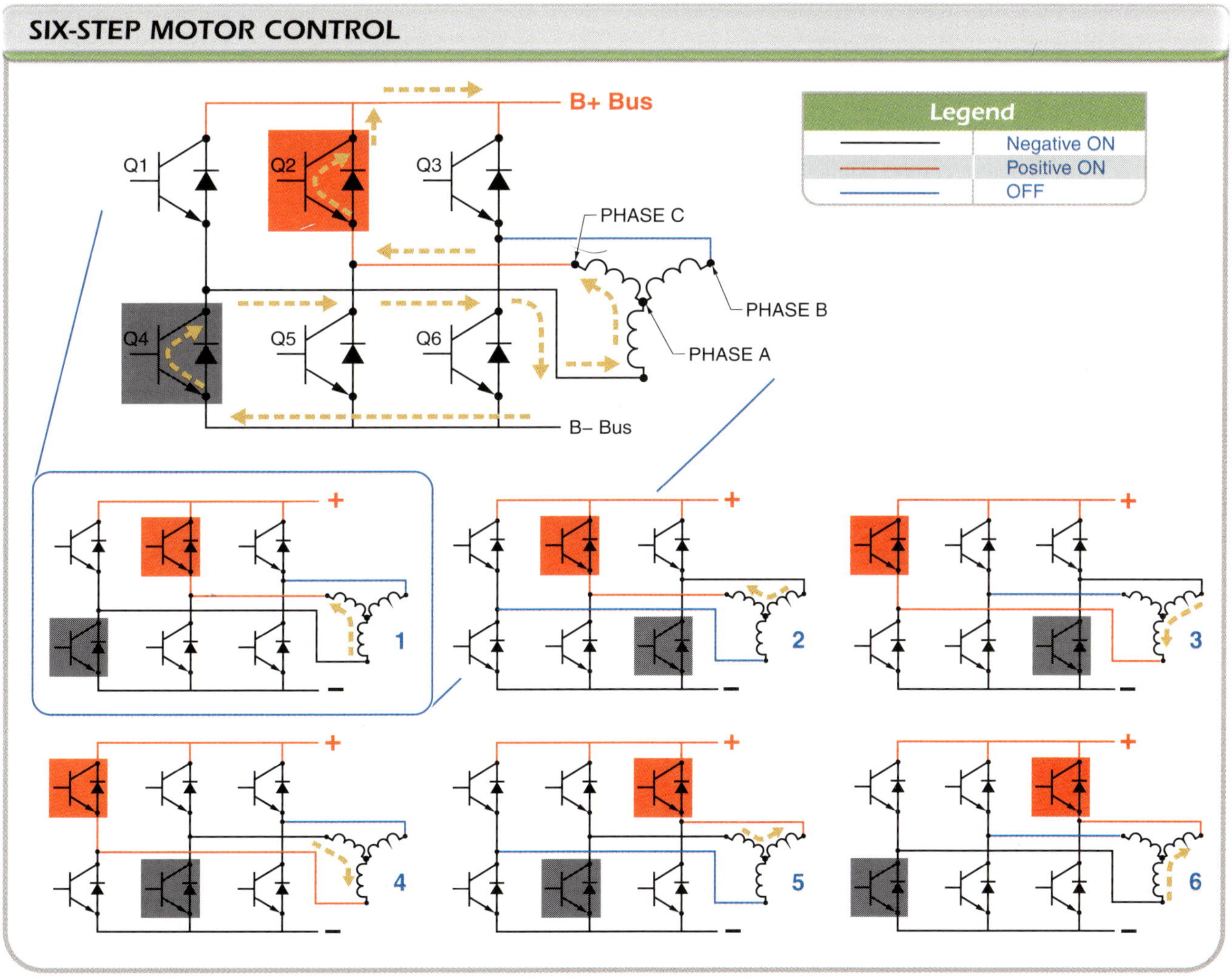

Figure 7-8. The controller uses one of the IGBTs of the previous triggering cycle in the next switching cycle.

In the illustration, the pair of IGBTs turned ON for the event is highlighted in red (positive ON), black (negative ON), and blue (OFF). One IGBT is turned ON and held ON for two sequential switching cycles. This control strategy is used to reduce electrical transients and reduce complexity in developing the overall IGBT switching plan in the software.

By following the six switching events, it can be seen that one IGBT remains turned ON from the previous switching cycle and is used in a next switching cycle. For example, in the first and second switching cycles, the IGBT highlighted in red remains ON for both cycles. The IGBT highlighted in black changes from one cycle to the next. During the first event, IGBT modules 2 and 4 are used to magnetize stator coil poles A and C. During the second event, IGBT modules 2 and 6 are used to magnetize stator coil poles B and C. There are a total of 6 events in one complete cycle.

The firing order of the IGBT modules is very similar to the firing order of an ICE. **See Figure 7-9.** It requires a total of 720° of crankshaft rotation to ignite all six cylinders in a four-stroke ICE engine. During each 360° cycle, a cylinder can be on the exhaust or compression stroke. The same concept can be applied for the IGBT motor drive in the power inverter. It requires 360°, or one full rotation of the rotor, to fire all 12 of the stator coils in a 3ϕ, 4-pole motor. To achieve the 360° rotation, all of the IGBTs must be used and the IGBT modules must work in pairs during the rotation, similar to the way companion cylinders operate in an ICE.

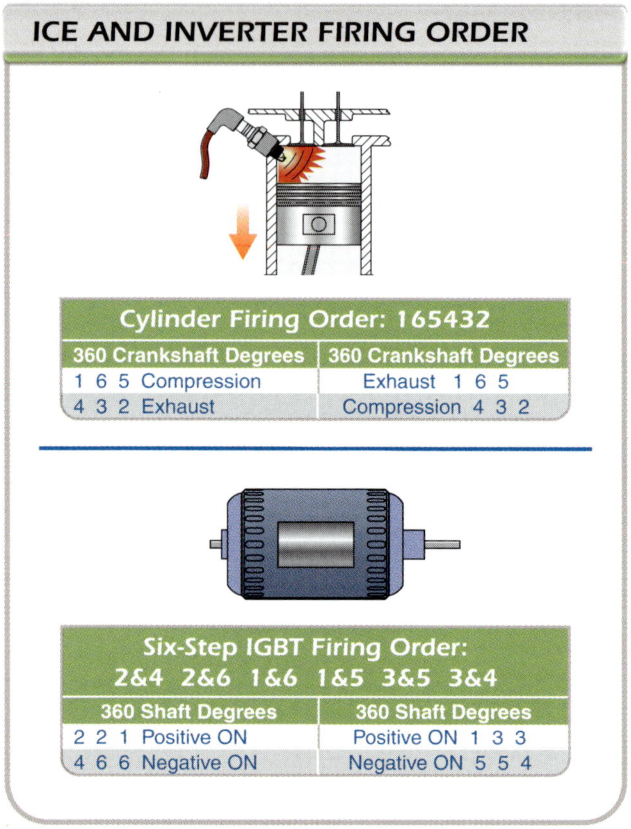

Figure 7-9. The firing order of an IGBT module is very similar to the firing order of an ICE.

Pulse Width Modulation (PWM)

Another popular control strategy is PWM sine wave control. The object of PWM sine wave control is to create an AC sine wave from a DC source. PWM provides smooth motor operation, especially at low speeds. Also, torque ripple is virtually eliminated with PWM sine wave control. *Torque ripple* is rough torque delivered by an electric machine caused by the use of waveforms that have aggressive changes in voltage, such as six-step and rectangular waveforms.

Pulse width modulation means that the duty cycle of a DC voltage signal is controlled from 0% to 100% to create a sine wave. **See Figure 7-10.** By changing the PWM duty cycle, the phase stator pole coil voltage and frequency are controlled. The IGBT power module section synthesizes AC power by triggering the gate drive circuitry at the proper intervals and on-time. More power is delivered when the pulse width is wider.

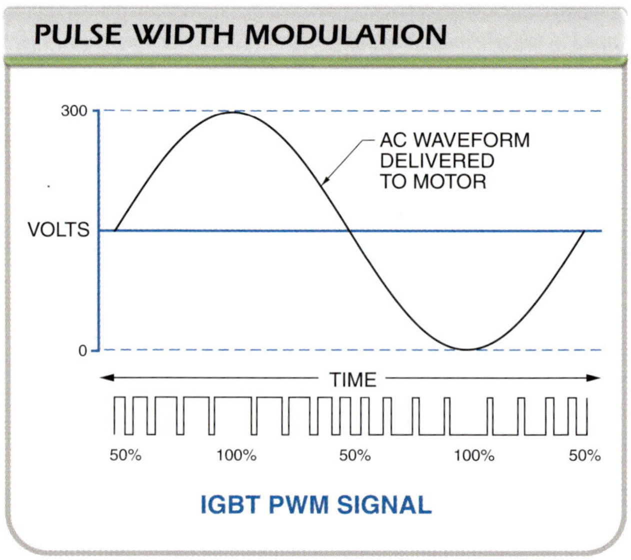

Figure 7-10. Pulse width modulation means that the duty cycle of the DC voltage signal is controlled from 0% to 100% to create a sine wave.

The process of producing a sine wave requires the use of a microprocessor that is designed for high mathematical computational throughput capacity for complex control functions. With contemporary microprocessors, creating PWM sine waves has become a much easier task. It is now the preferred method of motor control in applications that require smooth motor operation.

The control of the sine wave is limited by the digital-to-analog resolution and microprocessor speed. The sine wave table and the generation of the sine wave must also operate with both the speed and torque requirements that are requested by the operator of a vehicle. The AC waveform that is synthesized by a sine table in the software and executed by the controller to the power inverter closely resembles the AC power that is delivered by the utility company.

PWM Sine-Wave Creation. An AC waveform is created by controlling a PWM gate signal to synthesize (create) three sine waves to support 3ϕ motor operation. **See Figure 7-11.** The IGBT gate control determines the PWM and frequency of the waveform in the inverter based on operator input. As the sine wave frequency increases, the motor rpm increases accordingly. Many AC induction/PM machine inverters are designed to generate a waveform that will drive the rotor at 300 Hz to 400 Hz. The three sine waves are used to drive six IGBTs or MOSFETs that are housed in one power electronics module located in the power inverter.

The use of two IGBTs per phase is the same technique used in an H-bridge. The frequency of one wave cycle is determined by the rate at which the PWM duty cycle is commanded by the controller. The PWM signal from the microcontroller or DSP is delivered to the gates of two IGBT modules. **See Figure 7-12.**

PWM SINE-WAVE CREATION

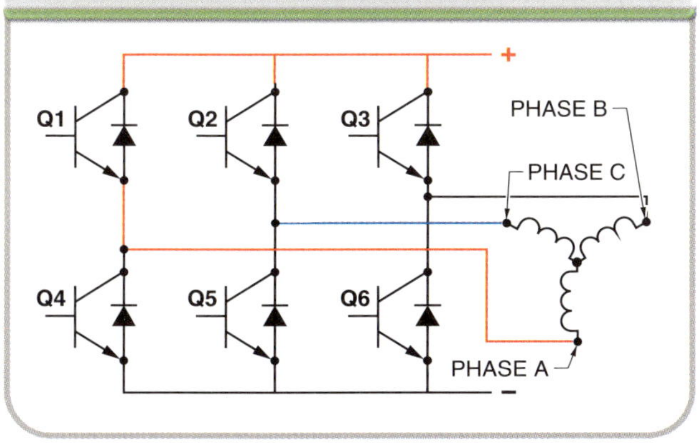

Figure 7-11. The IGBT gate control determines the PWM and frequency of the waveform in the inverter.

CURRENT FLOW

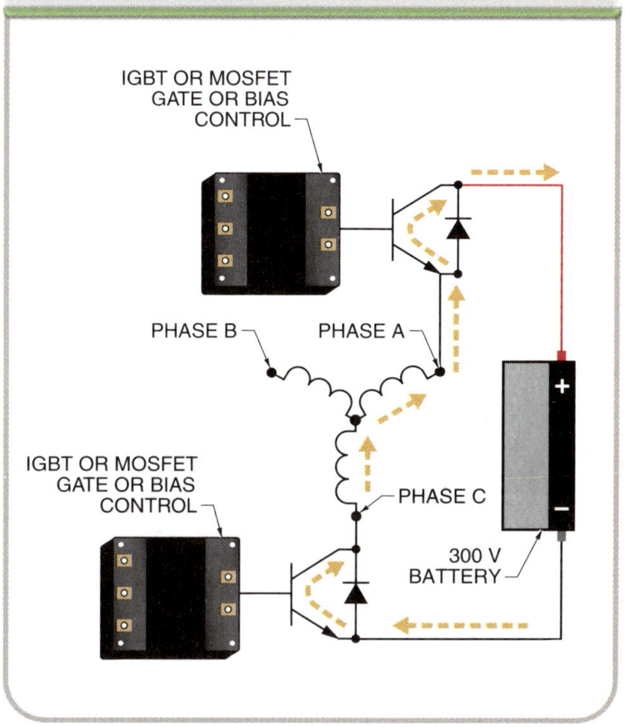

Figure 7-12. The PWM signal from the microcontroller or DSP is delivered to the gates of two IGBT modules.

The controller must trigger the inverter IGBT gate drive section at the proper time and with the correct voltage amplitudes and frequencies to generate the proper torque and speed. The electrical current exits the battery system and enters the phase C IGBT. The controller must send the proper signal to the IGBT to allow the current to flow through to the phase C coil. The electrical current then flows through the phase C coil, the phase A coil, the phase A IGBT, and finally back to the battery pack. This process is repeated for all of the electric machine motor phases. This is one of a possible six stator phase coil modes in which the inverter can be triggered.

For comparison, the PWM control strategy used for generating the sine wave is the same strategy used for controlling the PWM of a fuel injector, exhaust gas recalculation (EGR) valve, and many other devices in an ICE vehicle. With the fuel injector, more fuel is delivered as PWM is increased. This results in more power from the engine. However, the fuel must be delivered in the correct volume and at the proper time to generate more engine power. This strategy is mirrored for the induction/PM machine. The machines must also receive the correct levels of AC electrical power in order to generate the necessary output power.

Creating a sine wave from a DC source is very computation intensive. It requires a microprocessor that can quickly execute instructions and computations. A software algorithm that references a location within the software code called a "sine table" is used to modify the sine wave.

With the algorithm and sine table, the sine wave can be controlled to any desired voltage or frequency level by triggering the IGBTs at precisely the correct time to create the sine waves. The sine wave voltage controls the torque output. Higher voltage results in higher torque and lower voltage results in lower torque. The frequency of the sine wave controls the motor speed. Higher frequency results in higher speed and lower frequency results in lower speed.

Switching Frequency. Most contemporary induction/PM machines operate on switching frequencies from 10 kHz to 20 kHz. One of the benefits of switching the power modules at a high frequency is to create a smoother motor operation. Rotor cogging, torque ripple, and audible noises can result as transistor switching frequencies decrease.

Faster switching increases inductive spiking (transient voltages) in the system that can damage the power transistors. **See Figure 7-13.** The overall energy levels of the transients also increase with faster switching. This increases IGBT temperature, creates the possibility of overstressing the IGBTs from elevated voltage and current levels, and increases stator temperature. A liquid cooling system is usually necessary to transfer the heat generated from the IGBTs during the power switching process.

PWM VOLTAGE SPIKES

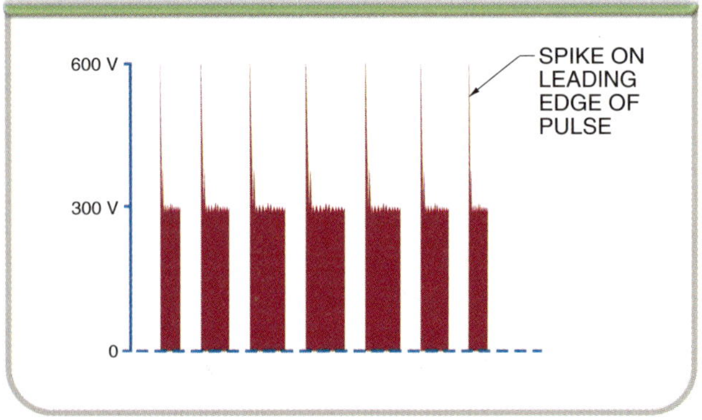

Figure 7-13. Faster switching increases inductive spiking (transient voltage) in the system that can damage the power transistors.

Audible noise is a concern with pulse width modulation control of electric machines. When electric machine switching frequency is below about 10 kHz, a person can hear this electrical switching noise. Because of this, designers usually specify a switching frequency of 10 kHz or higher.

The stator field frequency (sine wave frequency) is not the same as the IGBT switching frequency. The IGBT motor control switching frequencies need typically 10 kHz to 20 kHz to generate a smooth PWM pattern. It is the number of PWM cycles within a given amount of time that generate the sine wave frequency. The more PWM triggering cycles within a one-second period, the more waveforms that can be generated, and vice versa. This ultimately permits the controller to generate the desired stator frequency to give the speed the operator desires.

PWM control supplies a lower average operating voltage than six-step control. This results in higher motor currents. PWM sine wave control prevents the motor terminal phase voltage from reaching its maximum value. The minimum pulse width imposed by limitations of the switching transistors prevents full voltage from being reached. This reduced waveform voltage level can be costly since the motor must be designed for a lower voltage.

A lower voltage requires higher current to deliver a given horsepower. This higher current affects the operation of power switching devices. In addition, higher levels of motor harmonics cause elevated heating in the motor. However, cooling will be adequate because the motor is running near its rated speed. This results in an overall benefit.

Power and Torque

An important aspect of waveform control is the power level of the electric machines. Electric machine control engineers typically design the machines to operate on more than one type of waveform to more effectively control torque and speeds. A sine wave is used to provide smooth motor performance at low speeds. Progressively changing the waveform from trapezoidal to six-step, and finally to rectangular can increase the average voltage applied to the machine.

Voltages at the motor are increased to increase horsepower. There is a linear relationship between horsepower and voltage. If the voltage increases, then the horsepower and torque increase. The power is calculated as follows:

$P_w = V \times A$

where

P_w = power (in W)
V = voltage (in V)
A = current (in A)

For example, if the current is 400 A and the voltage is either 150 V or 300 V, the power is calculated as follows:

$P_w = V \times A$
$P_w = 150 \times 400$
$P_w = \mathbf{60{,}000\ W\ (60\ kW)}$

$P_w = V \times A$
$P_w = 300 \times 400$
$P_w = \mathbf{120{,}000\ W\ (120\ kW)}$

When the power in watts or kilowatts is known, the power in horsepower can be calculated as follows:

$$P_{HP} = \frac{P_w}{746}$$

or

$$P_{HP} = \frac{P_{kw}}{0.746}$$

where

P_{HP} = power (in HP)
P_w = power (in W)
P_{kw} = power (in kW)

For example, if the power is 60,000 W (60 kW) or 120,000 W (120 kW), the power in horsepower is calculated as follows:

$$P_{HP} = \frac{P_w}{746}$$

$$P_{HP} = \frac{60{,}000}{746}$$

$$P_{HP} = \mathbf{80.43\ HP}$$

$$P_{HP} = \frac{P_w}{746}$$

$$P_{HP} = \frac{120{,}000}{746}$$

$$P_{HP} = \mathbf{160.86\ HP}$$

This shows that when the voltage is doubled, the power also doubles. This is true whether the power is measured in watts, kilowatts, or horsepower. Furthermore, if the horsepower and rpm of the electric machine are known, then torque can also be calculated as follows:

$$T = \frac{P_{HP} \times 5252}{rpm}$$

where

T = torque (in lb-ft)
P_{HP} = power (in HP)

For example, if the power is 80.43 HP at 4000 rpm, the torque is calculated as follows:

$$T = \frac{P_{HP} \times 5252}{rpm}$$

$$T = \frac{80.43 \times 5252}{4000}$$

$$T = \mathbf{105.6\ lb\text{-}ft}$$

The motor torque profile must also be tailored to the application. Motor and controller manufacturers stress that the entire electric machine and power inverter package must be designed for a particular application. For example, a hybrid and electric power system package may be designed to generate electrical power for propulsion and also blend its electrical power with an ICE.

The Toyota Prius inverter is located underhood.

AC induction machines were designed by Nikola Tesla to operate on sine waves. However, electric machines can operate on other types of waveforms, such as square waves or stepped waves.

For an automotive technician, there will seldom be a need to calculate power, speed, and torque. However, as with an ICE, it is very important that a technician understand the concepts of cause and effect as it applies to electric propulsion systems. Moreover, a technician should understand how voltage levels affect electric machine operation and how an underperforming battery pack, hybrid generator, high-voltage cable, or connection can lead to a customer complaint about power output.

Rotor Frequency

The key to generating magnetic fields in an AC induction machine is by introducing an AC sine wave (or other waveform) into the stator pole coils. The polarity of a coil alternates any time a coil of wire is energized and then de-energized. In addition, if the amplitude of the electrical current that is transferred through a coil of wire is quickly increased and decreased, the polarity alternates. This causes the stator pole coil magnetic fields to expand and collapse.

The stator is similar to the primary winding of a transformer and the rotor is similar to the secondary winding. There is slip between the stator rotating magnetic field and the magnetic field generated in the rotor. Because of the slip, the rotor field always lags the stator field during drive operation. In other words, the frequency of the rotor always lags the frequency of the stator. The stator field must first induce a voltage (mutual induction) into the rotor. An *inductor* is an electrical device that stores energy in the form of a magnetic field. In its simplest form, an inductor consists of a wire loop or coil. In an inductor, the current lags 90 electrical degrees behind the voltage.

Rotor frequency is important in propulsion torque development and in explaining how high levels of regenerative current are developed. The polarity of the sine wave that is impressed upon the rotor changes twice as often as the stator field. When the generated voltages and current polarities in the stator are reversed during an interval in the sine wave cycle, the flux field is reversed approximately twice as often in the rotor bars. **See Figure 7-14.** This means that the rotor frequency is approximately twice the frequency of the stator field. The stator and rotor field frequencies are related as follows:

$f_r = f_s \times (2 - s)$

where

f_r = frequency of current in rotor conductor (in rpm)

f_s = frequency in stator coils (in rpm)

s = slip (decimal fraction)

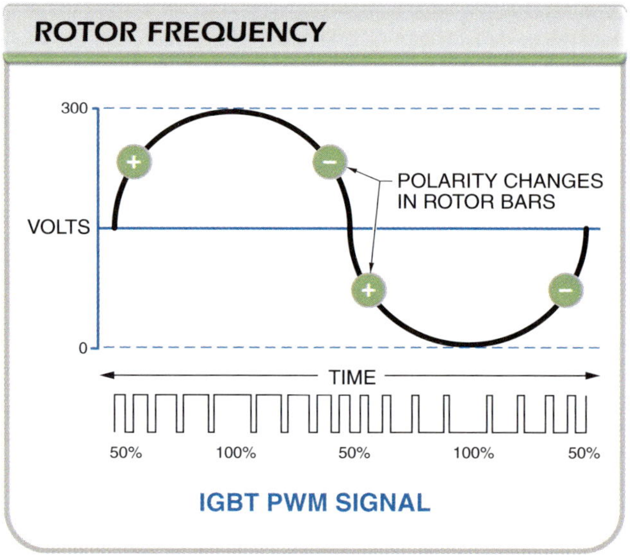

Figure 7-14. When the generated voltages and current polarities in the stator are reversed during an interval in the sine wave cycle, the flux field is reversed approximately twice as often in the rotor bars.

For example, if the frequency in the stator coils is 300 rpm and the slip is 8%, the frequency of current in the rotor conductor is calculated as follows:

$f_r = f_s \times (2 - s)$

$f_r = 300 \times (2 - 0.08)$

$f_r = 300 \times 1.92$

$f_r = \mathbf{576\ rpm}$

Maximum Motor Speeds

Although a power inverter can supply variable waveforms and frequency ranges to an electric machine, an electric motor has speed limitations that are determined by its mechanical design. The maximum synchronous speed is calculated as follows:

$$S = \frac{(F)(120)}{P}$$

where

S = synchronous speed (in rpm)

F = frequency (in Hz)

P = poles per phase

For example, with a 4-pole motor operated with a 400 Hz maximum frequency, the maximum motor speed is calculated as follows:

$$S = \frac{(F)(120)}{P}$$

$$S = \frac{(400)(120)}{4}$$

$$S = \mathbf{12{,}000\ rpm}$$

DC Brush-Type Motors

Controlling a DC brush-type electric motor requires that the electrical current from the battery system be transferred to a controller and then to the motor. The electric current is guided through brushes and a commutator. This generates the necessary magnetic field interaction between the field coils and armature to produce rotation of the armature. **See Figure 7-15.** The interaction of the brushes and armature limits the efficiency of DC machines.

DC BRUSH-TYPE MOTORS

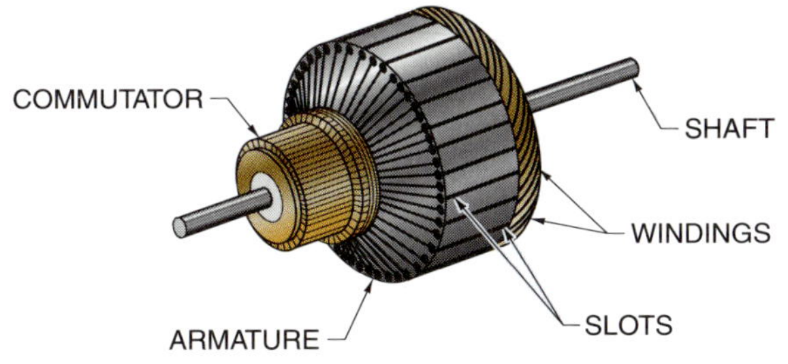

Figure 7-15. Electric current is guided through brushes and a commutator. This generates the necessary magnetic field interaction between the field coils and armature to produce rotation of the armature.

The brushes are placed at a constant angle with respect to the commutator bars to provide the magnetic field orientations that produce optimum torque. This fixed angle is sufficient until the speed of the armature reaches a point where arcing begins to occur because of the interaction of armature and battery currents at these higher armature speeds. In order to eliminate arcing, the angle needs to change.

This is analogous to the spark advance system for an ICE. An ICE needs more spark advance at higher rpms so that the ignition can ignite the gases with enough lead time to expand these gases to maximum pressure at the precise time. This is done to optimize cylinder power and generate maximum pressure to drive the piston downward. Without advancing ignition timing, engine power peaks at a much lower rpm.

This same principle can be applied to a DC brush-type electric machine. There is a point at which the mechanical angle of the brush and commutator must change (advance) if the rpm of the armature is to continue to increase. Advancing the brush set is analogous to advancing ignition timing on an ICE. Additional torque cannot be generated if the brush set angle does not change to advance the magnetic fields.

Although a PWM electronic motor controller can be used to control a DC brush-type electric machine, the relationship between the brush and armature limits the efficiency. Efficiency is also limited by the mechanical contact between the brushes and the armature. Brushes wear and must eventually be replaced and a commutator requires regular maintenance. In addition, the mass and construction of the armature limits the maximum rpm to speeds much lower than that of an AC induction electric machine.

In a DC brush-type electric machine, the voltage and current determine the strength of the magnetic field interaction between the armature and field coils. The brushes and commutator control the direction of current flow in a similar way as a sine wave controls the direction of electrical current in an AC machine. However, in an AC electric machine system, the sine wave frequency and amplitude can be infinitely controlled without contacting components to provide a wider rpm and torque range than with a DC brush-type electric machine.

INCREASING ELECTRIC MACHINE VOLTAGE

It is sometimes necessary to increase electric machine power and torque without changing the dimensions or mass of the propulsion system. This may also require increasing the horsepower and torque without increasing the battery-pack specific power and specific energy ratings. One of the most effective and efficient methods to accomplish this change is to increase the design battery pack voltage to increase the applied voltage to the motor. Overmodulation and voltage boosting are options to increase the voltage from an existing battery pack.

Overmodulation

Six-step and sine wave electric machine waveforms are commonly used in motor control. Other waveforms include trapezoidal and rectangular waveforms. **See Figure 7-16.** These waveforms are used for overmodulation. *Overmodulation* is a technique used to increase the available voltage to a motor by using waveforms other than sine waves. Different waveforms deliver different average voltages to the electric machine. An advantage of overmodulation is that a waveform can be selected in order to maintain the bus voltage. This results in an increase in horsepower and torque, and allows higher operating speeds.

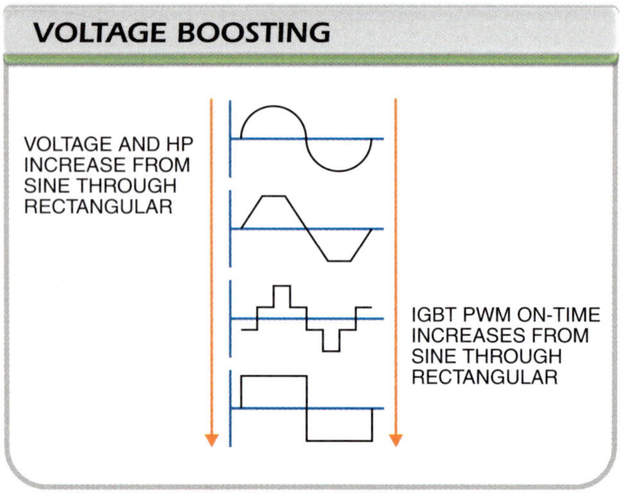

Figure 7-16. Common waveforms used in motor control include sine, six-step, trapezoidal, and rectangular waveforms.

The higher average motor voltage of alternate waveforms has other advantages. If the electric machines are used for electric assist at higher vehicle loads and speeds, the battery-pack terminal voltage may drop. The vehicle is more efficient when this voltage is kept high. Higher voltage decreases the motor electrical current and permits the battery state of charge to remain high during most operating ranges.

For example, if the battery-pack voltage or system voltage is low, the power inverter controller can switch from a sine wave to a six-step, trapezoidal, or rectangular wave to raise the voltage level at the electric machine. In addition, when the vehicle operator requests very high levels of torque and speed, such as with wide open throttle or when beginning a steep grade, the power inverter controller can immediately shift from sine through rectangular waveforms to ensure the electric machine maintains a high level of voltage.

There are disadvantages to overmodulation. As the control switching strategy moves from sine through rectangular waveforms, the electric machine harmonics, transient voltage, and current spikes increase significantly. The method of switching during overmodulation is termed "hard switching" due to the nature of turning the IGBTs ON and OFF at high voltage and current levels. The results of hard switching include increased motor temperatures, decreased system efficiency due to harmonic and IGBT switching losses, and increased torque ripple. Overmodulation provides more torque and speed at the expense of system efficiency.

Boost Converters

Overmodulation can be used along with a boost converter system. A *boost converter* is a special DC-DC converter that is used within a power inverter to increase voltage from the battery pack to the electric machine. **See Figure 7-17.** A boost converter system can reduce the system cost and weight of a battery pack. Boosting voltage with the power inverter is especially beneficial if the engineers are constrained by transaxle design, weight, and dimensional requirements. Boost converter systems are used in Lexus RX400h, Toyota Prius Generation II, and Toyota Highlander hybrids.

By boosting voltage, a higher level of performance can be achieved with minimal impact to mechanical, weight, and/or dimension aspects of an engine. There is almost a linear relationship between horsepower and torque and system voltage. Therefore, boosting the voltage can scale horsepower and torque.

To boost voltage in the Toyota and Lexus systems, a reactor is included within the power inverter. A *reactor* is a power inductor within a power inverter system that can boost battery-pack voltage to a higher voltage level. The reactor is part of the DC-DC boost converter of the power inverter. It is designed to operate in a similar manner as an ignition coil primary winding to boost voltage. For example, the voltage is boosted to a maximum of about 400 VDC in the Toyota Prius and to about 650 VDC in the Toyota Highlander and Lexus RX400h.

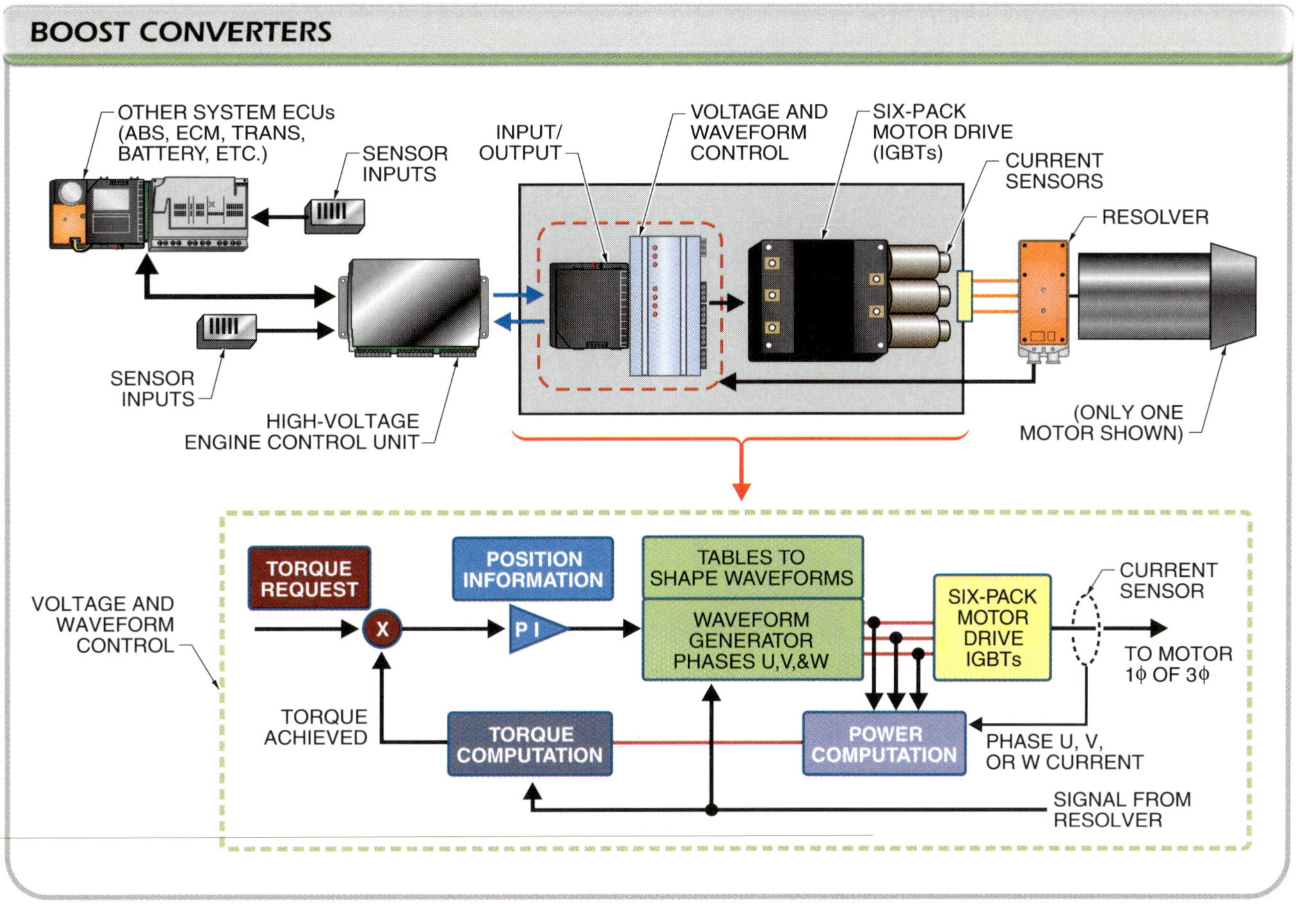

Figure 7-17. A boost converter is a special DC-DC converter that is used to increase voltage to an electric machine.

As with the primary ignition coil circuit, the reactor generates a magnetic field through electromagnetic induction to generate higher voltage. The boost converter is placed within the power inverter system between the battery pack and power inverter. **See Figure 7-18.**

Boost Converter Operation

A DC-DC boost converter system is a combination of controllers and other elements within the electronic control unit of the power inverter. The boost converter provides boosted electrical power to the electric propulsion motor. Inputs from sensors provide information to control the output of the boost converter in the power inverter. The voltage and waveform control section of the power inverter monitors and controls voltage to the motor(s). This determines the electric machine horsepower and torque output. The amount of voltage boost is variable and depends on the commanded horsepower and torque levels.

Figure 7-18. The boost converter is placed within the power inverter system between the battery pack and power inverter.

Boost Control Sensing Circuits. The power inverter system must maintain control over the boost converter to provide the optimum level of electrical power to match the amount of torque requested by the operator. The control system monitors prereactor and postreactor voltage to control the torque. A hardware circuit illustrates the main power electronics blocks that make up the boost converter system circuit and other input/output systems. **See Figure 7-19.** Sensor VB measures battery-pack voltage. Sensor VH measures the boosted voltage to control and record the level of boost. Sensor VL measures the unboosted voltage to calculate the difference between the existing voltage level and the desired boost voltage level.

Resistor R1 and capacitor C1 make a filter circuit to filter out electrical transient noise. A *capacitor* is an electrical device that stores energy in the form of an electrostatic field. In its simplest form, a capacitor consists of two conducting plates separated by an insulating material called a dielectric. R1 is a bleed resistor for C1.

Transistors Q1 and Q2 make up the half-bridge transistor circuit used to control the boost and regenerative electrical power. Diodes D1 and D2 are used to steer current during reactor charge and discharge cycles. Capacitor C2 limits bus voltage fluctuations from the power fluctuation of a single motor. Too much bus fluctuation affects the operation of other motors that share the high voltage bus. Resistor R2 is a bleed resistor for C2 and may also serve to filter the bus.

Reactor Charging. The first step in boosting voltage is charging the reactor. **See Figure 7-20.** A six-pack motor drive circuit provides overall switching to the motor and controls motor power output. A *six-pack motor drive* is a power transistor configuration that uses six IGBTs to control the torque and speed of an electric machine. The reactor must be charged to generate electromagnetism in order to expand the magnetic field energy. This is analogous to supplying electric current to the primary winding of an ignition coil to expand the coil's magnetic field.

BOOST CONVERTER CONTROLS

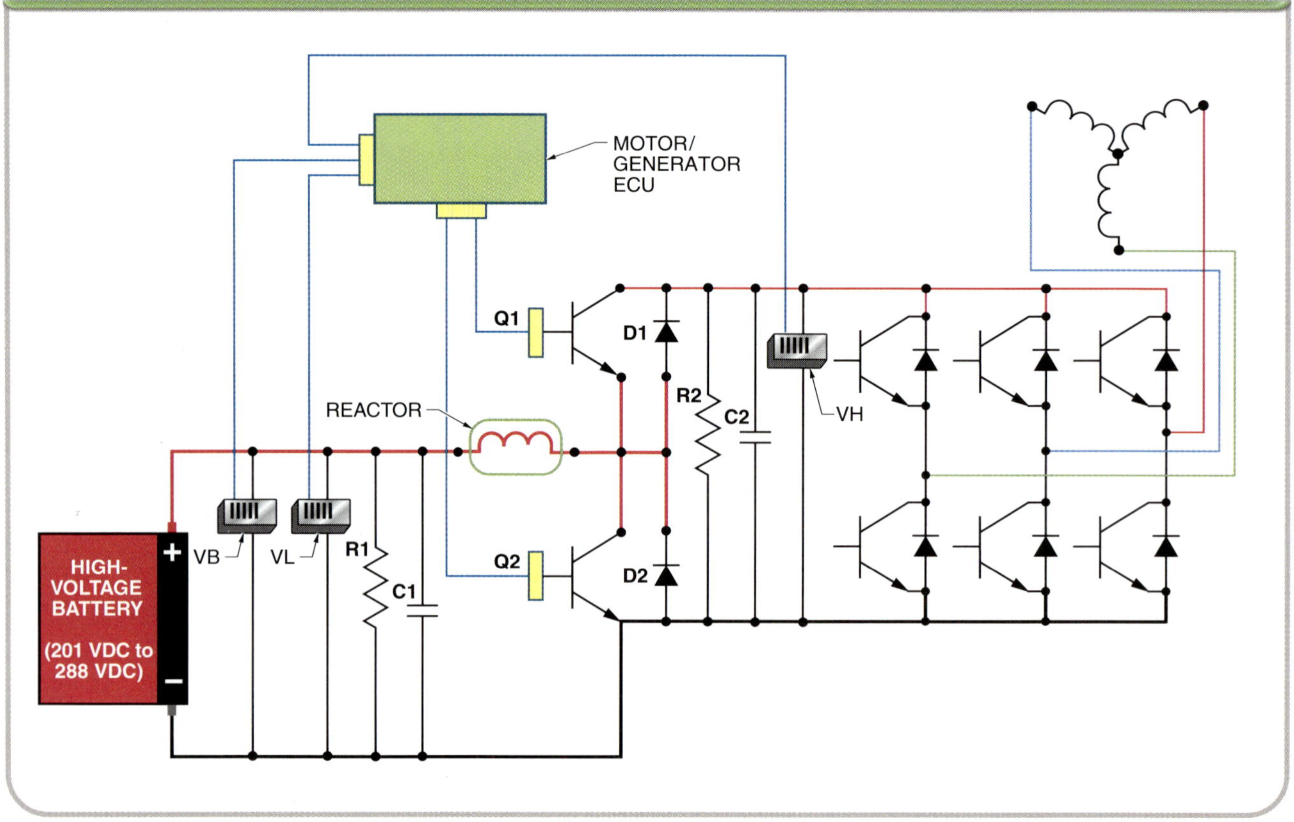

Figure 7-19. The control system monitors prereactor and postreactor voltage to control the torque.

REACTOR CHARGING

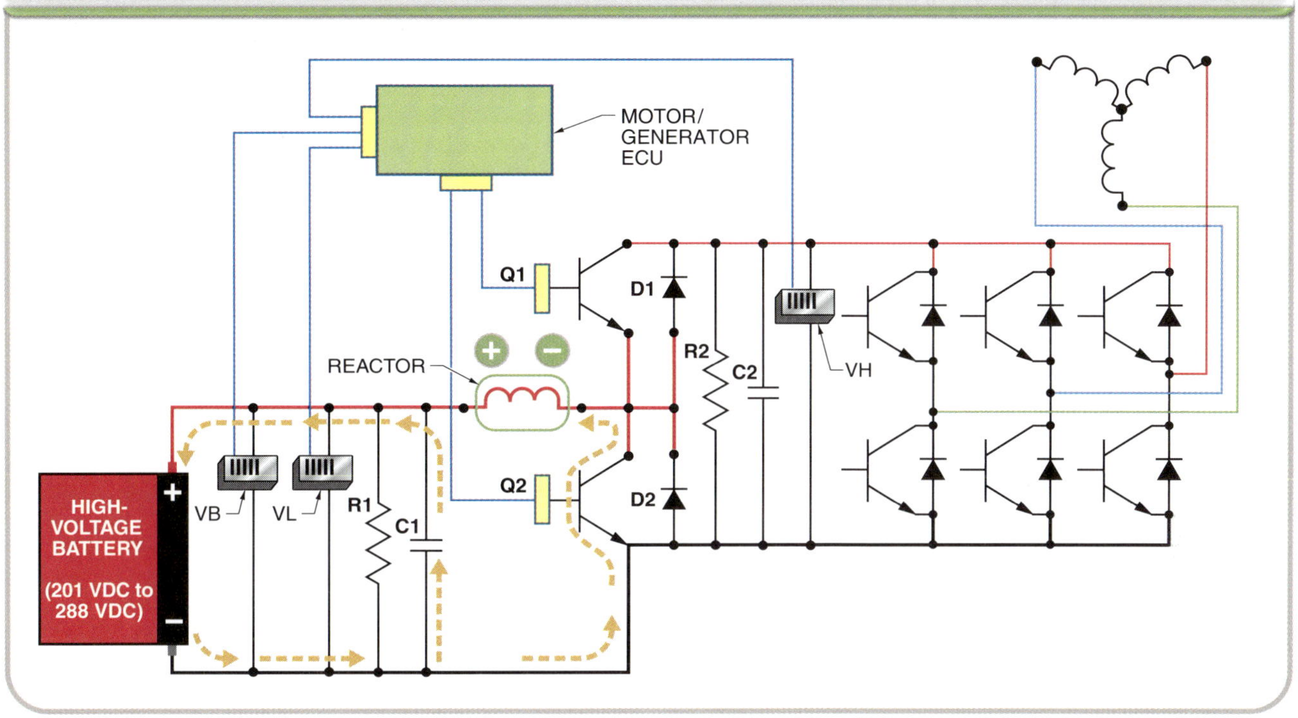

Figure 7-20. A six-pack motor drive circuit provides overall switching to the motor and controls motor power output.

The control circuit commands transistor Q2 to turn on to generate the voltage level required to provide the power level (torque) requested by the operator. Sensors VB and VL compare the battery-pack voltage to the reactor low-side voltage and monitor the starting and charging voltage levels. The sensors report this data to the power inverter controller. The PWM controller commands transistor Q2 on-time. Electrical current flow is from battery negative through Q2, through the reactor, and back to battery positive.

The reactor magnetic field builds and expands. The reactor charge voltage and the capacitor C1 charge voltage are both equal to the battery source. The reactor polarity is shown in the illustration.

Reactor Discharging. Discharging (or collapsing) the reactor magnetic field provides the boosted voltage to the power inverter six-pack motor drive. Discharging the reactor provides voltage levels about 2.25 to 2.5 times higher than the battery pack. This voltage increase results in increased electric machine horsepower and torque. Controlling reactor power is complex. The strategy involves the controller (software algorithms, calibrations, etc.), reactor, and the six-pack motor drive. **See Figure 7-21.**

The Toyota Prius boost converter is located inside the inverter enclosure.

Hybrid electric vehicles can use low-rolling resistance tires or inflation pressure can be increased slightly to help increase fuel economy.

REACTOR DISCHARGING

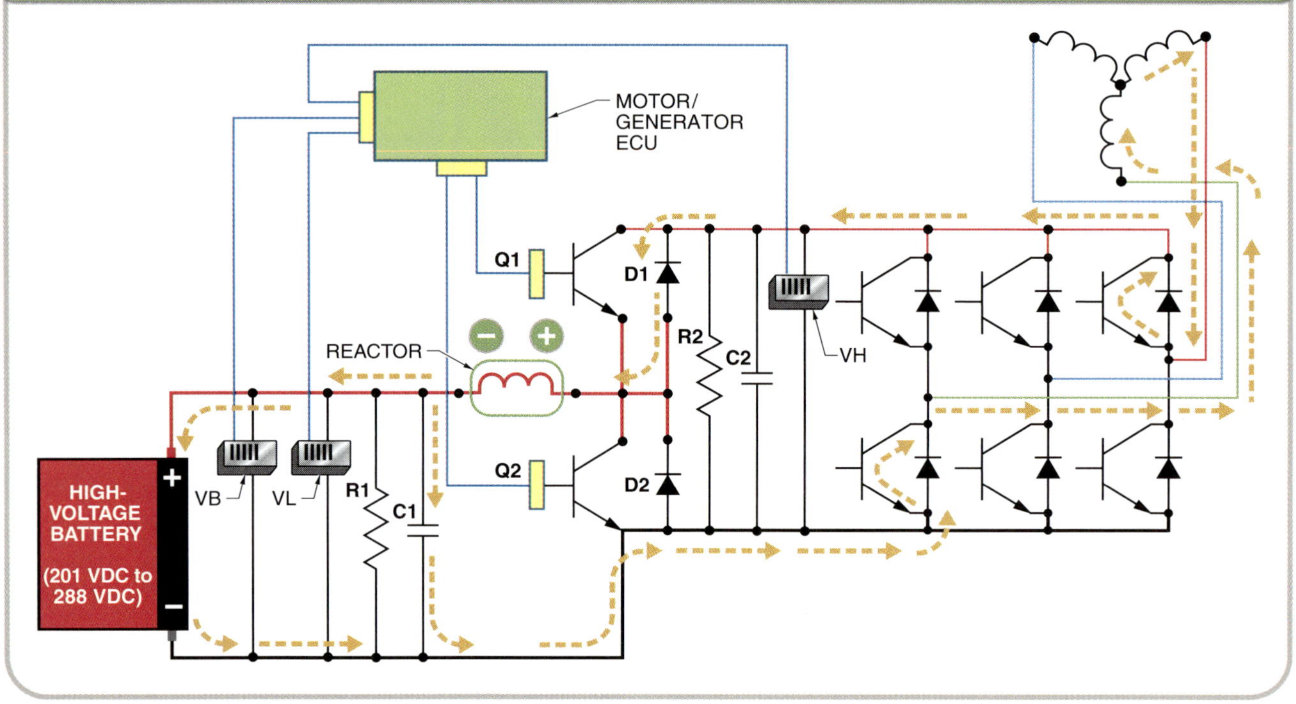

Figure 7-21. Controlling reactor power involves the controller, reactor, and six-pack motor drive.

During discharge, the polarity reverses on the reactor leads. The reactor energy (field collapse) is dissipated through the battery pack to the power inverter IGBTs and electric machine. The electrical current returns to the reactor through diode D1 to complete the circuit. Power inverter IGBTs control reactor energy delivery to the electric machine by using PWM on the gate circuits based on the desired positive torque level.

Diode D2 does not permit electrical current to flow directly back to the reactor. It steers electrical current into the power inverter motor drive IGBTs. The controller measures the voltage with VL and VH to ensure that the desired reactor voltage output level was achieved. It compares this to other control inputs , such as current sensors, resolvers, and throttle position sensors, to further verify that desired torque was achieved.

REGENERATIVE BRAKING

Regenerative braking (regen) is built into induction/PM machines with power inverter control. The function of regen is to recover energy and recharge the battery system by capturing the vehicle's kinetic energy. The vehicle driveline is loaded with negative torque that emulates the effects of vehicle braking. To change the operation of an electric machine from motor to generator, the system requires minimal mechanical or electrical changes. A majority of the changes are in the control software algorithms.

To use the electric machine as a generator, the rotor rpm must be faster than the stator frequency. This can be accomplished by reducing the stator frequency to a value less than rotor rpm, or by driving the rotor faster than the stator frequency during coasting or braking. During this negative slip situation, the battery pack or generator continues to supply electrical current to the stator phases as the rotor speed is leading (rotating faster than) the stator frequency. With the battery pack or generator continuing to supply power to the stator, the stator continues to supply power to excite the rotor.

The boost reactor in a Toyota Camry hybrid can produce approximately 21 kW of electrical power.

With the rotor speeding faster than the stator, the rotor bar magnetic fields cut across the stator coils and induce voltage back into the stator coils. There is a difference in voltage potential between the power generated by the electric machine (motor/generator) and the battery-pack terminal voltage level. For example, if the battery system is providing 300 V to excite the stator and the rotor produces 400 V in regen mode, then the rotor has 100 V higher potential than the battery system. The higher rotor potential causes current to flow through the IGBT flyback diodes and finally to the battery system. **See Figure 7-22.**

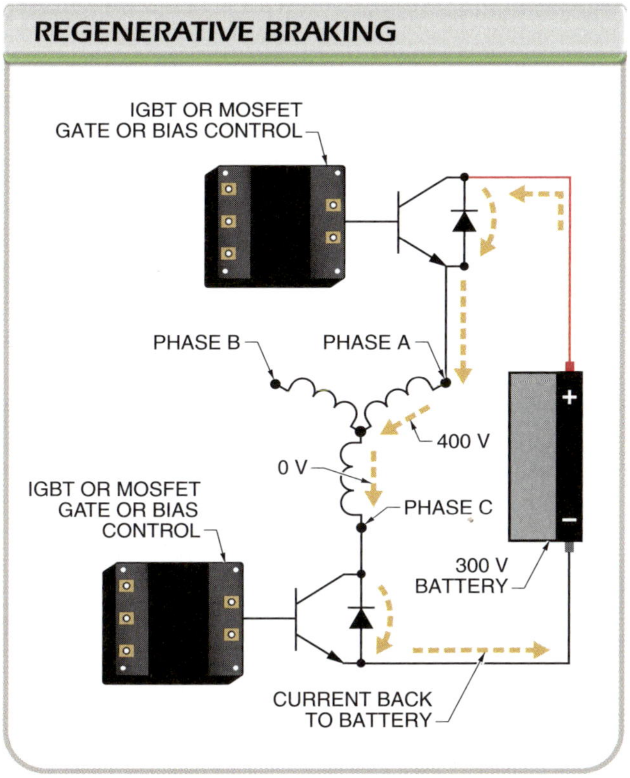

Figure 7-22. Higher rotor potential causes current to flow through the IGBT flyback diodes and finally to the battery system.

The sum of power concept is that current flows from a higher voltage to a lower voltage. The higher voltage produced by the rotor provides electrical current to the lower voltage battery system. This is similar to turning OFF a charging system indicator lamp by placing positive 12 V on both sides of the bulb. The light turns OFF because the electrical potential on both sides of the bulb is identical and no electrical current would flow.

Toyota Motor Company

Inverters come in different sizes and with different packaging.

During regen, the induction/PM controller must communicate with the battery system controller to calculate the necessary levels of regen power. If regen were uncontrolled, excessive levels of electrical current would be delivered to the battery. This would cause gassing of the battery module electrolyte, and battery failure would be inevitable. This also occurs in an ICE vehicle when the voltage regulator fails and output voltage is uncontrolled.

Induction Machines and Regen

When an induction machine is in regen mode, the battery system controller requests electrical current from the electric machine. The propulsion system controller decreases the stator field frequency so that the rotor speed is faster than the stator frequency. The rotor field induces voltage in the stator pole coils, which generates current in the stator. At the same time, the propulsion system controller begins to decrease the voltage on the gate signal to the IGBTs. This decreases the voltage from the stator field and permits electrical current to flow through the flyback diodes back to the battery system.

The IM controller needs approximately one line of software code and a regen table to command the stator field from positive to negative slip. No internal or external hardware additions are necessary. By modulating the IGBTs from 0% to 100%, the current transferred between the stator and the rotor can be infinitely controlled. During regen, more IGBT on-time means less regen current to the battery pack and less IGBT on-time means more regen current to the battery pack.

In general, battery-pack current is about 20% of rotor output. **See Figure 7-23.** For example, if 60 A is required to recharge a battery pack, then 20% of that value, or 12 A, must be supplied to the rotor. This 12 A supply is transferred by the stator to the rotor. The rotor then develops 60 A that is reflected back to recharge the battery pack.

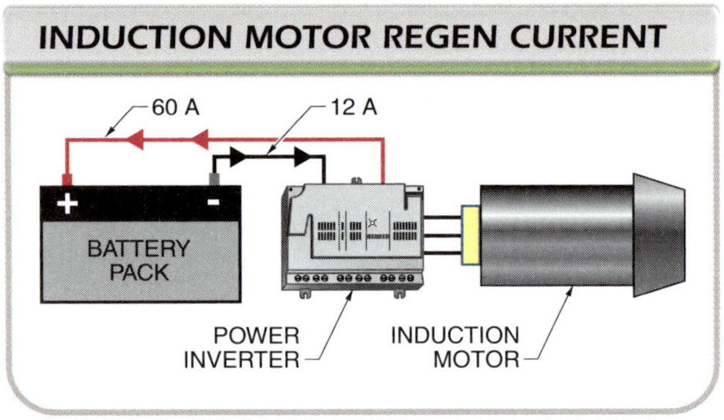

Figure 7-23. Battery-pack current is about 20% of rotor output.

If, however, the battery system is fully charged and regen is enabled during a vehicle coast down, the battery controller would communicate to the propulsion system controller that no electrical current is necessary. Therefore, the propulsion system controller would command the slip to about 0%. To do this, the controller commands the IGBT to an on-state at a voltage that will place the battery system and stator field at a similar voltage potential. This directs the stator field frequency at or near the same speed as the rotor and reduces the motor output torque to a negligible level.

PM Machines and Regen

For PM machine regen, the battery system does not need to excite the stator pole coils in order to generate output power. A PM electric machine has permanent magnets mounted in or on the rotor to provide continuous magnetic flux. A PM machine constantly generates output power. Therefore, it requires that the phase retard angle of the stator and the voltage to the stator field be controlled to control the output.

Regen is capable of generating hundreds of amps at voltages higher than the battery system voltage. This is a major concern that must be addressed when designing a PM regen system. Regen output would be very high if the control system were to fail while the PM machine is operating at high speed. The concern is that voltage could be over 600 V and it may be enough to damage power inverter components and battery-pack modules.

The battery system controller must work with the propulsion controller. If either controller fails to operate, the control strategy is designed to turn on all IGBT power switches of the power inverter. This short circuits the stator and confines the high voltage to the PM machine stator. This may damage the stator, but it will protect all power electronic devices and the battery pack.

Boost Converters and Regen

The DC-DC boost converter and reactor system function during regenerative braking mode. When a boost converter is being used, the electric machine operates on a voltage higher than the battery-pack voltage. The reactor provides a voltage level to the electric machine that permits it to operate at voltage levels of about 2.25 to 2.5 times higher than the battery pack. The result is that during regenerative braking, power is controlled by the reactor and IGBT PWM to reduce the electric machine voltage to a level high enough to recharge the battery pack but not high enough to damage it. **See Figure 7-24.**

During regen, the electric machine stator generates electrical power. Transistor Q1 controls and modulates the reactor so that electrical energy to the battery pack from the electric machine is controlled to a commanded level. With Q1 modulating the reactor energy to control electric machine power, the output voltage from the electric machines is controlled to levels that properly maintain battery-pack state of charge.

Regenerative electric machine power is controlled by the reactor voltage amplitude and polarities and by IGBT PWM signals. The reactor can either reduce or enhance these regenerative electrical currents to the battery pack. This is coupled with electric machine stator field control to provide the overall regenerative braking strategy. Control software determines Q1 PWM to achieve the desired system voltage level to maintain desired battery-pack state of charge.

> A boost converter can be used to increase motor power and torque without increasing the physical size of the electric transmission.

BOOST CONVERTERS AND REGEN

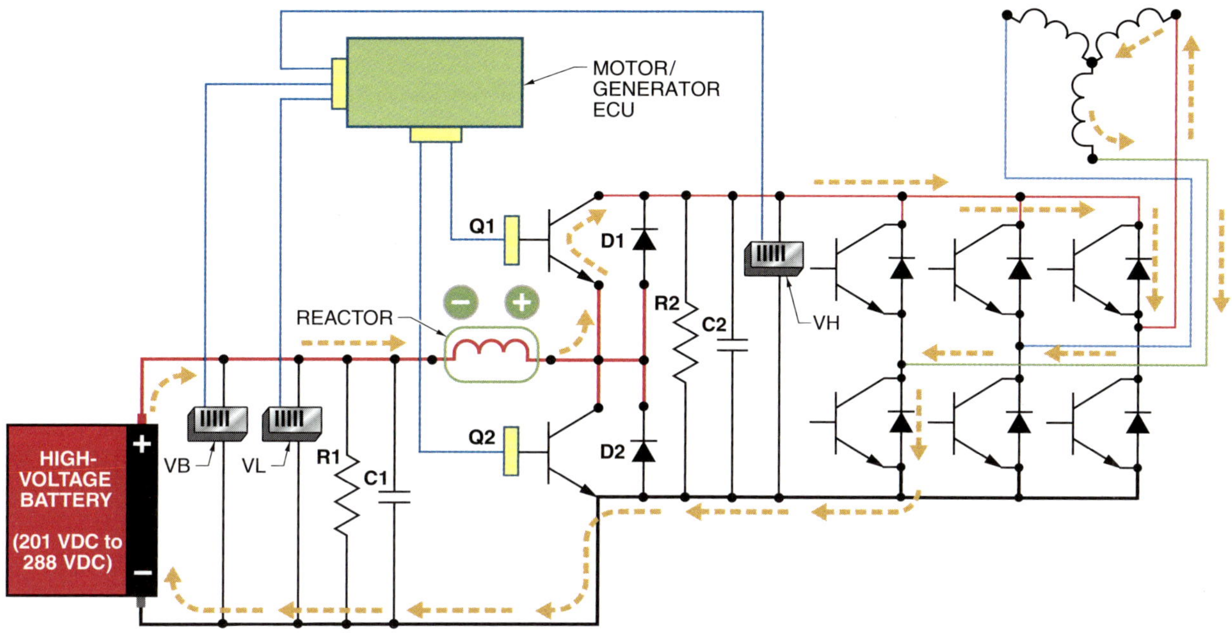

Figure 7-24. Electric machine power must be controlled by the reactor and IGBT PWM to reduce the voltage to a level high enough to recharge the battery pack but not too high so as to damage it.

Refer to the *Hybrid Electric Vehicle Technology* CD-ROM for a Quick Quiz® on the chapter contents.

POINTS TO REMEMBER

- Power inverters may be located within the engine compartment or in other locations in the vehicle.
- In an automotive application, power inverters synthesize (create) AC from a DC source (battery pack).
- Power inverters control electric machine torque through slip or vector control.
- Power inverters control electric machine torque and speed by using signals from several sensing circuits such as throttle position, speed (rpm), phase current, and brake switch sensors.
- Most modern power inverters use an insulated gate bipolar transistor (IGBT) as the means to deliver and control electrical power the electric machine.
- A power inverter typically uses six IGBTs in a six-pack design to deliver electrical power to the electric machine.
- An IGBT six-pack could be thought of as six large power switches.
- Waveforms are created with pulse width modulation (PWM) control.
- The four basic waveforms used to control an electric machine are the sine, trapezoidal, six-step, and rectangular waveforms.
- All induction electric machines are sine wave machines but can be operated on other waveforms during specified conditions.

- During regenerative braking, the IGBTs modulate electrical power to control electrical current that is steered through the flyback diodes and eventually is transferred to the battery pack.
- Electric machine horsepower, torque, and speed (rpm) can be increased by boosting supply voltage to the electric machine.
- Battery-pack voltage can be boosted up to a maximum of 500 VDC in some power inverter systems.

Refer to the *Hybrid Electric Vehicle Technology* CD-ROM for the Review Questions in Microsoft® Word format.

REVIEW QUESTIONS

1. Explain why an AC machine rotor operates at a different frequency than the stator.
2. What are the benefits of increasing voltage to an electric machine?
3. Describe the purpose of using overmodulation for electric machine control.
4. What is the sum of power concept?

7 Power Inverter Systems

HYBRID Electric Vehicle TECHNOLOGY

Trade Test

Name ______________________________ Date ______________

______________ **1.** The power inverter in a Honda Civic Hybrid can be found ___.
A. behind the left kick panel
B. behind the glove box
C. behind the rear seat
D. in the engine compartment

______________ **2.** The power inverter in a Toyota Prius can be found ___.
A. behind the left kick panel
B. behind the glove box
C. behind the rear seat
D. in the engine compartment

______________ **3.** One type of power transistor typically used in a power inverter is the ___.
A. MOSFET
B. FET
C. MITFET
D. IGFET

______________ **4.** In a basic power inverter system, ___ transistors are used to drive a 3ϕ electric machine.
A. 3
B. 6
C. 12
D. 24

______________ **5.** There are ___ basic waveforms for electric machine control.
A. 2
B. 3
C. 4
D. 6

______________ **6.** ___ is used to create an electric machine sine wave.
A. PWM
B. Triangle modulation
C. Flux vectoring
D. Unijunction angling

_____________________ 7. The sum of power concept is based on ___ difference between the rotor and stator.
A. voltage
B. current
C. triggering
D. control

_____________________ 8. The power inverter senses the ___ to control the amount of regenerative power generated during braking or vehicle coast.
A. IGBT temperature
B. vehicle speed
C. battery-pack state of charge
D. electric machine field vector

_____________________ 9. Voltage boosting of the power inverter is used by ___.
A. Toyota and Lexus
B. GM and Honda
C. Toyota and GM
D. Ford and Lexus

_____________________ 10. One of the primary functions of the flyback diode in an IGBT package is to steer electrical current during ___ mode.
A. advanced torque
B. regenerative
C. reverse
D. machine start

_____________________ 11. One of the reasons that a power inverter control system uses different waveforms to power an electric machine is to ___.
A. decrease maximum speed
B. decrease slip angles
C. maintain bus voltage level
D. none of the above

_____________________ 12. If the power inverter controller were to request only an increase in speed (or rpm), the ___ must increase.
A. waveform amplitude
B. number of rotor bars
C. number of magnets
D. stator frequency

_____________________ 13. ___ IGBTs are required to turn on two phases of a wye-connected stator.
A. Two
B. Three
C. Six
D. Eight

_______________ **14.** The boost converter is primarily used in the power inverter to ___.
A. increase horsepower and torque
B. decrease battery-pack size
C. reduce back EMF
D. all of the above

_______________ **15.** The power inverter controller must ___ to the IGBT to increase the voltage level to the electric machine.
A. increase frequency
B. increase PWM
C. decrease frequency
D. decrease PWM

_______________ **16.** Increasing the voltage to the electric machine also increases the ___.
A. horsepower
B. torque
C. speed
D. all of the above

_______________ **17.** A six-pack motor drive has ___ IGBTs.
A. 3
B. 6
C. 12
D. 36

_______________ **18.** An IGBT contains ___ transistor technologies.
A. MOSFET and bipolar
B. bipolar and unijunction
C. bipolar and gate turn-off
D. thyristor and bipolar

_______________ **19.** The sum of power concept is very similar to the function of a(n) ___.
A. alternator and generator rotor
B. battery and cranking motor
C. charging system indicator bulb
D. high-voltage vehicle light

_______________ **20.** To increase the electrical current to the electric machine stator coils, a power inverter must command sine wave amplitude to ___.
A. increase
B. decrease
C. remain unchanged
D. shut off for at least 3 seconds

_______________ **21.** The ___ control algorithm achieves torque control by controlling the phase electrical current in the stator coils and by controlling the slip frequency.
A. torque ripple
B. overmodulation
C. slip
D. reactive power

____________________ **22.** ___ is a technique used to increase the available voltage to a motor by using waveforms other than sine waves.

A. Overmodulation
B. Voltage boosting
C. Boost converting
D. Voltage bucking

____________________ **23.** Reactive power is generated during ___ mode when the magnetic fields of the rotor cut across the stator pole windings.

A. electric-assist propulsion
B. regenerative braking
C. capacitance charging
D. boost converter

Electric Propulsion Sensing Systems

8

Electric machines can generate electrical power to propel a vehicle or recharge a battery-pack system. The power inverter controls the power that the battery pack transfers to the electric machine. It also controls the energy the electric machine generates and transfers back to the battery pack. However, the power inverter must sense how much energy to deliver for vehicle traction or battery-pack recharging.

Sensing systems provide information to the power inverter so that the electric energy can be controlled. Common types of sensors used as the core sensor inputs in the propulsion system include motor speed, motor current, and throttle position sensors. There are many other sensors on a vehicle related to safety or convenience systems, but they are not unique in hybrid vehicles.

CHAPTER OBJECTIVES

- List the types of speed (or rpm) sensors used with electric machines.
- Describe the operation of resolver, encoder, and proximity sensor systems.
- Explain why at least two current sensors must be used in a 3ϕ power inverter.
- Describe the design of the throttle position sensors.
- List several other vehicle sensors that are used as inputs to control an electric propulsion system.

KEY TERMS

resolver: A sensor that provides two sine waves to a motor controller to indicate absolute rotor position and speed.

encoder: A sensor that provides digital, square-wave outputs that the electric machine motor controller counts to determine the approximate rotor position and speed.

proximity sensor: A sensor that provides a digital, square-wave output to the electric machine motor controller by sensing the position of the commutation plate to determine the approximate rotor position and speed.

current sensor: A sensor that provides a motor controller feedback information on phase current.

throttle position sensor (TPS): A sensor located on the engine throttle body that is used to monitor the position of the throttle.

pedal position sensor (PPS): A sensor located on the accelerator pedal assembly that provides an input to an electric machine motor controller so that it can sense the desired acceleration rate and level of power desired by the vehicle operator.

MOTOR SPEED AND POSITION SENSING

An induction/PM machine uses a speed-sensing or rpm-sensing device to provide an input signal to the controller. The controller uses the input signal to calculate the proper slip percentage or phase advance/retard to control the machine.

The signal output of an rpm sensor can be a square or sine wave. With some technologies, the signal is conditioned to a square wave. Within some rpm sensors, the signal may need to be amplified to a specific amplitude so that it can be interpreted by the controller and converted to an rpm value.

With PM electric machines, precise rotor positioning is also required. Rotor positioning is critical because the magnets in the PM rotor are mounted in fixed locations. The magnet positions are synchronized to specific points every revolution. The controller must know the exact position of the rotor so that it can synchronize the magnetic fields of the rotor and stator to produce the target output speed and torque.

Any number of designs and methods can be used to achieve motor rpm and position sensing. Electric machine speed and position sensing is typically accomplished with resolvers, encoders, or Hall-effect sensors. Speed sensing devices are usually connected directly to the rotor shaft with threaded, key, or press fit fasteners. The sensors are designed with slot poles or teeth to generate a signal. Some sensors are triggered by a reluctor or commutation ring.

Resolvers

A common type of speed sensor is a resolver. A *resolver* is a sensor that provides two sine waves to a motor controller to indicate absolute rotor position and speed. A resolver can detect rotational direction and rotor shaft rpm by varying sine wave amplitude and relative positions. In addition, it can detect the angular position of the rotor by comparing the time and amplitude between sine and cosine angles. The time varies with rotational speed.

For most electric machine and power inverter manufacturers, resolvers are the preferred sensors in contemporary electric propulsion systems. **See Figure 8-1.** They are used in GM, Ford, Honda, Lexus, and Toyota hybrids. Other manufacturers also plan to use the resolver for future hybrid and fuel-cell applications.

A resolver is also an absolute position sensor. Absolute sensing means that the sensor provides data on exact rotor position. It provides two AC waveform output signals and the controller tracks each sine wave to calculate the precise position of the rotor.

A resolver can be designed with any number of poles to produce consistent, smooth AC waveform outputs. **See Figure 8-2.** It must work with the rotor to report the precise rotor magnet position to the power inverter controller. The resolver generates a sine wave signal by using a sensor assembly and reluctor lobes. The number of reluctor lobes determines the resolution of the signal generated by the resolver and transmitted to the power inverter controller. More reluctor lobes mean a higher resolution signal because smaller rotations are detectable.

RESOLVERS

Figure 8-1. Resolvers are the preferred sensors in contemporary electric propulsion systems for most electric machine and power inverter manufacturers.

The power inverter controller can also decode rotor position even when the rotor is stationary. When the system is powered with the ignition key ON, the oscillator pulses the exciter winding. If a rotor lobe is aligned with the resolver pole(s), a corresponding sine and cosine AC waveform result. The power inverter controller senses the exact rotor position by sampling the position and amplitude of the resolver sine and cosine AC waveform outputs. The waveforms are then transmitted to the R/D converter.

The R/D converter decodes the AC signals and then transmits this data to the power inverter controller. Controller software can interpret the exact position of the rotor. The power inverter controller must know the exact position of the rotor, even when the rotor is stationary, in order to energize the proper stator coils and generate rotor motion.

RESOLVER POLES

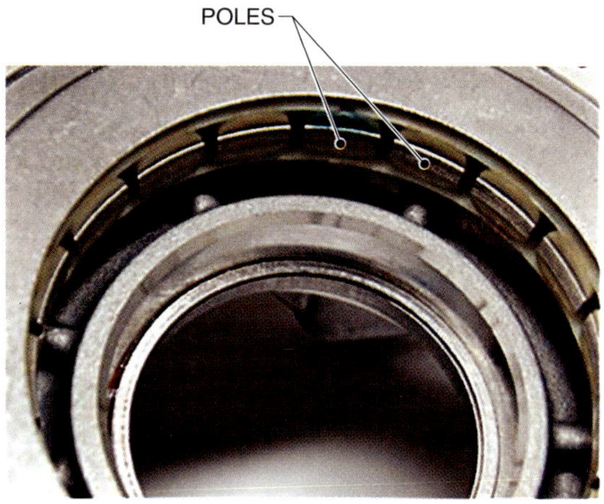

Figure 8-2. A resolver can be designed with any number of poles.

Resolver Operation

The two main components of a resolver sensing system are the sensor and the reluctor with lobes. **See Figure 8-3.** The circuit contains the fundamental elements that allow the resolver to generate analog, sine-wave outputs. Reluctor designs may include two, three, or four lobes.

The oscillator generates a pulsed signal and is connected to the exciter winding to create an AC sine-wave output that is to be used as a reference signal for measuring the sine and cosine outputs. Each pole tooth of the resolver is wound with exciter windings. Also, each pole has the identical number of exciter winding turns that will produce a consistent reference AC sine-wave amplitude output.

Each pole tooth of the resolver is wound with sine and cosine windings. Each pole is wound with the same number of exciter turns to establish a reference waveform. The number of sine and cosine windings varies with each pole to produce the varying sine waves that provide absolute motor rotor position. The different number of windings between the sine and cosine windings produces a different amplitude output at the same angle point for the respective poles. A resolver-to-digital (R/D) converter converts AC sine-wave inputs to parallel data output that is transmitted to the power inverter controller in bits. The controller decodes and translates the bits into exact rotor position.

RESOLVER OPERATION

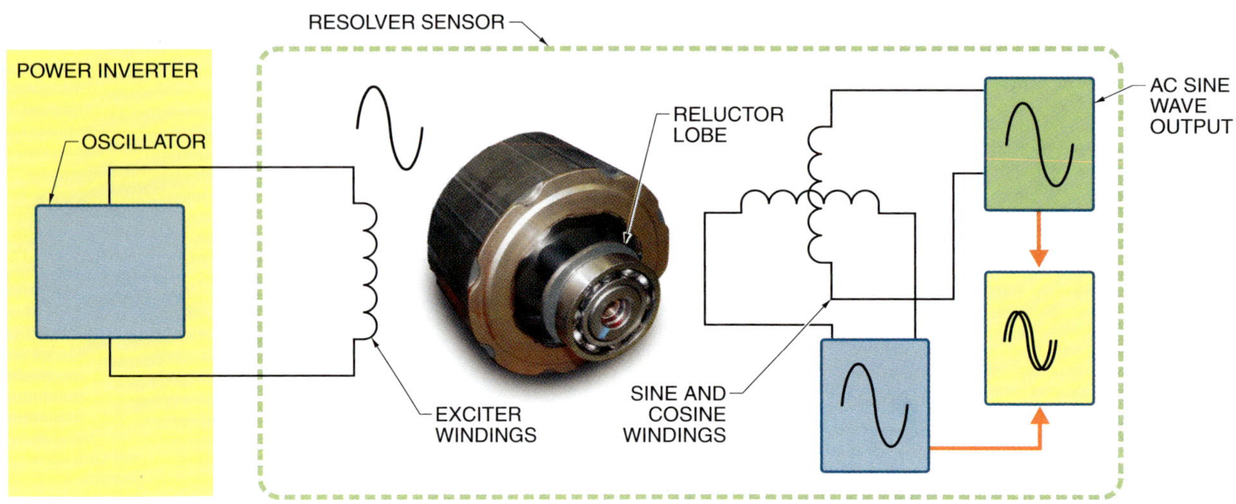

Figure 8-3. The two main components of a resolver sensing system are the sensor and the reluctor with lobes.

The wiring of a resolver is very similar to that of a 12 V series-wound cranking motor. The field coils of a 12 V cranking motor are actually one long ribbon of copper that is wound into individual coils and then fastened into the housing of the cranking motor. The result is that the coils appear to be individual circuits but are part of one series circuit.

All resolver pole windings begin as individual, long wires wrapped around each sensor pole. This results in three electrically isolated series circuits used to generate three separate AC sine-wave output signals. **See Figure 8-4.** The AC waveform outputs of the three windings do not overlap in most areas of the curves, and each waveform either leads or lags other waveforms.

There are three sine waves generated by the resolver circuit. Two sine wave signals are generated when the resolver rotates and one fundamental (excitation) sine wave is generated by the oscillator and excitation winding. The fundamental is constant and transferred from an oscillator in the power inverter to each sensing pole on the resolver.

Sine and cosine waveforms are generated when the excitation AC sine wave cuts across the resolver windings. The resolver integrated circuit, located in the power inverter, calculates the angular position by measuring and comparing the amplitude and position of the sine and cosine output waveforms.

There are six wires that connect the reluctor with the electronic control unit (ECU). Two of the six wires are the sine angle output. Two of the wires are the cosine angle output. The remaining two wires are the excitation input. **See Figure 8-5.**

The R/D converter decodes the sine and cosine waveforms at each degree (or smaller increments) of the 360° rotor rotational cycle. This provides the power inverter controller with a precise angular location of the rotor. The R/D converter can be programmed to provide parallel data in 10-bit (0 to 1024) or 12-bit (0 to 4096) formats.

With this data, the controller can calculate an accurate rotor angular position. Accurate position information allows the power inverter controller to trigger the six-pack motor drive IGBTs at precise intervals to efficiently control the interaction of the stator coil magnetic fields and rotor permanent magnet fields.

Tamagawa-Seiki Ltd. developed the use of resolver technology for hybrid electric vehicles. Resolvers can be used to obtain more precise speed and position information for better control.

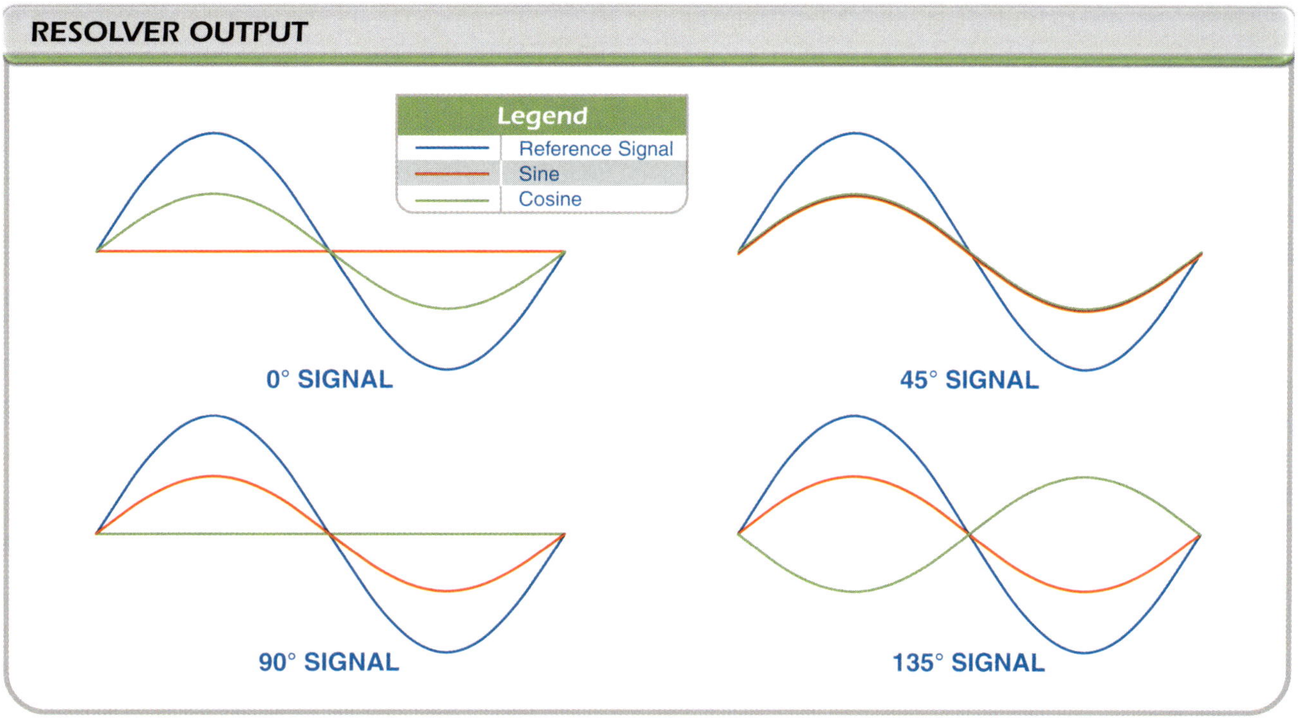

Figure 8-4. A resolver is used to generate three separate AC sine-wave output signals.

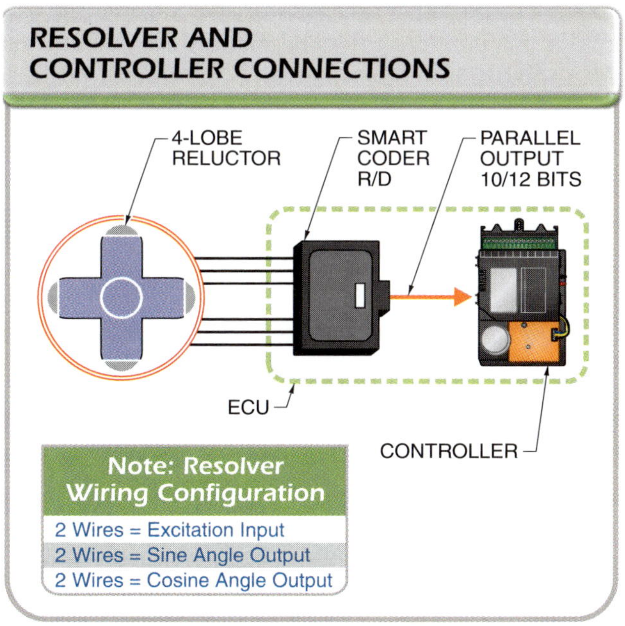

Figure 8-5. There are six wires connecting the reluctor to the ECU.

A resolver can generate sine waves because each pole has a different number of wire turns. As the reluctor rotates and aligns with the resolver poles, the sine-wave peak voltages from the two resolver windings increase significantly. The voltages increase because the steel rotor lobes provide a concentrated magnetic path for the pole winding magnetic flux when the lobes align with the resolver poles.

This concentrated magnetic flux path intensifies the electromagnetic induction of the signal and increases the voltage output of each winding. The output voltages generated from the sine and cosine windings are mostly asymmetrical and do not overlap relative to the fundamental sine waveform generated by the oscillator and exciter winding.

Encoders

An *encoder* is a sensor that provides digital, square-wave outputs that the electric machine motor controller counts to determine the approximate rotor position and speed. They provide position information by counting the number of leading and trailing edges of a square wave. An encoder cannot sense any positioning information that could be tracked between the leading and trailing square-wave edges. Therefore, the controller cannot sense the absolute position of the rotor because of the time gaps between the leading and trailing square-wave edges. Encoders are not designed to provide a high level of precision.

Magnetoresistive, Hall-effect, and optical sensing technologies are typical technologies used on a number of automotive systems including encoder designs for hybrid vehicle systems. For example, magnetoresistive sensor and optical encoders are often used in engine crankshaft positioning systems. Hall-effect sensors have been used for years in distributors and as crankshaft sensors. The diagnostic methods used for troubleshooting an induction/PM encoder are identical to those currently used for ICE engine sensors.

An encoder can be mounted on the side of an electric transaxle assembly or the shaft of an electric machine rotor. **See Figure 8-6.** The encoder generates a digital output pattern to show how the two channels are typically offset. This provides frequency and sensor rotational direction data to the controller.

ENCODERS

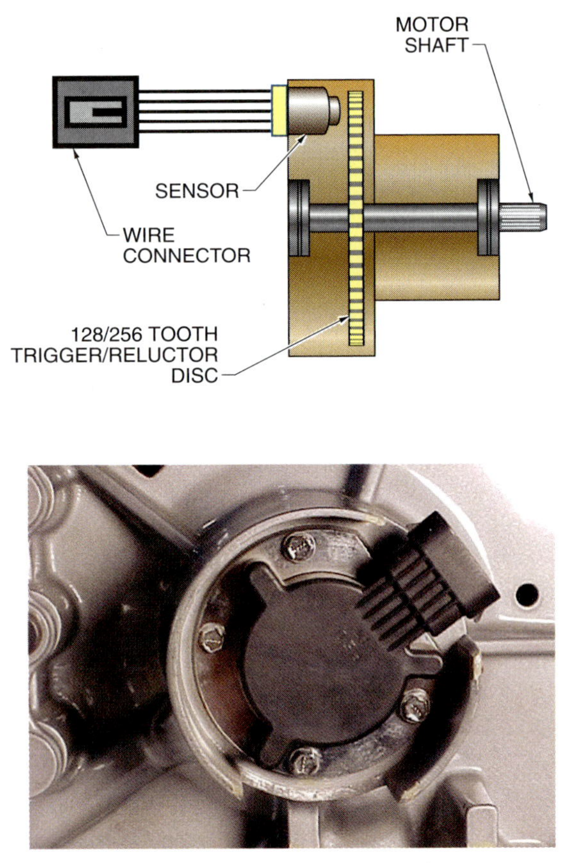

Figure 8-6. An encoder can be mounted on the side of the electric transaxle assembly.

An encoder can also generate a sine wave by using slots of different widths. This is known as asymmetrical digital design. The sine wave is synthesized when the sensor rotates and the asymmetrical slots produce incremental increases and decreases in output voltage. These voltage changes can be engineered to form an AC sine wave and decoded by a controller system. However, this adds extra functions to the controller and increases complexity. This can also affect accuracy because it is a digital sensing signal and not an analog signal.

Encoders are rarely used in modern electric, hybrid, or fuel-cell vehicles because they do not provide absolute rotor position. Absolute rotor position is especially important for PM electric machine control.

Proximity Commutation Sensor Systems

In the Honda Hybrid system, a commutation plate is indexed and bolted to the engine crankshaft. The commutation plate has 12 staggered metal tabs on its circumference. The tabs provide synchronous triggering points to communicate the positions of the 12 rotor magnets to the power inverter controller.

Sensor systems may use three Hall-effect or magnetoresistive proximity sensors. A *proximity sensor* is a sensor that provides a digital, square-wave output to an electric machine motor controller by sensing the position of the commutation plate to determine the approximate rotor position and speed. **See Figure 8-7.** This sensing system more closely resembles an encoder system than a resolver system because the output to the controller is digital.

Current sensors are small components that can be placed on a circuit board.

PROXIMITY COMMUTATION SENSOR SYSTEMS

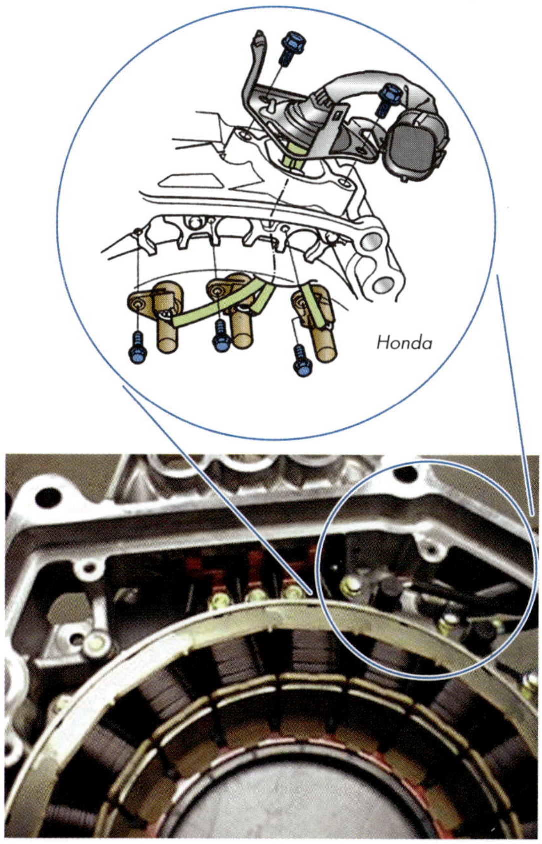

Figure 8-7. The proximity commutation sensor system may use three Hall-effect or magnetoresistive proximity sensors.

MOTOR LOAD SENSING

Hybrid vehicles need to control the electrical power that is generated or consumed by induction/PM machines. Most induction/PM controllers require an input signal from the power inverter giving the electrical current levels.

Electrical current measurement is a direct indicator of the amount of power generated or consumed and can be used to calculate the torque. By measuring the electrical current consumed or generated by the motor or generator, the controller can calculate how much current must be added or subtracted to meet the speed and torque requirements requested by the operator, and regen requirements of the battery system. It can also calculate the amount of regen that can be supplied.

Current Sensors

The controller uses a current sensor to measure the level of electrical current used to propel a vehicle or the level of regen power generated by the induction/PM machine. A *current sensor* is a sensor that provides a motor controller feedback information on phase current. With this information, the motor controller can determine if the correct level of electrical current has been delivered to the electric machine during propulsion mode or from the electric machine to the power inverter during regenerative braking.

A current sensor typically is a Hall-effect sensor that senses the magnetic field around the high-voltage cables that supply power to electric machines. **See Figure 8-8.** Although shunt-type sensors have been used in the past to sense motor current, Hall-effect sensors are more common in contemporary vehicles.

MOTOR LOAD SENSING

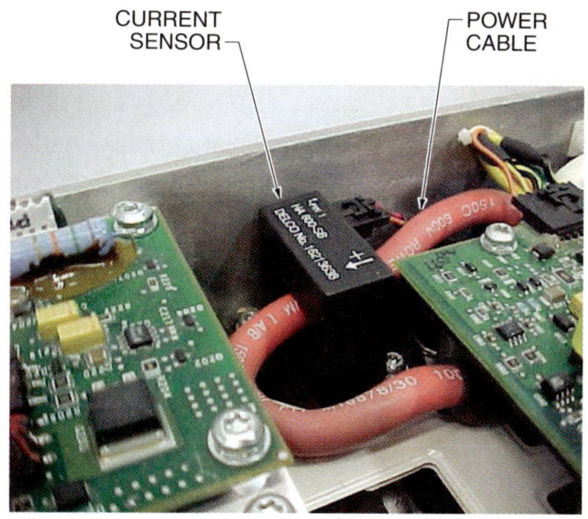

Figure 8-8. A current sensor typically is a Hall-effect sensor that senses the magnetic field around high-voltage cables.

Phase-current sensing systems typically can use two or three sensors in a 3ϕ electric machine. Three sensors are not required because the system only uses two of the three phases of the stator at any one time. Therefore, at least one sensor will always sense current through the stator phases and pole coils. However, three sensors are often used in more contemporary power inverter systems because if one sensor fails or its signal becomes skewed, the controller can ignore its signal and rely on the remaining two sensors.

Three-phase current is alternating current. At some point in the alternating cycle, the voltage and current are zero. Therefore, systems that use current sensors typically cannot operate on only one sensor. In such cases, the electric propulsion system will not function.

The power system controller uses a current sensor signal much like a powertrain controller uses a manifold absolute pressure (MAP) or mass airflow (MAF) sensor in an ICE fuel/emission control system. Both MAP and MAF sensors sense engine load. The MAP sensor uses engine pressure (vacuum) and the MAF sensor uses a hot wire to sense changes to incoming air mass. The current sensor also measures electrical loads and whether the electrical power is consumed (propulsion mode) or generated (regenerative braking mode).

ACCELERATOR PEDAL AND THROTTLE POSITION SENSING

There are systems in hybrid vehicles that use throttle position sensors and pedal position sensors, also called accelerator position sensors, as components of the overall motor/generator torque control strategy. A *throttle position sensor (TPS)* is a sensor located on the engine throttle body that is used to monitor the position of the throttle. A *pedal position sensor (PPS)* is a sensor located on the accelerator pedal assembly that provides an input to an electric machine motor controller so that it can sense the desired acceleration rate and level of power desired by the vehicle operator. A PPS provides throttle angle information to the controller in the form of a voltage signal.

Throttle Position Sensors

A common TPS system design is to have two or three Hall-effect sensors placed within the throttle sensing assembly. **See Figure 8-9.** By using more than one sensor, the controller can constantly cross-check the dynamic operation of each sensor by comparing one sensor with another. Each sensor is normally designed with a different voltage slope. This means that each sensor has a different voltage value at a given throttle angle. The power inverter controller can compare the sensors to determine if one of them is not reporting the same throttle position as the other(s). If the sensor voltages do not correlate properly, the controller sets a diagnostic code for the suspect sensor.

THROTTLE POSITION SENSING

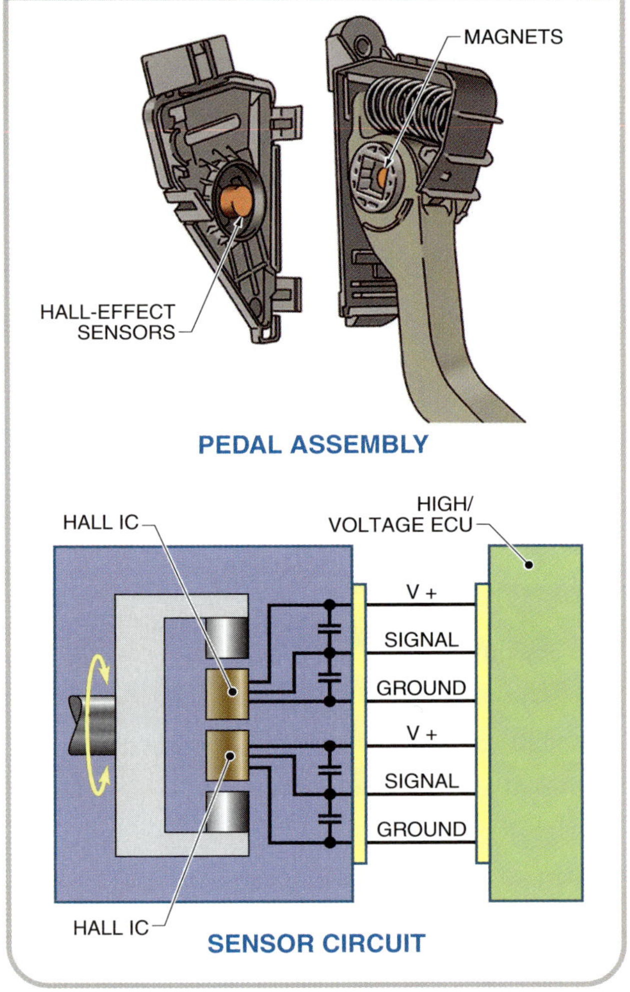

Figure 8-9. A TPS normally uses multiple sensors.

If only one sensor were used and it malfunctioned, the power inverter would command the electric machine to a maximum torque level. Therefore, more than one sensor is used to ensure safety and accuracy. If there is a discrepancy between any one sensor and the others, the controller no longer uses the malfunctioning sensor information. It uses only the remaining sensor(s). The controller lights the malfunction indicator lamp and a system fault code is stored in the controller memory.

The operation of a TPS should be very familiar to an automotive technician because all ICE engines with electronic feedback controls use a TPS. The TPS sensor measures throttle angle by resistive or electronic circuit technology. The sensor can be part of the throttle pedal assembly or located in the motor compartment as a separate device linked to the throttle. Older TPS systems used potentiometers instead of Hall-effect sensors. These are rarely used any more.

OTHER CONTROLLER INPUTS

There are other inputs that provide valuable information to the controller for propulsion control that are not unique to hybrid vehicles. **See Figure 8-10.** The sensors discussed up to this point are the most critical to the electric propulsion system. However, they are not the only inputs controllers use to control the propulsion system. There are other important inputs to a hybrid controller that control various propulsion and accessory functions such as switches related to safety or convenience systems. These switches include, but are not limited to, stop light, cruise control, coolant pump relay, 12 V power, airbag sensor, and transponder key ECU switches.

TORSIONAL HARMONIC DAMPENING

Electric propulsion systems provide significant flexibility in overall powertrain operating modes. Power electronic systems are infinitely variable in their ability to control power. Sensor systems play a significant role in power control. This allows the design engineer many different ways to blend electric propulsion with ICE systems. For example, an electric propulsion system can be used to optimize torsional harmonic dampening in ICE systems operation. Vehicle occupants can feel engine driveline power impulses caused by cylinders that are not producing power during certain modes of operation.

In 1994, the General Motors Impact (later to be named EV1) set the land speed record for electric vehicles at 189 mph.

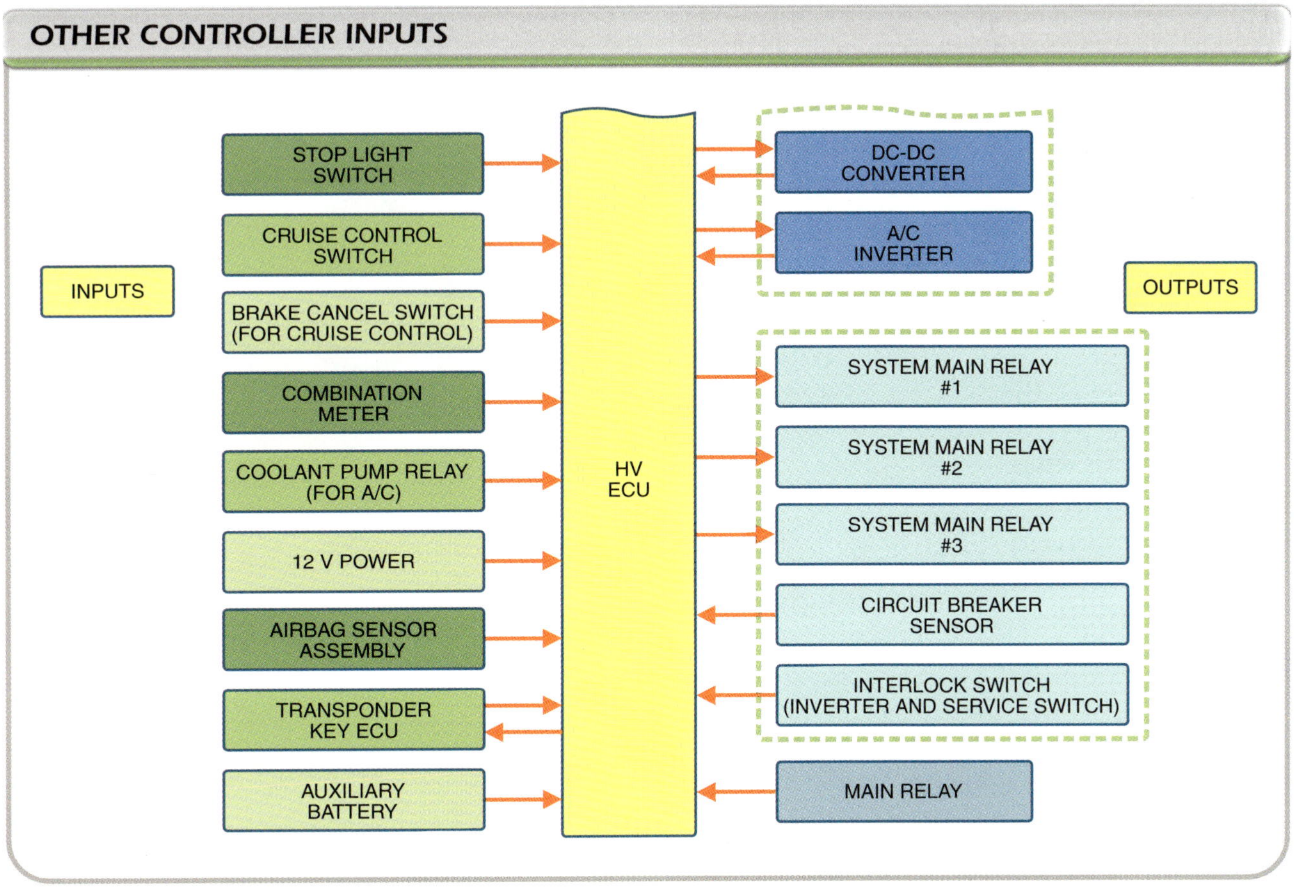

Figure 8-10. There are other important inputs to a hybrid controller that affect the hybrid system, such as switches related to safety or convenience systems.

An electric machine can be used to dampen the power impulses that cause engine cylinder torsional harmonics. Engine controller sensing systems are used to monitor engine rpm and firing sequence. The hybrid and power inverter controllers even out the power impulses by adding torque and rpm with the electric machine(s). **See Figure 8-11.** The controller response is based on the inputs from the engine rpm, TPS, MAP/MAF current sensors, and other sensor inputs.

Engine speed fluctuations of the pistons result in sinusoidal crankshaft speed changes and powertrain rpm variations. The controller monitors the engine rpm as a linear value where the rpm is increasing or decreasing at a specific rate. Based on that rate, and combined with other sensor inputs, the controller calculates the specific times when the driveline impulses are to occur. The controller applies the torque input to the driveline at the proper time to eliminate crankshaft speed reductions when the pistons are in a spectrum of the rotating cycle that is not producing power. This results in a smoother power transfer to the wheels.

The wiring for battery sensors is visible near the battery pack.

TORSIONAL HARMONIC DAMPENING

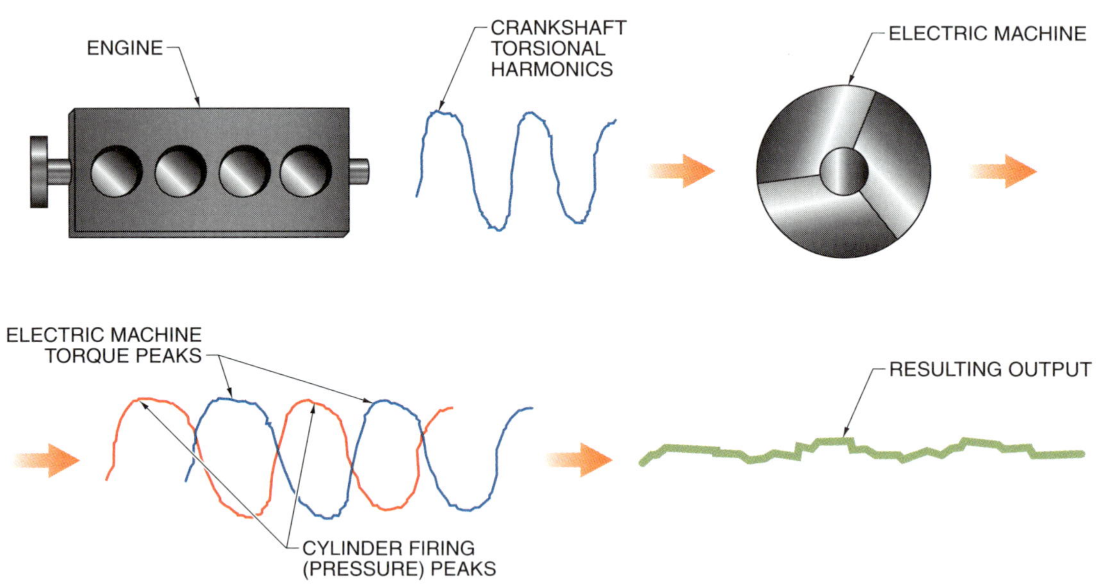

Figure 8-11. The hybrid and power inverter controllers even out power impulses by adding torque and rpm to the electric machine(s).

Refer to the *Hybrid Electric Vehicle Technology* CD-ROM for a Quick Quiz® on the chapter contents.

POINTS TO REMEMBER

- Two types of sensors used to sense electric machine speed (rpm) are the encoder and resolver.
- Encoders provide a digital, square wave to the controller. The motor controller counts the square waves to determine the approximate electric machine rotor position and speed.
- Resolvers provide two sinusoidal waveforms to the controller. The motor controller calculates the absolute rotor position and speed by comparing the amplitude and waveform position.
- Proximity sensors provide a digital, square-wave output to the controller by sensing the position of the commutation plate. The motor controller uses the edge of the square wave to determine the approximate electric machine rotor position and speed.
- Current sensors provide feedback information on phase currents. The motor controller determines if the correct level of electrical current has been delivered to the electric machine during propulsion mode or during regenerative braking mode.
- A minimum of two current sensors are necessary to control a 3ϕ electric machine. Three sensors can be used in a system.
- A current sensor is analogous to the MAP or MAF sensor in an internal combustion engine (ICE) system, which senses engine intake manifold pressure (vacuum) so that the engine controller can calculate engine load.
- A pedal position sensor (PPS) provides the throttle angle to the controller. The controller determines the desired acceleration rate and level of electric machine power desired by the vehicle operator.
- A PPS is analogous to the throttle position sensor (TPS) in an ICE control system.
- A hybrid controller uses many inputs to control the electric machine operation. These inputs are determined by the vehicle platform and architecture.

Refer to the *Hybrid Electric Vehicle Technology* CD-ROM for the Review Questions in Microsoft® Word format.

REVIEW QUESTIONS

1. What type of sensor is used in a PM electric machine to measure its speed and position?
2. Why are three current sensors used to sense electric machine phase current?
3. What is the result if an electric machine has only one current sensor input?
4. What type of waveform output is provided by an encoder?
5. How would adding more lobes to a resolver affect the accuracy of the sensor?

8 Electric Propulsion Sensing Systems

HYBRID Electric Vehicle TECHNOLOGY

Trade Test

Name ______________________ Date ______________

______ **1.** The two types of sensors used to sense motor speed are ___ sensors.
- A. reflective and encoder
- B. Hall and magnetosense
- C. proximity and current
- D. encoder and resolver

______ **2.** The Toyota Prius uses a(n) ___ sensor.
- A. encoder
- B. resolver
- C. magnetoresistive proximity
- D. Hall-effect

______ **3.** The Honda Civic Hybrid uses a(n) ___ sensor.
- A. encoder
- B. resolver
- C. magnetoresistive proximity
- D. Hall-effect

______ **4.** An induction/PM controller needs at least ___ internal power-inverter current sensor(s) to safely measure 3ϕ current.
- A. one
- B. two
- C. three
- D. six

______ **5.** A resolver creates a ___ waveform.
- A. square
- B. sine
- C. trapezoidal
- D. six-step

______ **6.** The Ford Escape Hybrid uses a(n) ___ sensor.
- A. resolver
- B. encoder
- C. magnetoresistive proximity
- D. magnetic

____________________ **7.** An induction/PM machine uses a ___-sensing or rpm-sensing device to provide an input signal to the controller, which uses the signal to calculate the proper slip percentage or phase advance/retard to control the machine.

A. speed
B. current
C. voltage
D. throttle

____________________ **8.** With a resolver, the ___ generates a pulsed AC sine-wave output that is used as the reference signal for measuring the sine and cosine outputs.

A. R/D converter
B. oscillator
C. reluctor lobe
D. encoder

____________________ **9.** A(n) ___ provides a digital, square-wave output in which the electric machine motor controller counts to determine the approximate rotor position and speed.

A. resolver
B. current sensor
C. four-lobe reluctor
D. encoder

DC-DC Converter Systems

9

The DC-DC converter is a critical component in the hybrid system because it replaces the traditional alternator/generator to provide low-voltage, and in some cases, high-voltage electrical power. Although the location of this converter can differ based on manufacturer preference, the operation of this power electronic device is similar (if not identical) regardless of the manufacturer. With the advent of hybrid and electric vehicle power systems with multiple voltage levels, a DC-DC converter plays the major role of controlling and supplying the power needs for each voltage level.

CHAPTER OBJECTIVES

- Describe the primary function of a DC-DC converter.
- Identify the locations of DC-DC converters.
- Explain the difference between step-down (buck) and step-up (boost) converters.
- List the components of a traditional vehicle that are replaced by a DC-DC converter in a hybrid electric vehicle.
- Describe the basic operating cycles of DC-DC converters.

KEY TERMS

DC-DC converter: An electronic device that converts high-voltage DC power to low-voltage DC power, or vice versa.

buck: A term used to describe DC-DC converter voltage that is stepped down, or reduced.

boost: A term used to describe DC-DC converter voltage that is stepped up, or increased.

forward-switching converter: A converter used to transform an input voltage into a regulated output voltage.

center-tapped inductor: An inductor coil with a secondary winding that has an electrical point in the center allowing it to be used as two electrical circuits.

full-wave rectifier: A diode system of at least two rectifier diodes that allows the conversion of positive and negative components of an AC sine wave into a DC output.

choke: An inductor used with an electric filtering system to smooth out varying electrical current.

DC-DC CONVERTERS

With AC power, power transformers can be used to step electrical power either up or down. The desired output can be acquired by designing a transformer winding with the proper turns ratio. A DC-DC converter is necessary because DC power cannot easily be transformed from a specified voltage level. A *DC-DC converter* is an electronic device that converts high-voltage DC power to low-voltage DC power, or vice versa.

In the past, DC-DC converters were used in truck applications where it was required to step 24 V DC down to 12 V DC to operate a car radio, communications transceivers, mobile phones, or other accessories. In other automotive applications, DC-DC converters were used to step down the 12 V battery to 3 V in order to operate a personal compact disc player, or to 5 V to operate a personal computer.

A higher power application may require a 300 V DC signal from a rectified 240 V AC source to be stepped down to 12 V DC or less. There were also special applications where a 12 V DC source needed to be stepped up to 650 V DC as part of a DC-AC sine-wave inverter. In each of these applications, there was a need to efficiently change DC energy from one voltage level to another.

In contemporary automotive systems, a DC-DC converter is necessary to provide electrical power at the specified power supply levels that the vehicle systems require. A 12 V battery system (14 V operating voltage) is used for ignition, fuel, emission, general electrical system, and standard accessory systems in most mid-level to high-end automobiles.

However, in HEVs that use a traditional 12 V electrical system for operations other than propulsion, the 12 V battery and electrical system are constantly supplied with electrical power from a source that can exceed the demand for electrical power. In addition, modern designs may not include the traditional 12 V cranking motor and generator/alternator systems.

DC-DC converters can provide electrical power at any voltage level needed. Traditional generators are no longer a component in most hybrid and electric power systems. Newer DC-DC converter systems replace the generator and supply the 14 V or 42 V system power to the circuits that require those voltage levels. With a DC-DC converter, a 12 V battery can supply power to the high-voltage battery pack at a higher voltage. The DC-DC converter is typically located within a power electronics package that includes other subsystems, such as the power inverter and microprocessor. **See Figure 9-1.**

Because of limited space, vehicle designers place DC-DC converters in various places. For example, the DC-DC converter in a Honda Civic Hybrid is housed near the battery pack. The DC-DC converter in a Ford Escape Hybrid is housed under the hood.

The DC-DC converter of the Ford Escape Hybrid steps down the 330 V DC battery source voltage to 14 V DC to operate the vehicle's electrical systems. The battery pack has two cables that connect to the DC-DC converter to provide electrical power to the 12 V systems. By housing the DC-DC converter close to the 12 V battery, voltage losses due to long lengths of power cabling are eliminated. This optimizes the transfer of electrical power. **See Figure 9-2.**

Although there are numerous DC-DC converter designs, the two most common designs are the buck-boost and forward-switching (isolated) converter.

Electric propulsion for boats is routinely found in the western coastal states of the United States. Electric propulsion is used to reduce noise and environmental pollution. The electrical storage system can be recharged from utility power or solar panels.

DC-DC CONVERTERS

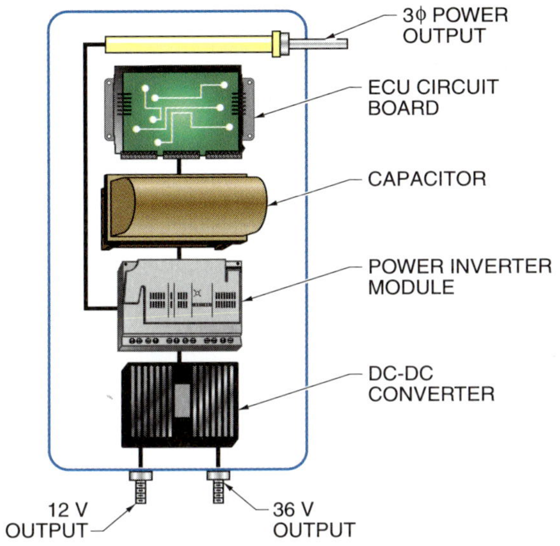

Figure 9-1. A DC-DC converter is typically located within a power electronics package that includes other subsystems, such as the power inverter and ECU circuit board.

CONVERTER LOCATION

Figure 9-2. The DC-DC converter in a Ford Escape Hybrid is housed under the hood.

Buck-Boost Converters

Buck-boost converters are used to buck (step down) or boost (step up) voltage. *Buck* is a term used to describe DC-DC converter voltage that is stepped down, or reduced. *Boost* is a term used to describe DC-DC converter voltage that is stepped up, or increased. **See Figure 9-3.**

The block diagram of a hybrid and electric power system DC-DC converter system shows how voltage can be stepped down or stepped up. The yellow center section shows two converter blocks. One block steps down (bucks) the voltage to 14 V and the other steps up (boosts) the voltage from 14 V. This type of DC-DC converter is a bidirectional DC-DC converter because it can buck or boost voltage as needed. The circuit components include a bipolar transistor (Q1), inductor (L1), diode (D1), capacitor (C1), and the essential wiring for the circuit.

When Q1 is turned ON in a buck-boost DC-DC converter, inductor L1 is connected directly across the source voltage and electrical current flows through the inductor. This stores energy in the magnetic field. No current can flow through D1 to the load because the diode is connected so that it is reverse biased and blocks electrical current. Capacitor C1 must supply the load current in this ON phase.

In addition, the pulse width modulated (PWM) control and logic, the transistor drive circuit, and all electrical loads are part of the circuit. The design of this converter permits the voltage to be stepped up or down depending on the control duty cycle commanded by the DC-DC converter controller. When Q1 is turned OFF, L1 is disconnected from the source. Through self-induction, L1 opposes any tendency for the electrical current to decrease, and it instantly reverses its voltage. This generates a voltage that forward biases D1 and electrical current flows into the load. The current also recharges C1.

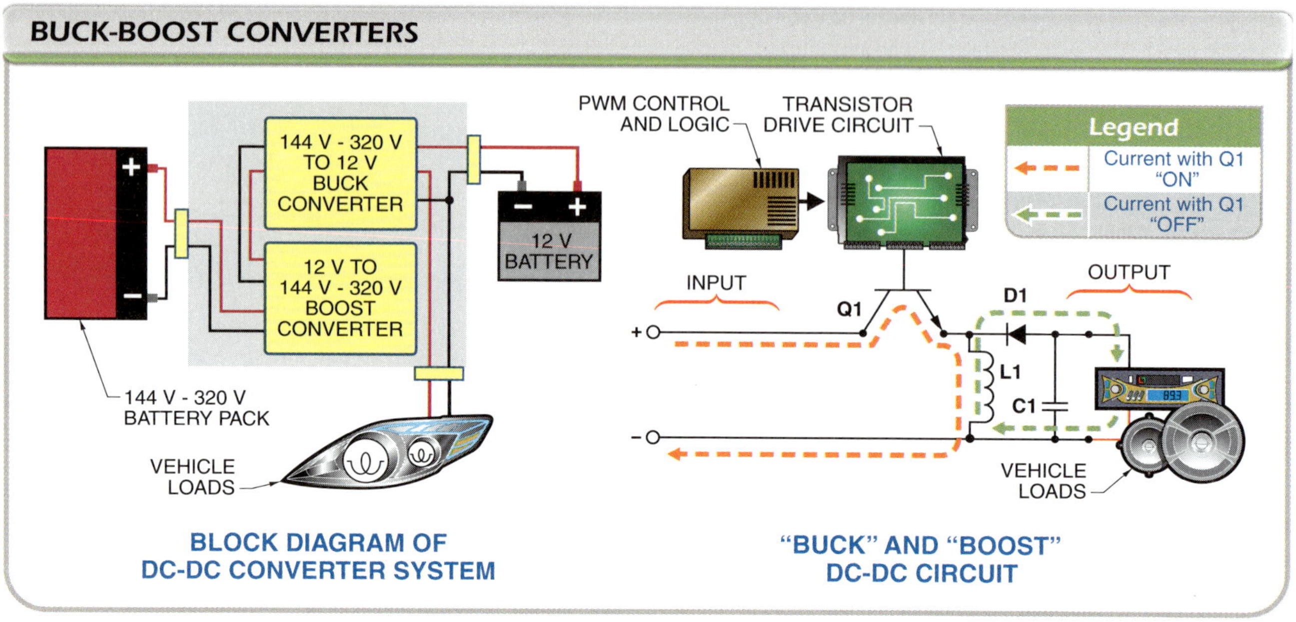

Figure 9-3. Buck-boost converters are used to buck (step-down) or boost (step-up) voltage.

Self-induction and mutual induction are important to understanding not only electric machine operation, but also any device that reduces or increases voltage levels using a coil. With this configuration of a DC-DC converter, the ratio between the output and input voltages is calculated with the following formula:

$$\frac{V_{out}}{V_{in}} = -\frac{T_{on}}{T_{off}}$$

where

V_{out} = voltage across the output (in V)
V_{in} = voltage applied to the input (in V)
T_{on} = transistor on-time (in sec)
T_{off} = transistor off-time (in sec)

The ratio of T_{on} to T_{off} is the duty cycle. For example, if the transistor on-time is 10 seconds and the transistor off-time is 5 seconds, the ratio between the output and input voltages is calculated as follows:

$$\frac{V_{out}}{V_{in}} = -\frac{T_{on}}{T_{off}}$$

$$\frac{V_{out}}{V_{in}} = -\frac{10}{5}$$

$$\frac{V_{out}}{V_{in}} = \mathbf{-2}$$

A buck-boost converter steps down the voltage when the duty cycle is less than 50% ($T_{on} < T_{off}$). It steps up the voltage when the duty cycle is greater than 50% ($T_{on} > T_{off}$). The output voltage is always reversed in polarity with respect to the input. Therefore, a buck-boost converter is also a voltage inverter that changes the polarity of the output voltage relative to the input voltage.

When the duty cycle is exactly 50%, V_{out} it is essentially identical to V_{in} except for the change in output polarity. Because a buck-boost converter reverses the voltage polarity, it may be used to generate a negative voltage to equipment that is operating from a single battery source. Because the power does not change when going through a buck-boost converter, the ratio between output and input current is the reciprocal of the voltage ratio.

Forward-Switching Converters

A *forward-switching converter* is a converter used to transform an input voltage into a regulated output voltage. The output may be lower or higher than the input voltage depending on its transformer turn ratio. A forward-switching converter uses two distinct phases for energy storage and delivery to the output. It uses a transformer (L1 primary winding and L2 secondary winding) to transfer the energy directly between the input and output in one step. **See Figure 9-4.** In the illustration, winding L1 is the primary and winding L2 is the secondary of the transformer.

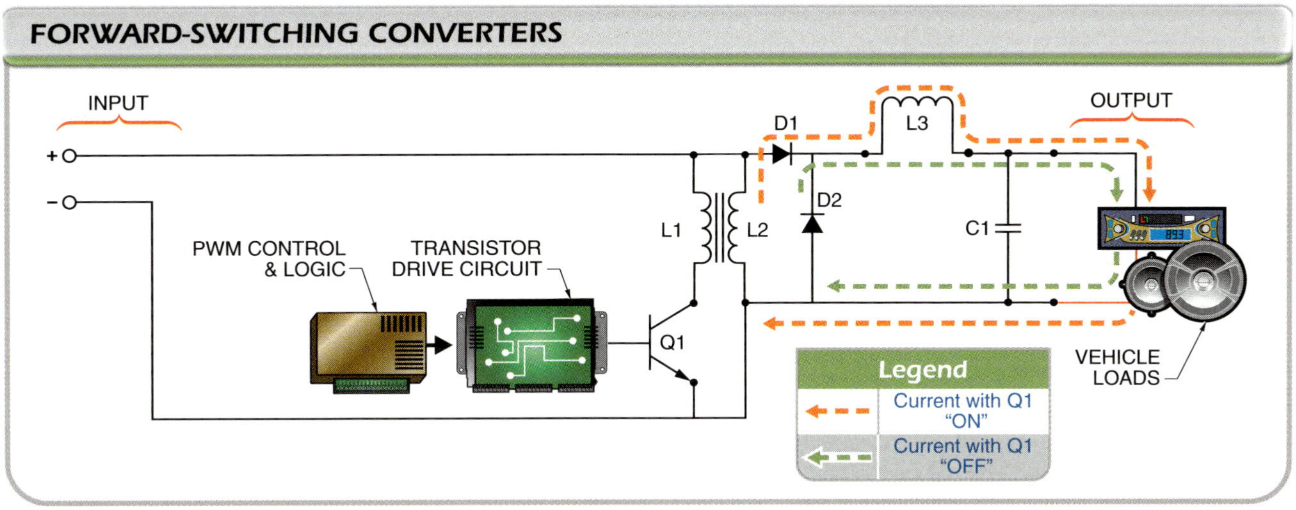

Figure 9-4. A forward-switching converter uses a transformer to transfer energy directly between the input and output in one step.

In this circuit, when Q1 is turned ON, the transformer primary current increases linearly at a rate approximately equal to $V_{in}/L_{primary}$, (L = transformer primary winding). During this time, a voltage equal to V_{in}/N (N = transformer turns ratio) is present at the transformer's secondary winding. The secondary electrical current that flows through rectifier diode D1 and output inductor L3 increases linearly at a rate approximately equal to L1 (V_{in}/N). This electrical current also flows into the output loads and the output capacitor C1.

As the voltage across C1 rises out of the PWM upper regulation tolerance window, Q1 turns OFF and terminates the electrical current flow from the voltage source. When Q1 is turned OFF, the secondary winding (L2) undergoes a voltage phase reversal and no longer provides electrical current through D1. However, electrical current continues to flow in L3 and the current is blocked by rectifier D2.

The L1 electrical current now decreases linearly at a rate approximately equal to V_{out} / L1. When the voltage across capacitor C1 falls out of the PWMs lower regulation tolerance window, D1 turns ON and repeats the cycle. The output voltage ratio of a forward switching converter is calculated as follows:

$$\frac{V_{out}}{V_{in}} = \frac{1}{N}\delta$$

where
V_{out} = voltage across output (in V)
V_{in} = voltage applied to input (in V)
N = transformer turns ratio
δ = duty cycle

A forward-switching DC-DC converter can also be used to step down the output voltage. To step down the output voltage, the primary winding is designed with more turns of wire than the secondary winding. Since the primary winding has more turns of wire than the secondary, the secondary output voltage is less than the input of the primary.

For example, if the primary winding produces 24 V and the turns ratio between the primary and secondary windings is 2:1, then the output of the secondary is 12 V. The only physical change between a buck or boost DC-DC forward (isolation) converter is the ratio of winding turns between the primary and secondary inductor coils.

This circuit is similar to an ignition coil and ignition module circuit where two coils make up the basic components of the automotive ignition coil. The automotive ignition coil has primary and secondary windings with a specified ratio of turns in the primary and secondary.

For example, if an ignition coil rated at 40,000 V output has a secondary winding with 10,000 turns and a primary winding with 100 turns, the ratio of secondary to primary is 10,000:100, or 100:1. The primary winding consists of a solid, heavy-gauge wire (around a #22 gauge) to carry the larger primary current. The secondary consists of a very fine wire (around a #41 gauge) to carry the smaller secondary current. If the primary of the ignition coil can produce 400 V of induced voltage when the ignition module turns OFF and stops the primary current, the secondary induced voltage will be 40,000 V (400 V primary × 100 turns ratio = 40,000 V secondary).

Also, the primary and secondary windings in an ignition coil do not come in contact with each other. The electrical energy can be isolated within a circuit and, therefore, not influence an adjacent or succeeding circuit. This isolation is used for circuit safety and to protect electronic components that are susceptible to electrically noisy circuits. In the forward/isolation transformer, the only transfer of energy between circuits is in the form of magnetic energy through mutual induction.

DC-DC CONVERTER CIRCUITS

Most automotive technicians do not analyze or diagnose a DC-DC converter at the circuit board level. However, it is important that a technician has a fundamental understanding of DC-DC converter operation. This understanding gives a technician a higher level of confidence when using electrical schematics for performing voltage tests at the converter wire connectors and terminals.

In contemporary hybrid vehicle designs, a DC-DC converter has replaced the traditional generator to provide electrical power. A DC-DC converter provides this electrical output without needing any rotating components or a belt drive system. There are a number of internal components that are necessary to ensure converter operation. The internal components include the PWM switching circuit, transformer windings, full-wave rectifier, and output filter. **See Figure 9-5.**

PWM Switching Circuit

Four transistors are used as switching pairs (Q1 and Q4; and Q2 and Q3) to switch inductor L1. **See Figure 9-6.** Switching these transistors in pairs results in a synthesized fundamental AC waveform as an input to L1. This switching is designed to resemble an AC sine wave that would typically be found in a commercial 120 V AC outlet.

The DC-DC converter designed for an automotive application does not use 120 V AC electrical power as an input to operate the converter. It typically receives electrical power from the hybrid high-voltage battery pack (about 144 V DC to 320 V DC). The electrical power from the battery pack is typically first routed through the power inverter as a "pass through" and then to the DC-DC converter. To permit this electrical routing, the DC-DC converter typically is mounted directly to the power inverter or mechanized as an integral system of the power inverter.

Transformer Windings

The inductor L1 functions as the primary winding of a transformer. The L2 winding functions as the secondary winding of a transformer and its operation is analogous to the secondary (coil) winding of an ignition coil. The diameter size of the L1 wire is larger because it must carry more electrical current to generate a strong magnetic field that collapses across the L2 inductor.

DC-DC CONVERTER DESIGN

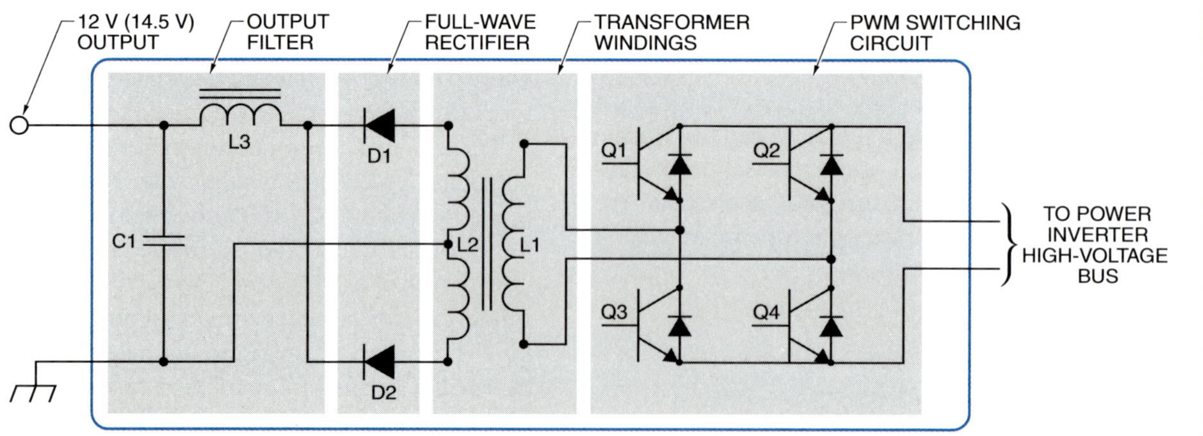

Figure 9-5. The internal components of a DC-DC converter include the PWM switching circuit, transformer windings, full-wave rectifier, and output filter.

PWM SWITCHING CIRCUIT

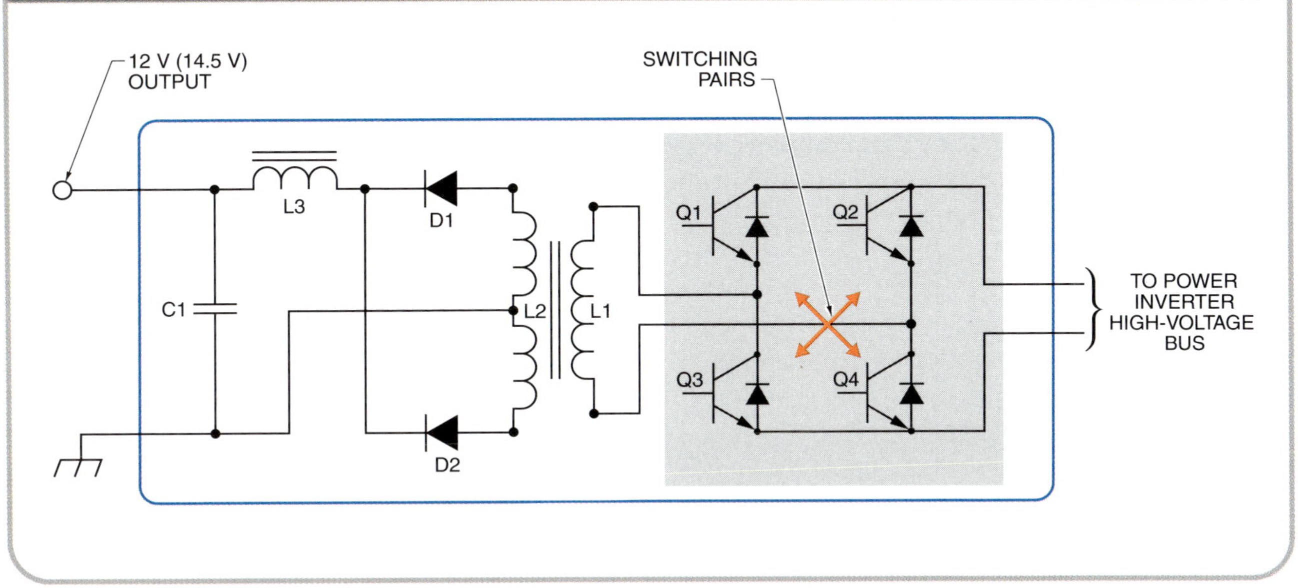

Figure 9-6. Four transistors are used as switching pairs (Q1 and Q4; and Q2 and Q3) to switch inductor L1.

The L2 inductor is designed as a center-tapped inductor. A *center-tapped inductor* is an inductor coil with a secondary winding that has an electrical point in the center allowing it to be used as two electrical circuits. **See Figure 9-7.** The wire at the center of the coil is connected to vehicle chassis ground. The wire functions as the ground reference for the L2 winding. This design permits the windings to provide alternating polarities on L2 with respect to the ground wire.

By connecting the leads of L1 to the switching pairs Q1/Q4 and Q2/Q3, the polarity of L1 can be alternately reversed so that its coil functions with alternating polarities to operate as a transformer. The result of this alternate switching on L1 is an AC output waveform that collapses across L2. This alternating switching pattern that provides an AC output waveform to L1 is meant to be similar to the waveform of the transformer L1 primary windings found in residential AC circuits.

Full-Wave Rectifier

A *full-wave rectifier* is a diode system of at least two rectifier diodes that allows the conversion of positive and negative components of an AC sine wave into a DC output. **See Figure 9-8.** The function of diodes D1 and D2 is to rectify (change) the AC output of L2 to a DC output. An automotive electrical system is designed to operate on DC voltage. Therefore, the DC-DC converter is designed to provide a DC output through D1 and D2. The output from the diodes goes to the output filter circuit as a pulsating DC power.

A DC-DC converter needs its own cooling system.

TRANSFORMER WINDINGS

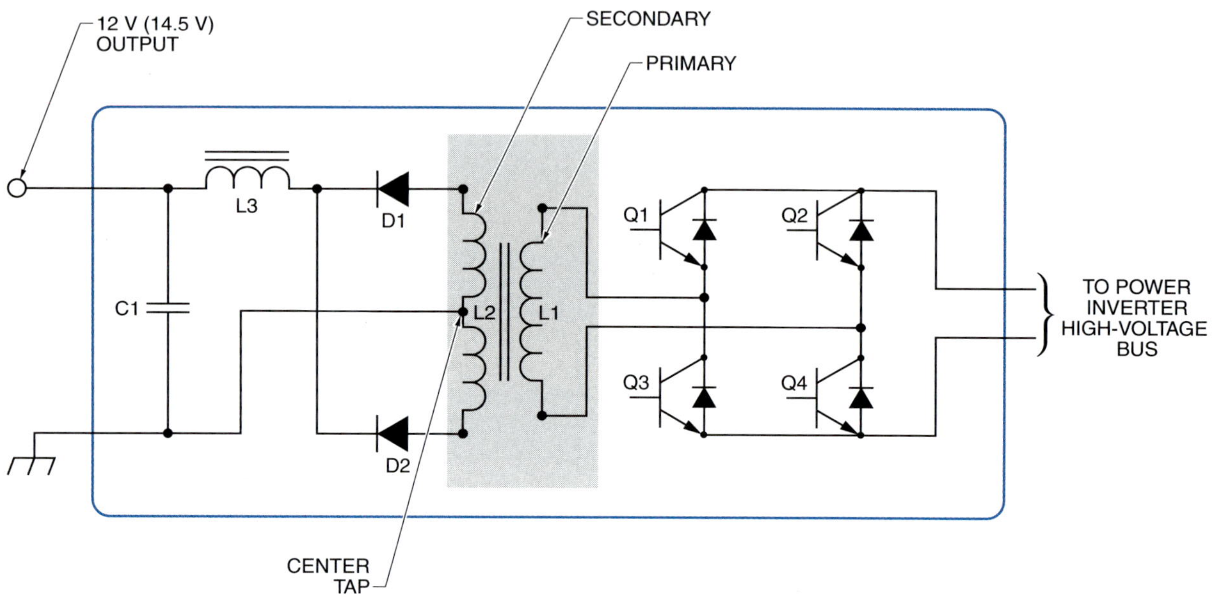

Figure 9-7. A center-tapped inductor has an electrical point at the center of the coil that allows the coil to be used as two separate electrical circuits.

FULL-WAVE RECTIFIER

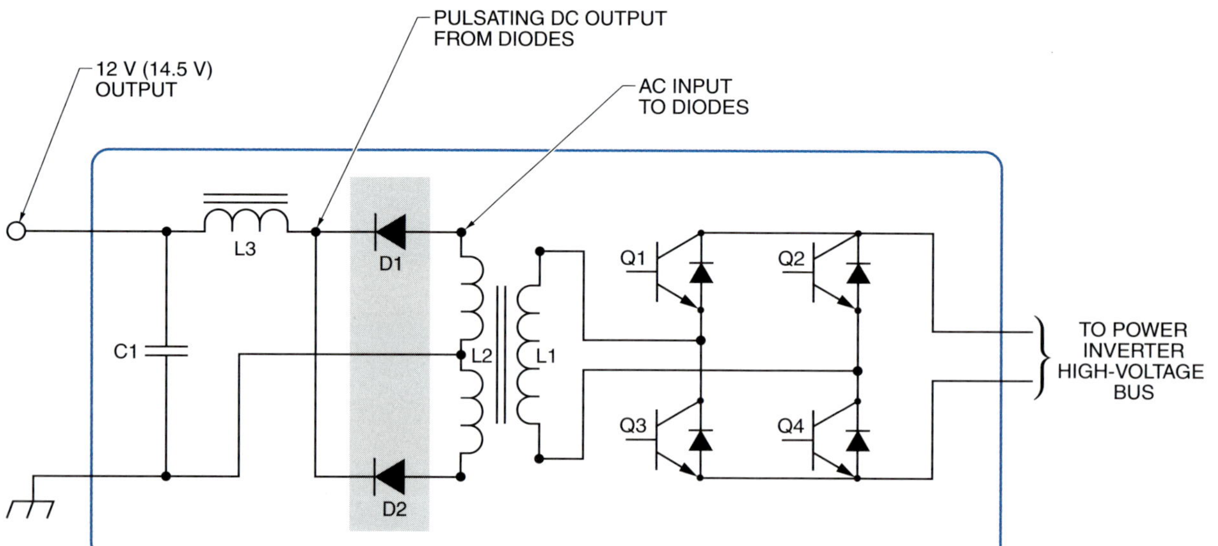

Figure 9-8. A full-wave rectifier converts both the positive and negative components of an alternating current sine-wave power into a DC output.

Output Filter

The purpose of the output filter is to smooth out the pulsating DC power from the rectifier circuit. Pulsating DC waveforms can cause electrical noise in the electrical system. This noise is created when electrical current is changing amplitude and/or direction. Pulsating DC also causes heating in components that receive the waveform because of the peak power values of the waveform.

The average value of a voltage waveform is 63.7% of the peak voltage value. This means that peak levels provide more electrical power. More power means more heat in electrical components. Therefore, electrical engineers design the circuits with filters to minimize the ripple fluctuations. This controls the output voltage to a level that permits safe circuit operation.

A filter consists of a capacitor and an inductor, or choke. A *choke* is an inductor used with an electric filtering system to smooth out varying electrical current. The filter electrically reduces waveform peaks and "fills in" waveform gaps, or low points, to form a more even voltage output. This smoothing of the waveforms results in less electrical noise due to smaller variations in current and less vehicle system circuit heat. **See Figure 9-9.**

In newer filter circuit designs, semiconductor devices, such as MOSFETs and IGBTs, are being used to replace the diodes. This is because the output of the DC-DC converter can be more easily controlled if a power semiconductor device is used. Infinite control of the output can be achieved by using these devices. Newer DC-DC converters also have the dual role of not only bucking the voltage, but also boosting the voltage. To achieve both the buck and boost output, power electronic devices such as MOSFETs or IGBTs are necessary.

An operator display provides information about the operation of a vehicle.

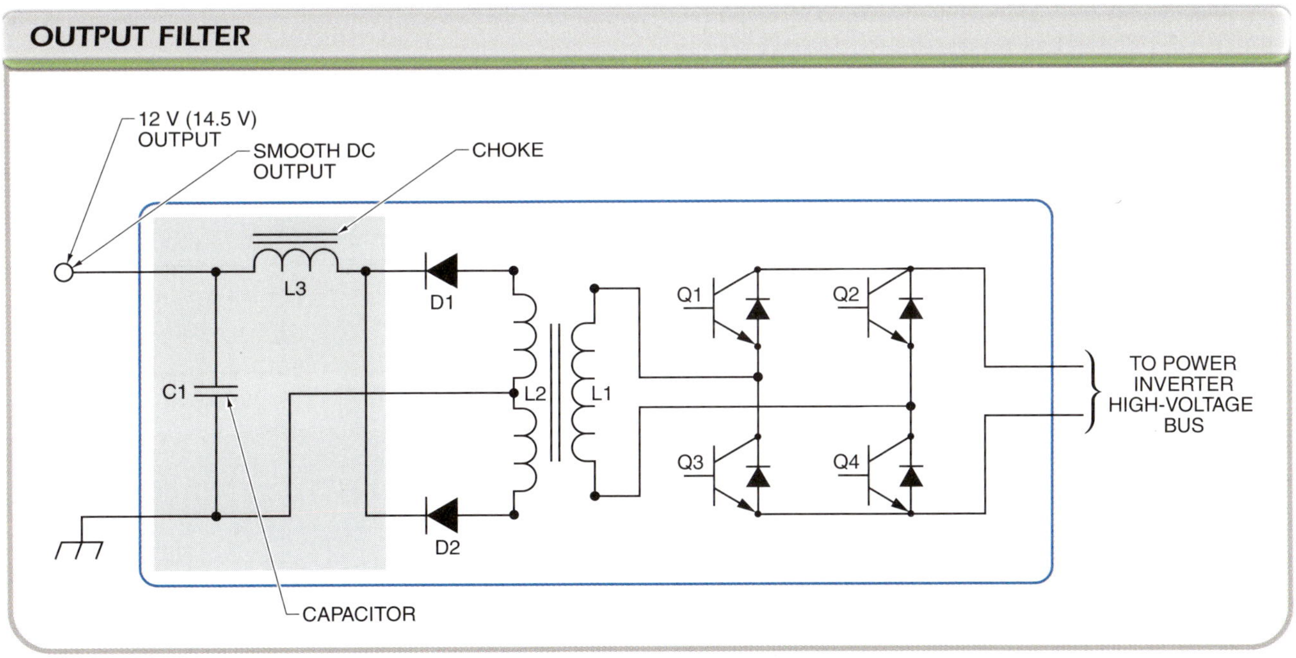

Figure 9-9. A filter consists of a capacitor and an inductor, or choke.

DC-DC Converter Operation

The DC-DC converter changes AC power to DC power in two half-cycle operations. The first half-cycle begins when Q1 and Q4 are turned ON to transfer current through the windings of L1. After a specified period, Q1 and Q4 are turned OFF. This causes the magnetic field to collapse and to induce voltage and current flow. **See Figure 9-10.** Note that the polarities generate a resulting voltage with positive and negative pulses (waveforms).

The second half-cycle begins when Q2 and Q3 are turned ON to transfer current through the windings of L1. After a specified period, Q2 and Q3 are turned OFF. Note that the polarities generate a resulting voltage with positive and negative pulses that are opposite of the first half-cycle.

When L1 is alternately switched ON/OFF by the transistor pairs, its magnetic field collapses during both the first and second half switching cycles. This switching function results in an alternating current waveform at L1.

When L1 is switched ON and OFF by Q1 and Q4, the magnetic field of L1 cuts across L2. When the magnetic field builds and collapses, L2 develops an induced voltage and current flow because of mutual induction. Rectifiers D1 and D2 control the output of L2 because L2 is a center-tapped transformer. In the first half-cycle, diode D1 is forward biased (turned ON) and D2 is reverse biased (turned OFF). Because D2 is not operational in this half-wave rectification cycle, half of the electrical power is delivered by the full-wave rectification cycle.

In the second half-cycle, diode D2 is forward biased (turned ON) and D1 is reverse biased (turned OFF). Because D1 is not operational in this half-wave rectification cycle, half of the electrical power is delivered by the full-wave rectifier. **See Figure 9-11.**

The L2 coil is center-tapped, so it functions as two separate electrical circuits connected to the full-wave rectifier circuit of D1 and D2. The result of the center-tapped coil is two separate outputs that generate positive voltage and current outputs. The first and second half-cycle output waveforms are added to give the total electrical power output of the converter. A typical half-cycle output frequency is 25 kHz and the combined output frequency of both cycles is 50 kHz. The filter smooths the output of the rectifier.

The U.S. Military has implemented hybrid electric vehicle development programs for use in combat operations that require stealth capability.

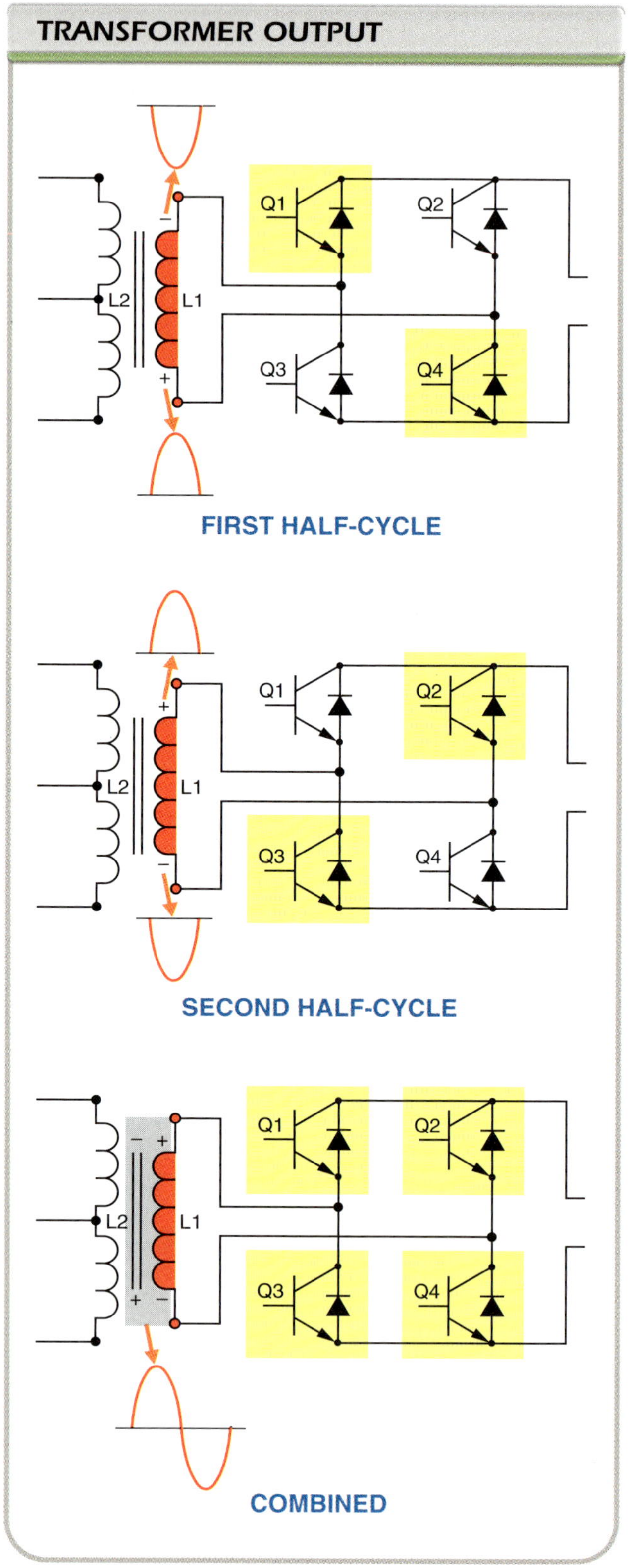

Figure 9-10. A DC-DC converter changes DC power from the high-voltage bus into AC power at the transformer in two half-cycle operations.

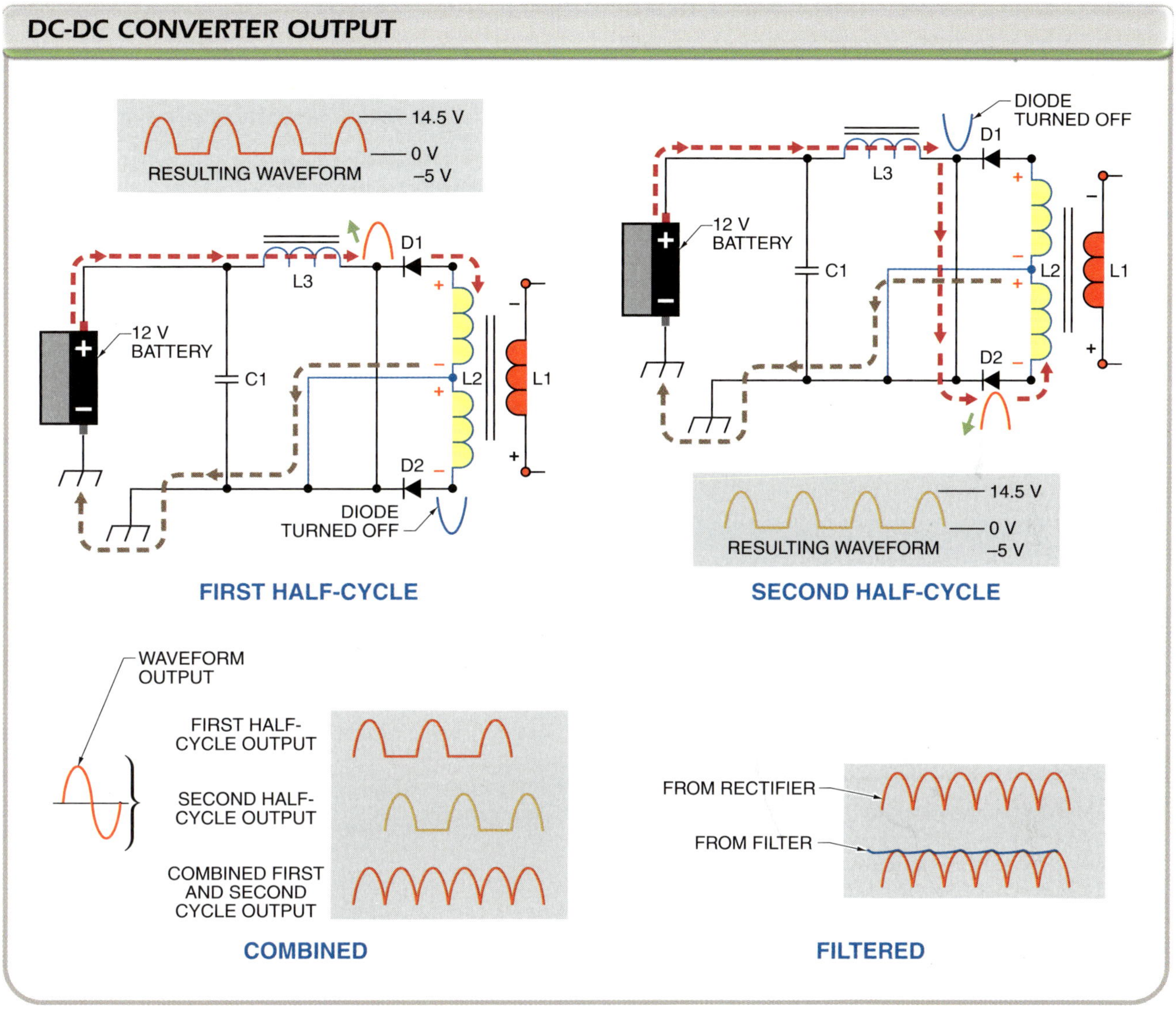

Figure 9-11. A DC-DC converter changes AC power at the transformer into DC power at the rectifier in two half-cycle operations.

Refer to the *Hybrid Electric Vehicle Technology* CD-ROM for a Quick Quiz® on the chapter contents.

POINTS TO REMEMBER

- A DC-DC converter system is used to replace the traditional generator/alternator on an electric, hybrid, or fuel-cell system.
- DC-DC converter operation can be designed to buck, or step down, battery-pack voltage to the normal vehicle operating voltage of about 14 V to operate vehicle systems and charge the 12 V auxiliary battery.
- DC-DC converter operation can be designed to boost, or step up, the 12 V auxiliary battery voltage to charge the battery-pack system.

- A traditional DC-DC buck converter system uses four power transistors to reduce high voltage to low voltage.
- An inductor coil with a center-tapped secondary winding is used to provide electrical power to the vehicle accessories and 12 V auxiliary battery.
- Two rectifier diodes are used to rectify the inductor output from AC to DC.
- The approximate switching frequency of a DC-DC converter in an automotive application is 25 kHz.
- An electrical filtering system is made up of chokes and capacitors.
- Filters are placed on the output of the DC-DC converter to smooth the ripple of the output signal to minimize electrical noise.

Refer to the *Hybrid Electric Vehicle Technology* CD-ROM for the Review Questions in Microsoft® Word format.

REVIEW QUESTIONS

1. What type of DC-DC converter steps up the output voltage relative to the input voltage?
2. What type of DC-DC converter steps up and steps down the output voltage relative to the input voltage?
3. Which hybrid component provides the electrical input to the DC-DC converter?
4. What is the minimum number of diodes necessary for a full-wave rectifier bridge to produce DC electrical power?
5. What is the purpose of the filter circuit in a DC-DC converter?

9 DC-DC Converter Systems

HYBRID Electric Vehicle TECHNOLOGY

Trade Test

Name ______________________ Date ______________

______ **1.** A DC-DC converter replaces the ___ of a traditional vehicle design.
A. cranking motor
B. voltage regulator
C. generator/alternator
D. accessory power inverter

______ **2.** The Honda Civic Hybrid DC-DC converter is located near the ___.
A. battery pack
B. engine
C. electric machine
D. 12 V battery

______ **3.** A buck converter uses ___ diodes.
A. 2
B. 4
C. 6
D. 8

______ **4.** The power input provided to a DC-DC converter comes from the ___.
A. power inverter transistor
B. high-voltage bus
C. 12 V battery
D. diodes

______ **5.** The DC-DC converter of the Toyota Prius is located in the ___.
A. battery pack
B. transmission
C. transaxle
D. power inverter

______ **6.** The DC-DC converter of the Ford Escape is located in the ___.
A. transmission
B. underhood/engine compartment
C. power inverter
D. battery pack

_____________________ **7.** A ___ DC-DC converter steps down, or reduces, high voltage to low voltage.
A. buck
B. boost
C. reactive
D. doubler

_____________________ **8.** The common components of a filter circuit include the ___.
A. battery pack and inductors
B. choke and capacitor
C. inductors and diodes
D. transistors and capacitors

HYBRID Electric Vehicle TECHNOLOGY

Transaxles, Gears, and Cooling Systems

10

Electric machines are capable of creating high levels of torque considering their size. However, electric machine torque does not provide proper levels of acceleration in most applications. Therefore, transaxle systems are coupled with the electric machine to provide gear reduction in order to increase torque. As with a traditional transaxle for an engine, a differential system is connected to the gear reduction stage of the transaxle as an output. In addition, torque production produces heat, and this heat must be removed from the electric machine and power inverter to ensure that these components operate at safe temperatures.

CHAPTER OBJECTIVES

- Describe the basic components and concepts of the Toyota Prius electric transaxle system.
- Explain why the system of the Toyota Prius is categorized as an electronically variable transmission (EVT) system.
- Explain the role that the transmission generator plays in cranking a Toyota Prius engine.
- Describe the component within the Toyota Prius transmission that controls the infinite gear ratio.
- Describe the construction of power electronics and an electric machine cooling system.

KEY TERMS

transaxle damper assembly: A device that permits electric machine power and torque to be transmitted from an engine to an electronically variable transmission. It mechanically links the engine to the input shaft of the transmission.

motor-generator one (MG1): An electric machine contained within the Toyota Prius electric transaxle that determines transaxle gear ratio, generates electrical power for the drive motor and battery-pack systems, and functions as the engine cranking motor.

motor-generator two (MG2): An electric machine contained within the Toyota Prius electric transaxle that serves as the propulsion motor and operates in conjunction with the engine system to provide vehicle propulsion and reverse propulsion mode.

TRANSAXLES AND GEARS

Most electric machines are coupled to a gearbox and differential to increase torque. **See Figure 10-1.** The torque of an electric machine can be increased to very high levels to provide fast acceleration. An electric machine can operate at a speed range of 6000 rpm to 16,000 rpm. Therefore, a multispeed transmission or a planetary gear set and clutch arrangement is not necessary. By adjusting the maximum electric machine rpm and sizing the correct gear reduction and differential ratios, an electric propulsion system can rival most ICE vehicles and deliver 140 HP of peak horsepower.

With the introduction of hybrid vehicles and the use of electric machines, varied transmission designs and system architectures are possible. In the Ford Escape, GM-Allison, Lexus, and Toyota hybrid models, electric machines are part of the transmission. Two electric machines are coupled to a planetary gear set. This provides a method for electrically blending engine and electric machine torque and speed that produces infinitely variable gear ratios. The result is reduced emissions, better efficiency, and reduced fuel consumption.

Toyota Prius Components

The Toyota Prius electric machine and planetary gear set system has a transaxle with electronically variable transmission (EVT) operating modes. Although the Prius transmission is only one of several EVT technologies, it can serve as a foundation to understand EVT technology.

The construction of the Toyota Prius EVT is much simpler than a traditional automatic transmission because it contains fewer components. It does not have a valve body system, hydraulic clutches, servos, bands, or high-pressure hydraulic pumps. However, the transmission does have several new components. These include a transaxle damper assembly, motor-generator one (MG1), motor-generator two (MG2), and a drive chain and gears. **See Figure 10-2.**

DIFFERENTIAL AND OUTPUT GEARS

Figure 10-1. Most electric machines are coupled to a gearbox and differential to increase torque.

TOYOTA PRIUS COMPONENTS

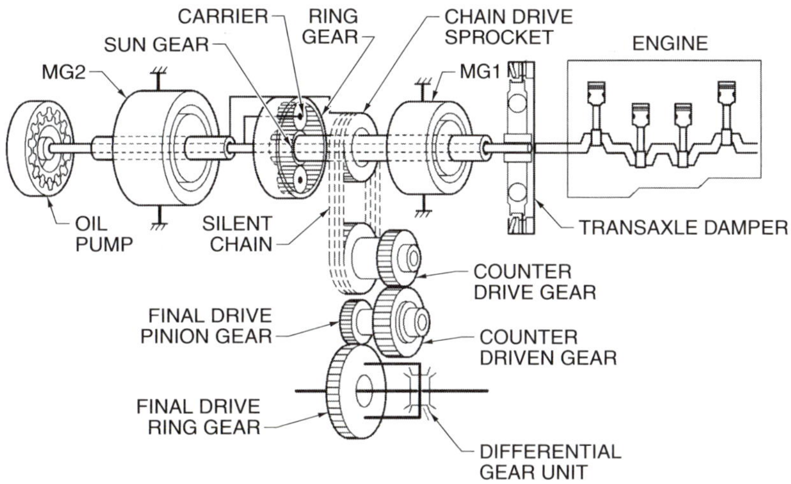

Toyota Motor Company

Figure 10-2. The Toyota Prius transmission includes a transaxle damper assembly, motor-generator one (MG1), motor-generator two (MG2), and a drive chain and gears.

Transaxle Damper Assembly. A *transaxle damper assembly* is a device that permits electric machine power and torque to be transmitted from an engine to an electronically variable transmission. It mechanically links the engine to the input shaft of the transmission. The electronically variable transmission reduces engine torque pulsations and driveline impulses for smoother drive system operation.

The transaxle damper assembly functions as a torsional dampening device to smooth engine torsional oscillations by using a system of springs and discs. It is also used to dampen torque fluctuations. It has four coil springs with low torsional characteristics to dampen the engine drive forces that are transmitted in the driveline. A mechanical absorption method is used that is designed with a dry, single-plate friction material. **See Figure 10-3.**

Motor-Generator One (MG1). *Motor-generator one (MG1)* is an electric machine contained within the Toyota Prius electric transaxle that determines transaxle gear ratio, generates electrical power for the drive motor and battery-pack systems, and functions as the engine cranking motor. MG1 is mechanically connected to the sun gear of the planetary gear set and determines the gear ratio.

MG1 splits, or blends, the torque of the engine and MG2 assists the battery pack in providing additional electrical power to the drive motor (MG2) during high load conditions. MG1 recharges the battery pack when the vehicle is in park, with the engine operating, and starts the engine by functioning as a cranking motor.

TRANSAXLE DAMPER ASSEMBLY

Figure 10-3. The transaxle damper assembly uses four coil springs with low torsional characteristics to dampen the engine drive forces that are transmitted in the driveline.

Motor-Generator Two (MG2). *Motor-generator two (MG2)* is an electric machine contained within the Toyota Prius electric transaxle that serves as the propulsion motor and operates in conjunction with the engine system to provide vehicle propulsion and reverse propulsion mode. MG2 is mechanically connected to the ring gear of the planetary gear set and provides additional torque to supplement engine torque. Propulsion torque is supplied in electric mode with the engine OFF if the hybrid controller has prohibited engine operation. MG2 functions as a generator in regenerative braking mode to recharge the battery pack if state of charge is low.

Drive Chain and Gears. The drive chain transmits engine and/or MG2 torque from the planetary gear set and ring gear to the counter gears, final drive gears, and drive axles. The power-splitting device is made up of a single planetary gear set. A planetary gear set consists of a ring gear, planet carrier and planet carrier gears, and a sun gear. It splits power and provides infinite gear ratios when blending engine and MG2 propulsion torque at the drive chain and drive axles.

The planetary ring gear is mechanically connected directly to the input drive chain gear to transmit torque directly to the drive chain. Counter gears are connected to the drive chain and provide gear reduction input to the final gears. The carrier is attached to the engine output shaft. The final gears are connected via gear teeth to the counter gears and provide a gear reduction to the differential assembly and axles. **See Figure 10-4.**

Engine power is transferred to the transmission input shaft with a transaxle damper assembly. The operation of the damper is similar to the clutch disc assembly for a manual transmission system. Most manufacturers of hybrid EVTs use an engine damper similar to the Toyota Prius damper.

Toyota Prius Cranking

The carrier for the planetary gear set is connected to the engine via the input shaft and transaxle damper assembly. The planetary gear set is driven by the planet carrier gears to crank the engine. The ring gear is connected to the drive wheels through the transmission gearing and differential.

In the 1960s, TRW and Ford Motor Company collaborated on one of the first electric transmission development programs.

DRIVE CHAIN AND GEARS

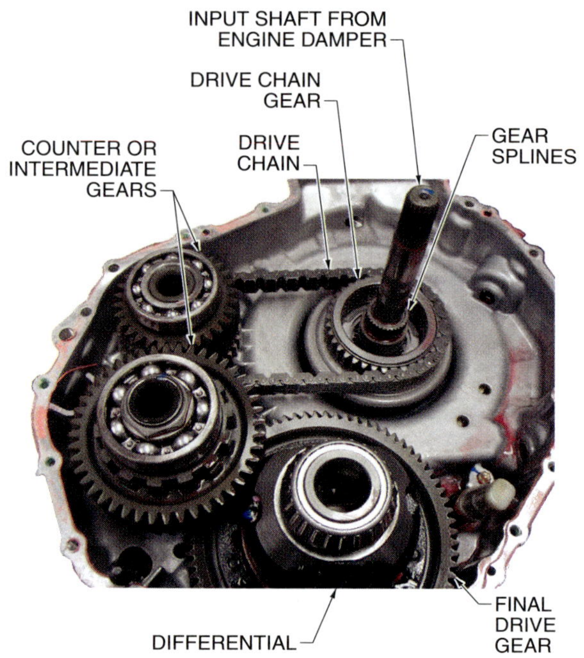

Figure 10-4. The drive chain transmits MG2 torque and/or engine speed from the ring gear of the planetary gear set to the counter gears, final drive, and drive axles.

The ring gear, which is connected to MG2, must be held stationary to function as a reaction member of the planetary gear set. This is accomplished by the drive chain holding the ring gear via the drive axles and tires or by transferring electrical power from the battery pack to the power inverter and MG2. The sun gear, which is connected directly to MG1, supplies torque and rotation to the planet carrier and planet gears. **See Figure 10-5.**

Electric machine output torque is increased by a factor equal to the planetary gear ratio. The planetary gear ratio during starting depends on the number of teeth on the sun gear and in the ring gear. The planetary gear ratio is calculated as follows:

$$R = \frac{N_S + N_R}{N_S}$$

where

R = planetary gear ratio

N_S = number of teeth on sun gear

N_R = number of teeth in ring gear

TOYOTA PRIUS CRANKING

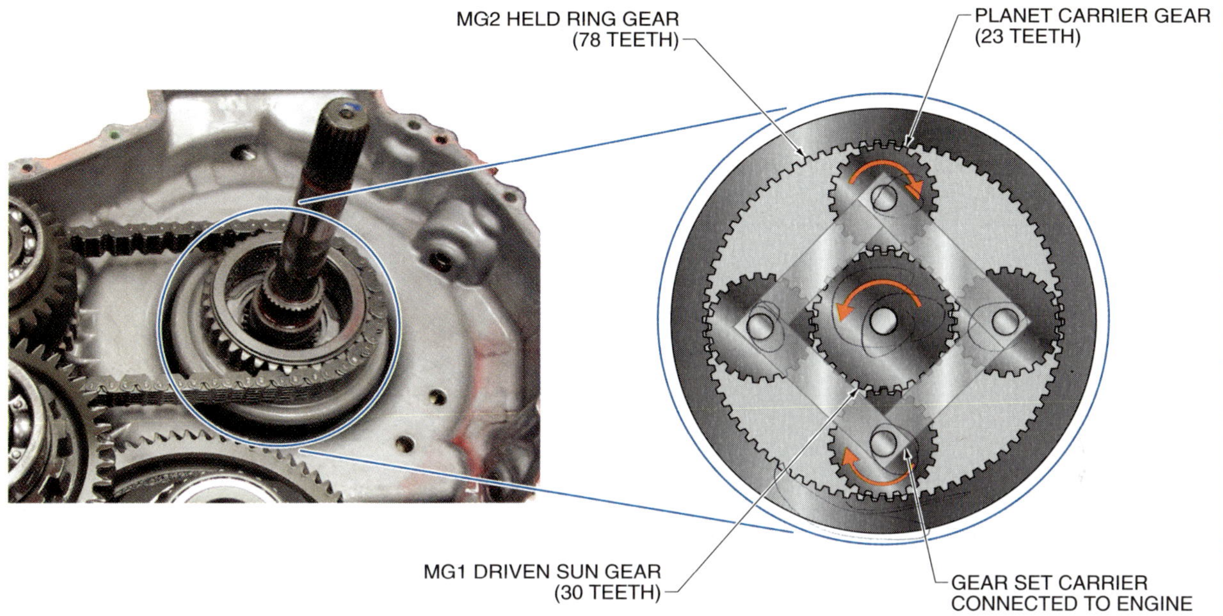

Figure 10-5. The planetary gear set is driven by the planet carrier gears to crank the engine.

For example, if there are 30 teeth on the sun gear and 78 teeth in the ring gear, the planetary gear ratio is calculated as follows:

$$R = \frac{N_S + N_R}{N_S}$$

$$R = \frac{30 + 78}{30}$$

$$R = \frac{108}{30}$$

$$R = \mathbf{3.6}$$

This means that the sun gear rotates 3.6 turns for every turn of the carrier. The torque output of the planetary gear is greater than electric machine output torque by the same factor. If the electric machine output torque is 50 N-m (37 lb-ft) and the output torque of the planetary gear is increased by a factor of 3.6, then the output of the planetary gear is 180 N-m (133 lb-ft).

Toyota Prius Propulsion

The torque band curves for an electric machine and an ICE are different. **See Figure 10-6.** The maximum torque for an electric machine is at the low end of the torque band graph, from 0 rpm to about 6000 rpm. The maximum torque for an ICE has a higher range, from about 4000 rpm to about 6000 rpm.

Electric machines used in automotive applications are constant torque machines at the lower end of the rpm range. An ICE vehicle has high torque at the upper end of the rpm range to deliver high horsepower and torque. An ICE vehicle must use multispeed transmissions that shift at specific points, while electric machines have no shift points. This means that an electric machine does not have to wait for an increase in rpm to regain its torque band area.

By connecting the planetary sun gear to MG1, infinite gear ratio combinations and a continuous blending of engine and MG2 torque to the final drive gears can be applied. This produces an EVT system. Blending engine and MG2 torque is accomplished by varying the planetary gear ratio, which varies the torque and speed of MG1. As the hybrid controller changes MG1 torque and/or speed, the torque and/or speed of the engine and MG2 also change(s). Hybrid controller calculations control the amount of MG1 torque/speed changes to maximize ICE and electric propulsion operating efficiencies. **See Figure 10-7.**

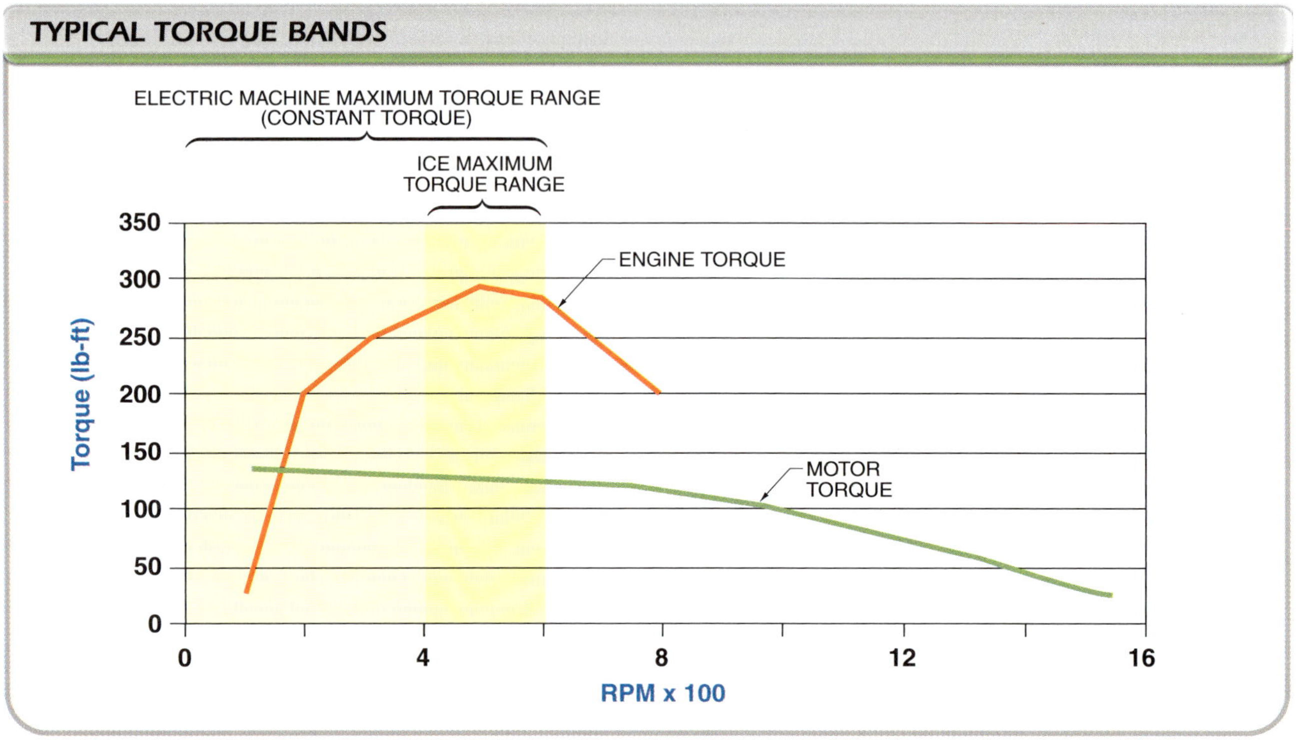

Figure 10-6. The torque band curves for an electric machine and an ICE are different.

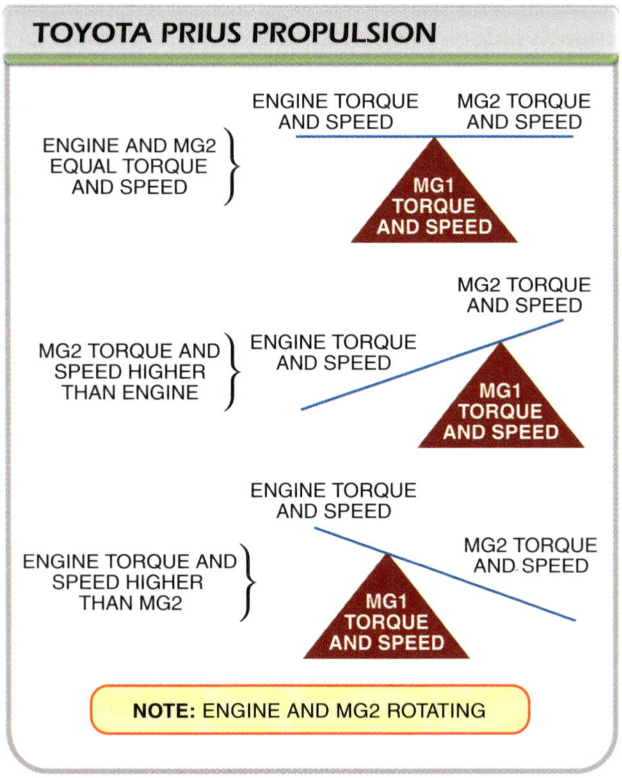

Figure 10-7. As a hybrid controller changes MG1 torque and/or speed, the torque and/or speed of the engine and MG2 also change(s).

Vehicles with high-voltage electric propulsion systems use special coolants with deionized water for the power electronics cooling system. This is to maintain electrical isolation between the vehicle chassis and high voltage.

The torque/speed relationships between the engine and MG1/MG2 transmission elements are important for understanding the power control strategy. As the controller adjusts the torque and speed of MG1, the percentage of the total torque applied by the engine or MG2 is increased or decreased. Therefore, the hybrid controller can adjust the powertrain system for maximum efficiency by continuously adjusting MG1 torque/speed. The balance between MG1, MG2, and the engine continually changes, or adjusts, the torque/speed.

Other EVT Systems

The Toyota Prius EVT is one system of many systems that uses the combination of engine and electric machine technologies for vehicle propulsion. The GM-Allison,

Lexus RX400h, Lexus GS450h, and Toyota Highlander are other examples that use EVT technology. Other hybrid products with variations of EVT technology will be designed in the future. It will be important for technicians to understand the operation of EVT systems to more thoroughly analyze and diagnose related vehicle operational problems.

MOTOR AND POWER INVERTER COOLING

Most electric machines and power inverters require cooling to keep operating temperatures within design parameters. These cooling systems differ only slightly from the systems found in ICE systems. Inverters must be cooled because the IGBT modules switch high voltage and current. Therefore, inverters must have coolant passages designed within the chassis to accommodate liquid cooling. **See Figure 10-8.**

Toyota Motor Company

The Lexus RX400h EVT consists of the lubrication pump, MG2, gear sets, and MG1.

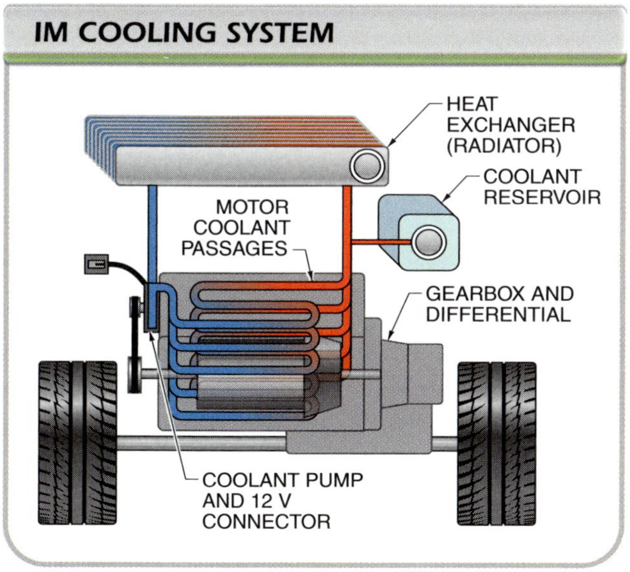

Figure 10-8. Most electric machines and power inverters require cooling to keep operating temperatures within design parameters.

Electric machines must also be cooled. High electrical current in stator coils can generate substantial heat that must be removed with liquid coolant. In an EV, there is no engine, so there are no belts to drive a coolant (water) pump. The cooling systems in EV, HEV, and fuel-cell power electronics systems use a low-voltage (12 V) electric coolant pump to circulate coolant through the system.

Along with the coolant pump, a standard heat exchanger (radiator) and coolant reservoir are used. The cooling system does not need to be pressurized because the temperatures do not approach that of an ICE system. The maximum temperature in a typical electric propulsion system is typically about 120°F to 140°F (50°C to 60°C).

Contemporary hybrid systems use a dedicated cooling system for the power inverter/DC-DC converter system and electric machines. Traditionally, these systems are equipped with a low-voltage (12 V) coolant pump and small radiator.

Refer to the *Hybrid Electric Vehicle Technology* CD-ROM for a Quick Quiz® on the chapter contents.

POINTS TO REMEMBER

- Differentials and gearbox combinations are used with the electric machine to increase torque.
- An electric machine develops maximum torque at low speeds, while an internal combustion engine develops maximum torque at high speeds.
- Electric machines used in automotive applications are controlled with constant torque control strategies.
- The Toyota Prius transaxle system is an example of a system that uses an electronically variable transmission (EVT) that provides infinitely variable gear ratio combinations.
- The Toyota Prius transaxle system uses a single planetary gear set that is used in combination with a high-voltage drive motor and generator.
- The Toyota Prius operates the high-voltage drive motor and the generator at specified, infinitely variable torque and rpm levels in order to operate the engine at its maximum efficiency.
- A high-voltage power electronic system and an electric machine (transaxle) system use a dedicated cooling system that includes a radiator, coolant reservoir, low-voltage (12 V) coolant pump, and coolant pump control system.

Refer to the *Hybrid Electric Vehicle Technology* CD-ROM for the Review Questions in Microsoft® Word format.

REVIEW QUESTIONS

1. What is the function of the transaxle damper assembly in a hybrid system?
2. Explain why an electric machine does not necessarily need a multispeed transmission.
3. What is the purpose of MG1 in a Toyota Prius?
4. What is the purpose of MG2 in a Toyota Prius?
5. Provide a brief explanation of how the Toyota Prius engine, MG1, and MG2 operate to produce an electronically variable transmission (EVT) system.

10 Transaxles, Gears, and Cooling Systems

HYBRID Electric Vehicle TECHNOLOGY

Trade Test

Name ______________________ Date ______________

______ **1.** The acronym given to the cranking motor in the Toyota Prius electric transaxle is ___.
A. GEN-1
B. MG1
C. IG-1
D. MG2

______ **2.** The acronym given to the drive motor in the Toyota Prius electric transaxle is ___.
A. MG1
B. DM-2
C. MG2
D. IG-1

______ **3.** The Toyota Prius electric transaxle contains ___ planetary gear set(s).
A. 1
B. 2
C. 3
D. 4

______ **4.** The Toyota Prius engine is connected to the electric transaxle through the ___.
A. torque converter
B. flywheel
C. flexible coupling
D. transaxle damper assembly

______ **5.** The coolant pump of the Toyota Prius electric propulsion system is driven by a(n) ___.
A. 12 V electric pump
B. 36 V electric pump
C. accessory drive belt
D. serpentine drive belt

______ **6.** The Toyota Prius generator MG1 is connected directly to the ___.
A. drive chain gear
B. main shaft
C. planet carrier
D. sun gear

_______________ **7.** The Toyota Prius generator MG2 is connected directly to the ___.
A. drive chain gear
B. main shaft
C. planet carrier
D. sun gear

_______________ **8.** ___ controls the infinite gear ratio in the Toyota Prius electric transaxle.
A. MG1
B. MG2
C. Speed control technology
D. Planet carrier technology

HYBRID
Electric Vehicle
TECHNOLOGY

Energy Management Hardware Systems

11

The energy management hardware system is made up of many different components. Each component has a particular function that provides the overall operating parameters for the energy management system. The hardware components, combined with the system controller, result in a system that stores high-voltage electrical energy in the vehicle. Understanding this system is a core building block in gaining overall knowledge of an electric, hybrid, or fuel-cell energy storage system.

CHAPTER OBJECTIVES

- Identify the typical location of the high-voltage system fuse.
- Explain why high-voltage wires and cables use shielding.
- Describe the operation of the high-voltage precharge circuit.
- Describe the operation of the battery-pack contactor system.
- Compare the types of active and passive thermal systems and describe their operation.
- Describe the operation of terminal-voltage sensing circuits.

KEY TERMS

electromagnetic interference (EMI): The magnetic field emanating from wire or cable that can cause operational disruptions with other electrical circuits.

shielding: A braided mesh of aluminum or copper that encases a cable or wire to negate magnetic field emissions, or to ensure that magnetic field emissions do not breach a wire or cable.

contactor: A large relay used to transfer electric current.

precharge circuit: A circuit with a resistor and contactor that is used to protect high-voltage bus capacitors by limiting electrical current into discharged bus capacitors when a high-voltage system is initialized.

thermal system: A cooling and heating system used to control the temperature of a battery pack.

passive cooling system: A forced-air cooling system that uses ambient air to cool a component or system.

active cooling system: A forced-air cooling system that uses an air conditioner to cool a component or system.

ELECTRICAL CONNECTIONS

Battery-pack connections consist of the electrical connections between individual batteries and between the battery and other components. The connections and components must be shielded from electromagnetic interference. Other electrical connections include fuses, current sensors, and control of the battery-pack system contactors that switch the current.

Battery Interconnection Cables

Battery interconnection cables are normally short lengths of #2 or #4 gauge, stranded-battery or welding cable, and are very similar to the cables used for automotive cranking systems. On each end of the cable is a mechanically crimped terminal (usually a ring terminal) to attach the cable to the battery terminal.

The size (gauge) of the cable is determined by the amount of electrical current that is expected to pass through the cable. The power law states that volts × amps = watts. It can be seen that as system voltage increases, the amount of electric current decreases to deliver the same amount of power. This means that the cable gauge can be reduced. If system voltage decreases, the amount of electric current increases to deliver the same amount of power. Therefore, the cable gauge needs to be increased. This is one of the many advantages of a higher voltage system.

Electric propulsion components, such as the battery pack, power inverter, auxiliary power module, and electric machines are connected with cables. The cables commonly have ring-type terminations. Copper busbars can be used to connect battery modules in a Toyota Prius. Copper busbars are less expensive than stranded battery cables. **See Figures 11-1.**

COPPER BUSBARS

Figure 11-1. Copper busbars can be used to connect battery modules.

Electromagnetic Interference

When there is a large amount of current flowing through a cable, there are magnetic fields surrounding them. An important concern with high voltage and current is the magnetic fields that cause electromagnetic interference. *Electromagnetic interference (EMI)* is the magnetic field emanating from wire or cable that can cause operational disruptions with other electrical circuits.

The EMI frequency of a power electronics system can reach the megahertz (MHz) range. These frequencies can be generated from the variable-frequency electric traction system or from any other system that constantly changes power levels and frequency ranges.

In a modern electric vehicle (EV) and HEV, a high-voltage cable typically is covered with braided aluminum or copper shielding to filter magnetic field emissions and prevent EMI. *Shielding* is a braided mesh of aluminum or copper that encases a cable or wire to negate magnetic field emissions, or to ensure that magnetic field emissions do not breach a wire or cable. Shielding is connected to the connector at each end of the cable to shunt electrical noise to ground. **See Figure 11-2.**

SHIELDING

Figure 11-2. Shielding is connected to the connector at each end of a cable to shunt electrical noise to ground.

It is especially important to protect low-voltage microcontroller/microprocessor circuits. The three separate elements that contribute to EMI levels are voltage, current, and circuit impedance. If any two of these three elements are high, then EMI will result and shielding is necessary.

External Cabling, Terminals, and Connections

High-voltage electric propulsion or power electronics systems require electrical hardware that will ensure efficient power transfer. The battery pack must transmit energy to the power inverter. The power is then transferred to the electric motor to produce electric propulsion. This means that there must be cabling and connections between the battery pack, power inverter, and electric machines to transfer the power.

Users, technicians, and engineers need to be protected from high voltage at the terminals. Fingerproof terminals are used to prevent anyone from inadvertently touching a high-voltage circuit. **See Figure 11-3.** A special tool or small instrument is needed to access the embedded high-voltage terminal.

FINGERPROOF TERMINALS

Figure 11-3. Fingerproof terminals are used to prevent anyone from inadvertently touching a high-voltage circuit with their hands or fingers.

High-voltage systems present a unique problem for cable terminal designs. They require the terminal to transmit the electrical power efficiently. This type of terminal is physically large and expensive. In many cases, the power inverter, auxiliary power module, and electric machines are located in areas that are populated with other electric devices or controllers. Therefore, the connectors must be able to shield against EMI in order for the surrounding devices to operate properly.

Alternative types of connectors are hardened to offer immunity from EMI as well as protection against water, humidity, and dirt. It is a normal procedure to replace the entire cable and connector as an assembly because repairs maintain integrity for EMI shielding. The connectors are very expensive and should only be used when the highest connection integrity is needed. **See Figure 11-4.**

HARDENED CONNECTORS

FEMALE CONNECTOR

MALE CONNECTOR

Figure 11-4. Alternative types of connectors offer immunity from EMI as well as protection against the elements such as water, humidity, and dirt.

> Fork lifts, golf carts, and small utility vehicles now typically use 3ϕ electric propulsion systems. These systems are significantly more efficient than traditional DC systems.

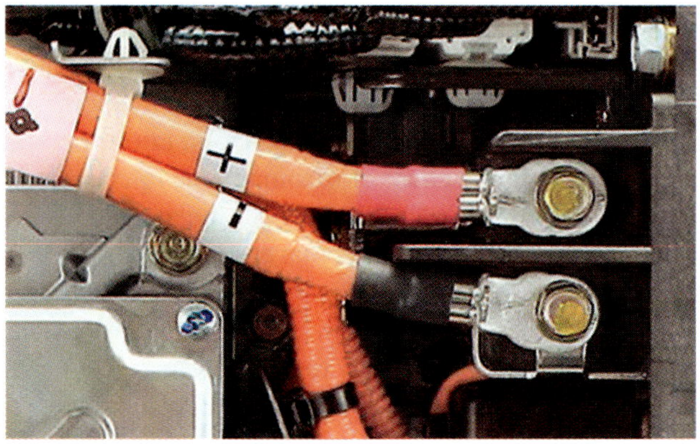

Battery cable terminal connections often use terminal rings for secure connections.

Battery-Pack Fuses

Safety standards require a fuse in the high-voltage system so that energy in the battery pack can be isolated in the event of a catastrophic condition. **See Figure 11-5.** Fuses are typically located within the battery-pack assembly or as part of the service/manual disconnect switch. Fuse ratings are typically 400 A to 600 A in an EV application, and about 200 A in a typical HEV application. Although battery-pack fuses do not need periodic maintenance, a technician should know where they are located.

Battery-Pack Current Sensing

The battery current sensor provides information to the battery controller on the electrical current flowing to or from the batteries in the high-voltage battery system. **See Figure 11-6.** The propulsion system controller uses this information to control power to the electric machine. This allows the controller to manage the speed, torque, and regenerative braking (generating) power. The current sensor is a shunt or Hall-effect sensor. Both sensors produce a small voltage signal that is transmitted to a signal conditioner (analog to digital conversion) and/or an amplifier before being sent to the controller.

One method for measuring electric current involves a shunt sensor that is connected in series with the batteries. The sensor contains an internal parallel circuit with a calibrated resistance that shunts, or diverts, a small amount of current and measures the voltage drop across the resistor. It transmits this data to a converter circuit in the battery controller where the voltage information is converted to binary coded information for the controller. The current is calculated from the known resistance and measured voltage drop.

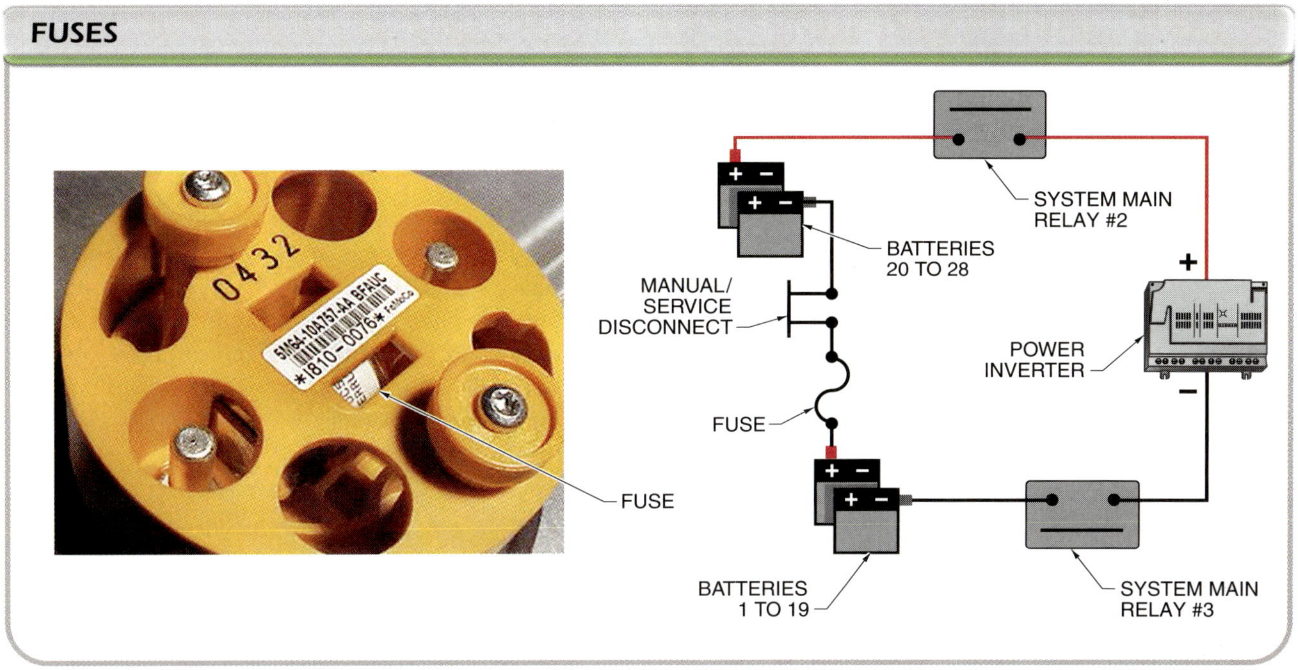

Figure 11-5. Safety standards require a fuse in a high-voltage system.

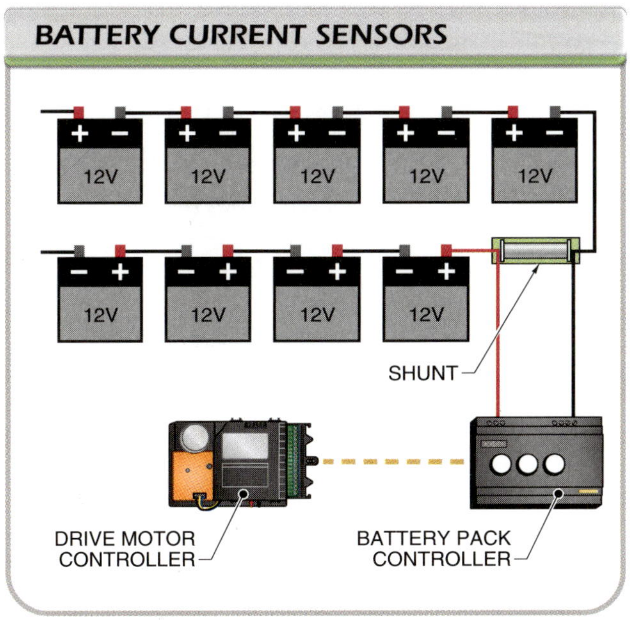

Figure 11-6. A battery current sensor provides information to the battery controller on electrical current flowing to or from the batteries.

The other method used to measure electrical current is with a Hall-effect sensor. Current flowing through a conductor creates a magnetic field around the conductor. Hall-effect sensors are placed around a conductor to measure the magnetic field. The strength of the magnetic field is proportional to the amount of electrical current. The output of the Hall-effect sensor is transmitted to the battery controller in the form of a linear, low-voltage signal.

Battery-Pack System Contactors

A *contactor* is a large relay used to transfer electric current. Battery-pack contactors are rated for use in high-power DC circuits. The contactors control power flow between the battery pack and power inverter. The power transferred from the battery pack is DC and is converted by the power inverter system to AC power for the electric machine motor. The contactors are controlled by the battery-pack controller. The controller uses a low-voltage signal, usually 12 V to actuate the contactors. The contactors switch high-voltage, such as a 300 V power source.

For safety reasons, most battery packs are constructed with two contactors. One contactor is electrically connected to the positive side of the battery pack and the other to the negative side. If one contactor fails to open, such as if it were welded closed, the other contactor can open the circuit. This isolates the battery energy within the battery pack. **See Figure 11-7.**

CONTACTORS

Toyota Motor Company

Figure 11-7. Contactors are used to isolate the battery energy within the battery pack.

The contactor must be mechanized to disconnect the power connection between the battery pack and the power inverter, even during propulsion. The drive motor is an inductive load. The internal contact plate and terminals of the contactor can disrupt high levels of energy. When power to an inductive load is disrupted, large voltage and current spikes are produced.

There are several methods of protecting the contactor internal plate and terminals against arcing. For use in these harsh electrical environments, some manufacturers use hermetically sealed gas or vacuum chambers to house the plate and terminals. In addition, plunger springs with very high spring rates mechanically force the plate from the terminals when de-energized.

Arc suppression devices can also be incorporated into the contactors. Arc suppression devices are connected across the contactor terminals to shunt the transient energy from the electrical terminal contacts. This minimizes the damaging effects of welded or pitted contacts from transient electrical energy. These contactors can have a peak rating of more than 300 V DC and 500 A DC. The continuous current rating can be more than 150 A DC.

The battery-pack controller has many diagnostic routines and it continually monitors the status of the contactors. It will request high-voltage information from other vehicle controllers to verify that high voltage has either been removed or applied to the propulsion system, depending on the specific command.

For example, if a contactor has failed with its contacts welded closed, any one of the numerous controllers would have reported this to the battery-pack controller. If a contactor fails, propulsion would be disabled after the operator parks the vehicle, diagnostic codes would be latched, and instrument displays would warn the driver of the existing system problem.

Battery-Pack Contactor Precharge System

A *precharge circuit* is a circuit with a resistor and contactor that is used to protect high-voltage bus capacitors by limiting electrical current into discharged bus capacitors when a high-voltage system is initialized. **See Figure 11-8.** The purpose of a precharge system is to ensure the safe operation of the high-voltage bus capacitors and contactor/relay contacts. It limits the amount of electrical current that flows through the contactors/relays and the high-voltage bus capacitors to minimize the possibility of capacitor damage.

PRECHARGE CIRCUIT

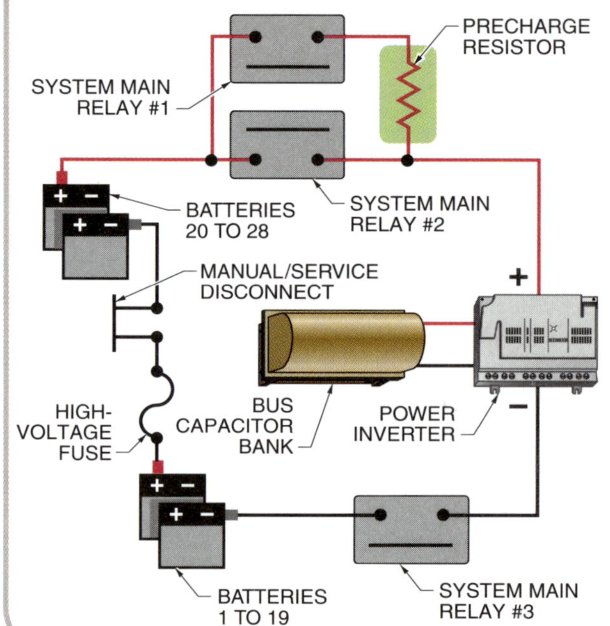

Figure 11-8. A precharge circuit consists of a resistor and contactor and is used to protect high-voltage bus capacitors.

The precharge circuit is typically located inside the power inverter enclosure. The precharge system consists of a precharge relay and resistor that are connected in parallel to one of the main system contactors/relays. The precharge relay is controlled by the battery-pack controller in conjunction with the two main high-voltage contactors that connect the battery pack to the power inverter.

Vehicle precharge and startup follows a standard sequence. When the vehicle system is OFF or at rest and the operator initiates vehicle power, the ignition or power system signal provides a wake-up signal to the hybrid controller. The hybrid controller sends a data message to the power inverter controller and battery-pack controller to initialize the precharge process.

The hybrid controller then commands the battery-pack controller to begin the precharge process. During the process, the hybrid power inverter and battery-pack controllers need to be involved. The power inverter tracks the capacitor voltage (charge) level and the battery-pack controller controls the contactors.

The battery-pack controller then sends a data message to the power inverter controller to begin the precharge process. The battery-pack controller closes system main relays #1 and #3. Current begins to flow from the battery pack through the relays and resistor to the power inverter to begin charging the high-voltage bus capacitors.

At the beginning of the precharge process, the capacitors are not charged and resistance is low (about 0 Ω). The precharge resistor limits the current to the power inverter and capacitor bank. If the resistor were not in the system, there would be nothing to limit battery-pack energy to the bus capacitors. As a result, the capacitors would overheat. This would damage the capacitors and may cause them to explode.

The time required to charge the capacitors to a safe level of about 90% of battery-pack voltage depends on the time constant of the precharge circuit. The time constant is calculated as follows:

$t = RC$

where

t = time constant (in sec)

R = resistance (in Ω)

C = total capacitance (in F)

For example, if a resistor is rated at 200 Ω and the capacitor is rated at 8100 μF (0.0081 F), the time constant is calculated as follows:

$t = RC$

$t = 200 \times 0.0081$

$t = \mathbf{1.62\ sec}$

The Ford Ranger electric truck was a rear wheel drive vehicle. The GM S-10 electric truck was an engine-powered rear wheel drive with electric propulsion for front wheel drive.

The power inverter monitors the capacitor voltage charge level. Once the 90% voltage level has been reached, the power inverter controller sends a message to the battery-pack controller to open system main relay #1 and close system main relay #2. This connects the battery pack directly to the power inverter and finishes charging the bus capacitors to a 100% voltage level.

For the capacitor to reach 90% of charge, the capacitor must charge at the battery voltage for an amount of time equal to three time constants. In this case, three time constants last 4.86 sec (3 × 1.62 = 4.86). This is the precharge time. The 4.86 sec precharge time begins the moment a wake-up signal is received. This normally occurs when the vehicle is unlocked and the doors are opened.

If there is a malfunction that does not permit the system to reach the 90% voltage level within the 4.86 sec time period, or the voltage level is reached too quickly, then a precharge failure is assumed and the battery-pack controller opens all of the relays. The battery-pack controller software may permit a total of three attempts to precharge. If the precharge process fails after the third attempt, the system disables the electric propulsion and latches a diagnostic code in the controller memory.

BATTERY-PACK THERMAL SYSTEMS

When hybrid vehicle sales began to increase, the design of thermal systems became important. A *thermal system* is a cooling and heating system used to control the temperature of a battery pack. Battery packs produce heat that needs to be removed from the confined battery storage area. The battery temperature must be monitored and controlled to ensure peak performance. Whether it is PbA, NiMH, or Li-ion, each technology benefits immensely from an onboard thermal system. Passive and/or active thermal systems will always be used in EV, HEV, or fuel-cell battery thermal systems.

Thermal System Operation

The thermal system is similar to a cocoon that maintains the necessary constant temperature environment for the battery-pack system in order to promote peak performance. **See Figure 11-9.** Typical NiMH and Li-ion battery packs generate high temperatures and rely on thermal systems for cooling to maintain peak performance. Because these battery technologies require cooling, both types would operate very nicely in cool climates. A typical lead-acid battery pack actually may require additional heat to operate at peak levels. Lead-acid batteries are poor performers in cold climates but thrive in warmer climates.

THERMAL SYSTEMS

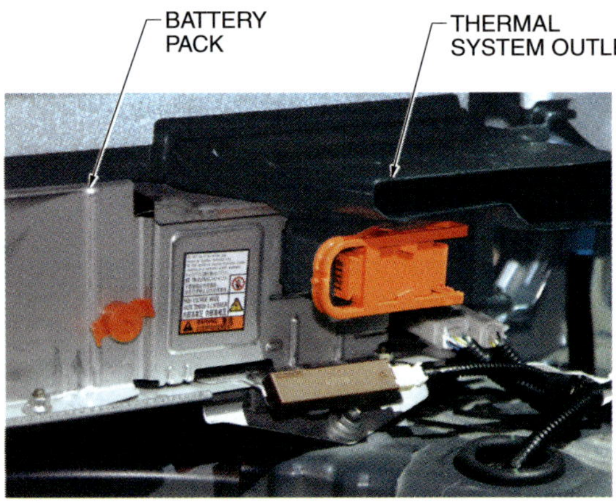

Figure 11-9. The thermal system is similar to a cocoon that maintains the necessary constant-temperature environment for the battery-pack system in order to promote peak performance.

The battery-pack thermal control system uses either a passive cooling system or an active cooling system. A *passive cooling system* is a forced-air cooling system that uses ambient air to cool a component or system. An *active cooling system* is a forced-air cooling system that uses an air conditioner to cool a component or system. Future designs will include a closed-loop cooling system to cool battery modules.

A typical battery-pack thermal system consists of a system made up of the following components:

- thermal grid or blanket
- cooling system using a dedicated air conditioning system or conditioned air from a passenger cabin
- electric fan with speed control for moving hot or cold air
- airflow and air temperature sensors
- electrically controlled air duct systems
- microcontroller for controlling the thermal system

The microcontroller can be integrated into the battery-pack controller functions. The cooling fans can vary in dimension, air movement specification, motor technology type, mass, and mounting location. Cooling fan motors are typically very quiet and durable.

In addition to cooling, battery-pack heating is also important if the vehicle is to be operated in very cold environments. For example, Ford offers an accessory for the Escape Hybrid that permits the vehicle operator to attach an external power-supply inverter to a connector on the vehicle. This supplies electricity to a heating grid system and raises the temperature of the batteries. As a result, battery capacity and performance are increased where the ambient temperature is below 0°F.

The thermal system must maintain tight temperature control to provide an environment for optimum performance. For example, the Toyota Prius system uses cool ambient air or conditioned air from the vehicle cabin and routes it through the battery pack. Honda Hybrids have systems similar to Toyotas. They also route cabin air through the battery pack to cool the power electronics and battery-pack modules.

The Ford Escape Hybrid uses a combination passive and active battery-pack thermal system. It has an air conditioning evaporator located in the left rear of the vehicle and is an integral part of the battery-pack cooling system. The passive cooling part of the system uses ambient air to cool the battery pack until the internal temperature increases to a point that ambient air can no longer provide adequate cooling.

At this point, the battery-pack controller sends a data message to the air conditioning system requesting additional active cooling. The battery-pack controller commands the air conditioning system to supply refrigerant to the evaporator and use cooler conditioned air from the evaporator. A series of ducts are designed to use ambient or conditioned air to cool the battery pack. Mode doors controlled by the battery-pack controller determine the flow and temperature of the air delivered to the battery pack to ensure the proper cooling level for maximum performance.

Airflow and Temperature Sensing

The battery-pack controller senses battery module voltage and current. In addition, it may receive the temperature of the battery modules and data on the airflow through the battery pack. Battery packs can be designed with one airflow sensor and four or more thermistor temperature sensors for critical modules. **See Figure 11-10.** The critical modules operate at temperatures greater than other modules because of their placement. Therefore, monitoring the temperatures of these specific battery modules is essential in controlling the health and performance of the battery pack.

Typically, thermistors are placed on the surface of modules that have been identified in testing as operating at the most elevated temperatures. It is assumed that the temperatures of the remaining modules are acceptable.

AIRFLOW AND TEMPERATURE SENSORS

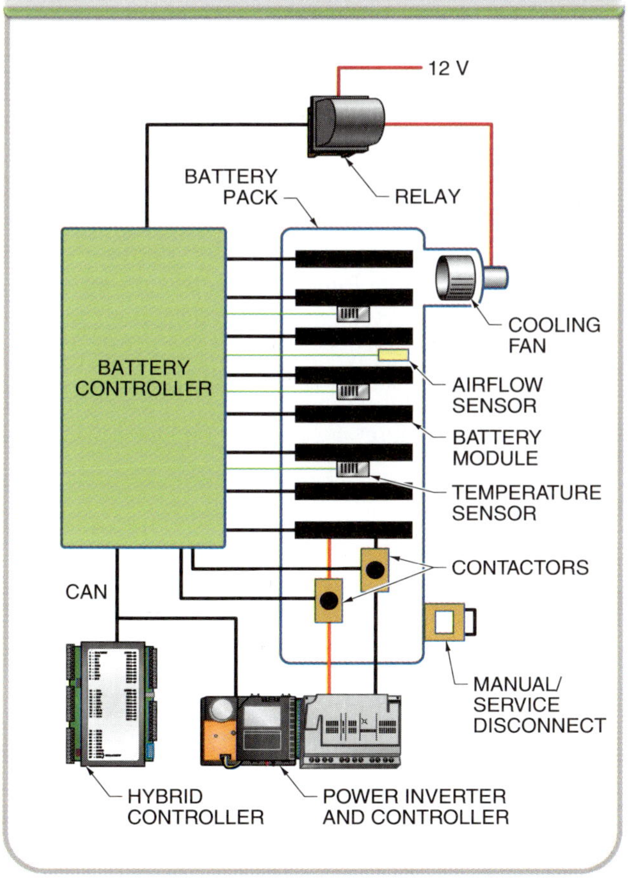

Figure 11-10. Battery packs can be designed with airflow sensors and thermistor temperature sensors for critical modules.

Many of the battery-pack diagnostic codes are based on battery-pack airflow and module temperatures. With a software algorithm that monitors battery module voltages, battery-pack current, airflow, and module temperatures, the battery-pack controller can easily determine the operating state and status of the battery pack, individual modules, and cooling system.

BATTERY-PACK CONTROLLERS

The function of the battery-pack controller is to control the energy entering and leaving the battery pack. It monitors many aspects of hybrid vehicle operation to control the battery pack. For example, it monitors the battery-pack shunt or Hall-effect sensors for the raw DC electrical current delivered to the power inverter. It also monitors the state-of-charge terminal voltage of each battery in the battery pack for optimal performance.

Battery-Pack Charging and Discharging

A typical function of the battery pack and its controller is to supply DC energy to the power inverter. The power inverter inverts the DC power to AC and powers the drive motor. The battery-pack controller constantly communicates electrical current and battery-pack status to the drive motor and hybrid controller. This is very critical in an EV application because various factors affect the distance the vehicle can travel, such as acceleration speed, hill grade, and time already spent in operation.

The current sensor provides a valuable input to the battery-pack controller to measure the incoming and outgoing battery-pack current. In addition, a wiring harness connects sensing leads to each battery module and the battery-pack controller measures the terminal voltage. **See Figure 11-11.** The battery-pack controller uses the terminal voltage measurements of each module and the electrical current measurement from the current sensor to determine the SOC.

The sensing leads are used to monitor battery condition. The battery discharges when it is used for propulsion or while at rest from self-discharge. The battery charges when the electric machine charges the battery pack during operation or when it is being charged from an external charger. With a 360 V battery pack, the controller could have 30 or 60 separate wires connected to it for battery-pack monitoring. This high number of separate wires leads to a fairly high incidence of connection failures along with connector and terminal maintenance issues.

TERMINAL VOLTAGE MEASUREMENT

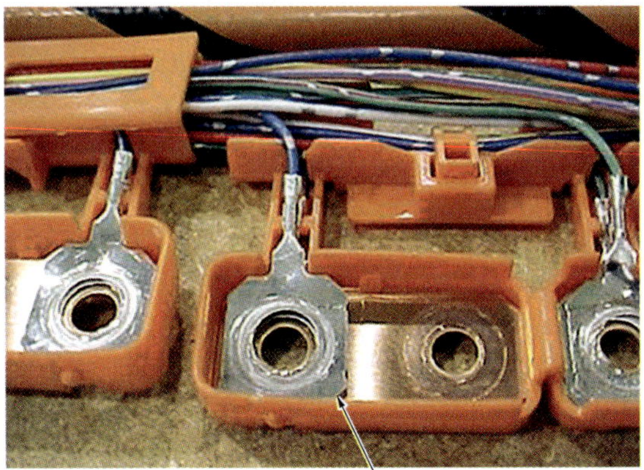

Figure 11-11. A wiring harness connects sensing leads to each battery module and the battery-pack controller measures the terminal voltage.

Electric Vehicles

HEVs typically do not have problems with low SOC in the battery pack. However, EVs definitely have problems. During a long commute or aggressive vehicle operation, an operator may face a nearly depleted battery pack and low SOC. This scenario could be disastrous and cause battery pack failure if the software algorithm is not developed correctly.

There is a risk of plate damage and cell-polarity reversal if the battery pack is depleted to the point at which it is "flat" (terminal voltages less than 11 volts). This can damage the battery cell plates. Therefore, communication between the battery-pack controller and the drive motor controllers is vital to prevent a depleted battery module.

The battery-pack controller can be designed to command the contactors to open. This avoids damage to the battery pack if voltage decreases to a point where damage can occur. One or two warning lights in the instrument cluster may be commanded ON to warn the operator of a pending and/or existing problem with the vehicle and to take immediate action.

Another possible problem could be that only one battery has a low SOC and all other batteries are in good health. In this situation, the battery-pack controller would command one or two warning lights ON in the instrument cluster to warn the operator.

A software algorithm could be written to avoid a possible battery failure in an EV. The battery-pack controller can be designed to inform the drive motor controller that the battery-pack SOC is low. It can also command reduced electric machine power and performance to conserve energy until the battery pack is charged. In response, the drive motor controller could prevent mild or hard accelerations to preserve the remaining energy in the battery pack. In most cases, reduced performance does not inhibit the operator's ability to cruise on the interstate. It may take longer to reach highway speed, but this mode conserves energy and increases the driving range.

Refer to the *Hybrid Electric Vehicle Technology* CD-ROM for a Quick Quiz® on the chapter contents.

POINTS TO REMEMBER

- Most battery-pack fuses are located in the service disconnect.
- A high-voltage cable contains a shielding system to minimize electromagnetic interference, with electronic components near the cable.
- High-voltage cable terminations can be accomplished by using ring terminals or connectors. The connectors may have fingerproof designs.
- High-voltage cable connectors must be designed to suppress electromagnetic interference from leaking into other components.
- Battery-pack systems typically use two high-voltage contactors (relays), but systems that use only one contactor are also used.

- Connections between battery modules, or battery module sections, can be made with cables or busbars.
- The battery-pack control system typically uses a Hall-effect current sensor to monitor electric current exiting and entering the battery pack.
- The precharge system is a circuit designed to protect the high-voltage bus capacitors by limiting electrical current when the high-voltage system is initialized.
- The precharge system is made of a precharge resistor and precharge relay. The precharge system is used with the high-voltage contactor(s) to safely charge the high-voltage bus capacitors.
- The time to charge the bus capacitors to a safe level of about 90% of battery-pack voltage depends on the storage size of the bus capacitor (in μF) and the resistance value of the precharge resistor (in Ω).
- A passive cooling system uses ambient air to cool a component or system.
- An active cooling system uses an air conditioner to cool a component or system.
- Battery-pack systems can use airflow and battery module sensors to monitor internal and external module and system temperatures.

Refer to the *Hybrid Electric Vehicle Technology* CD-ROM for the Review Questions in Microsoft® Word format.

REVIEW QUESTIONS

1. Why is shielding used around cable or wire?
2. How many contactors are typically used in a high-voltage hybrid system?
3. What is the primary reason a precharge system is used in a high-voltage hybrid system?
4. What is the difference between a passive and active battery-pack cooling system?
5. Why are temperature sensors used to monitor specific battery modules?

Energy Management Hardware Systems

Trade Test

Name ______________________________ Date ______________

______ **1.** The typical HEV high-voltage fuse rating is approximately ___ A.
A. 100
B. 200
C. 500
D. 1000

______ **2.** In most manufacturer systems, the high-voltage fuse is located in the ___.
A. power inverter
B. manual or service disconnect
C. battery pack
D. DC-DC converter

______ **3.** A Toyota Prius uses ___ to connect battery modules.
A. busbars
B. cables
C. welded tabs
D. looped membranes

______ **4.** If the precharge process fails, the high-voltage system will ___.
A. charge the battery pack
B. disable propulsion
C. enable limp-home mode
D. discharge the battery pack

______ **5.** The battery-pack controller monitors the voltage of each battery module, or pair of battery modules, as one of the inputs to determine the ___.
A. state-of-charge
B. power inverter resistance
C. current-sensor learn cycle
D. propulsion system torque level

______ **6.** The primary reason for the precharge system is to protect the ___.
A. power inverter transistors
B. bus capacitors
C. battery-pack controller
D. DC-DC converter transformer

_______________ 7. During regenerative braking, the controller recovers energy that was lost during ___.
A. coasting
B. acceleration
C. braking
D. cornering

_______________ 8. To monitor the temperature of battery-pack modules, thermistor sensors are placed or mounted ___.
A. in front of the cooling fan
B. behind the cooling fan
C. on or in critical battery modules
D. at each end of the battery pack

_______________ 9. The braided shielding around each high-voltage cable is used to eliminate or reduce the effects of ___.
A. ESD
B. RFI
C. PWM
D. EMI

_______________ 10. The two categories of battery-pack thermal (cooling) systems are ___.
A. open and closed flow
B. forward and reverse flow
C. passive and active
D. over and under airflow

Battery Construction and Technologies

12

A battery module is a core component of an electric propulsion storage system. It is critical that a technician learn about its function in order to perform vehicle analysis, diagnosis, and repair. Battery technologies have evolved very slowly over the past 100 years until the last few years. New technologies include advanced lead-acid batteries and newer nickel-metal hydride and lithium-ion batteries. Valve regulated lead-acid and absorbent glass mat battery technologies are commonly used in modern hybrid vehicles. In addition, battery ratings are important in the design of hybrid and electric vehicle systems as well as the operating modes of the vehicle.

CHAPTER OBJECTIVES

- Define state of charge (SOC).
- Explain the difference between flooded and absorbent glass mat batteries.
- Describe the discharging and charging cycle of a lead-acid battery.
- Describe gas recombination in a battery.
- Define capacity rate and amp-hour.
- Define energy density and power density and explain their importance.
- Define specific energy and specific power and explain their importance.

KEY TERMS

flooded battery: A lead-acid battery in which all plates in the battery cells are immersed in a liquid electrolyte solution of sulfuric acid and water.

valve regulated lead-acid (VRLA) battery: A battery that has a pressure regulation valve located inside to control the internal pressure.

absorbent glass mat (AGM) technology: Battery-cell technology that uses electrolyte stored in a mat placed between battery plates instead of plates immersed in a liquid electrolyte solution.

ion: An atom or molecule that is either positively or negatively charged.

state of charge (SOC): A measure of the remaining capacity of a battery relative to the capacity at full charge, usually expressed as a percentage.

self-discharge rate: The rate at which a battery will discharge without being connected to an external load and is due to the natural chemical reactions in the battery or micro shorting in a cell plate.

capacity rate (C-rate): The rate at which the capacity of a battery is discharged until its terminal voltage reaches a specified level. The C-rate is the amount of electrical current that can be supplied.

amp-hour (Ah): A unit of measure calculated by multiplying the amount of electrical current a battery will deliver (in amps) by the time in which it will deliver the current (in hours) until the battery terminal voltage reaches a specified level.

energy density: The amount of energy a battery stores per unit volume at a specified discharge rate, usually expressed in watt-hours per liter (Wh/L).

power density: The amount of power a battery can deliver per unit volume at a specified state of charge, usually expressed in watts per liter (W/L).

specific energy: The amount of energy a battery stores per unit mass at a specified discharge rate, usually expressed in watt-hours per kilogram (Wh/kg). This is a measure of the number of watts the battery can deliver per hour based on its mass. Specific energy is analogous to the reserve capacity (RC) rating.

specific power: The amount of power a battery can deliver per unit mass at a specified state of charge, usually expressed in watts per kilogram (W/kg). This is a measure of the instantaneous power a battery can deliver based on its mass. Specific power is analogous to the cold cranking amp (CCA) rating.

BATTERY OPERATING PRINCIPLES

The predominant technology for making batteries has been the 12 V flooded battery technology. A *flooded battery* is a lead-acid battery in which all plates in the battery cells are immersed in a liquid electrolyte solution of sulfuric acid and water. This battery technology is slowly phasing out in favor of newer valve regulated lead-acid, absorbent glass mat (VRLA AGM) technology.

A *valve regulated lead-acid (VRLA) battery* is a battery that has a pressure regulation valve located inside to control the internal pressure. *Absorbent glass mat (AGM) technology* is battery-cell technology that uses electrolyte stored in a mat placed between battery plates instead of plates immersed in a liquid electrolyte solution. Although there are a number of battery configurations, most battery manufacturers have adopted the Society of Automotive Engineers (SAE) standard for PbA batteries. It should be noted, however, that advanced battery designs do not necessarily adhere to the SAE standard in terms of mass and dimensions.

The weight of a battery in a hybrid and electric vehicle can range from 35 lb to 45 lb, which is within the mass range of a traditional automotive battery. A hybrid or electric vehicle may need up to 26 PbA or NiMH batteries. The total weight of these battery systems can range from about 150 lb for a hybrid vehicle using NiMH technology to about 1300 lb for a pure electric vehicle.

Flooded Battery Technology

The design of flooded batteries is fairly simple. Negative plates, made of lead or lead alloy, are sandwiched between positive plates of a lead alloy of calcium or antimony. **See Figure 12-1.** Sheets of nonconductive, microporous separators are placed between the plates to keep them from touching. The plates are pasted with a lead sulfate active material that provides a large surface area to store electrochemical energy. The positive cell plates for each cell are welded together and attached to an intercell connector or the positive battery terminal post. The negative cell plates for each cell are welded together and attached to an intercell connector or the negative battery terminal post.

Plate assemblies are placed in a case. A cover with a vent cap/flame arrestor and hydrometer hole is integrated into the battery top and case. Typically, the battery cover and case are bonded together with adhesive to provide a leakproof seal. The container is then filled with a diluted sulfuric acid electrolyte.

FLOODED BATTERIES

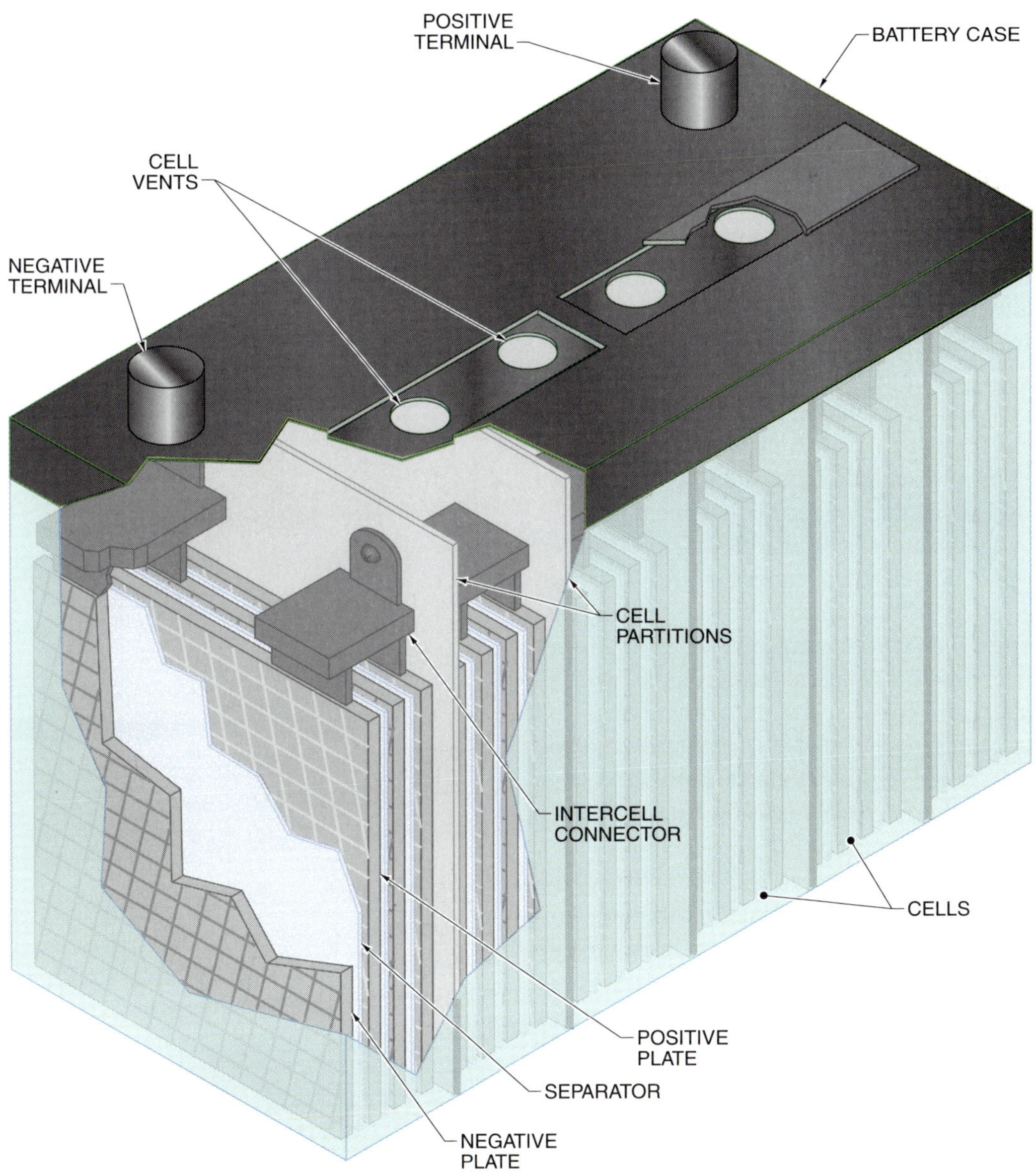

Figure 12-1. Negative plates, made of lead or lead alloy, are sandwiched between positive plates of a lead alloy of calcium or antimony.

Discharging and Charging. A chemical reaction takes place when a battery is discharging and delivering current. Each sulfuric acid molecule in the electrolyte dissociates into two hydrogen (H) ions that carry a positive charge and one sulfate (SO_4) ion that carries a negative charge. An *ion* is an atom or molecule that is either positively or negatively charged. A positively charged ion has a deficiency of electrons and a negatively charged ion has an excess of electrons.

During discharge, SO_4 ions go to the spongy-lead negative plates and react with the lead (Pb) in the plates. The SO_4 and Pb combine to form lead sulfate ($PbSO_4$). During the reaction, the SO_4 ions relinquish their negative charge and create an excess of electrons on the negative plate. The H ions go to the positive plate and react with the oxygen of the lead peroxide (PbO_2) in the plates. The H and PbO_2 combine to form water (H_2O). During the reaction, they take electrons from the positive plate. The lead material contained within the lead peroxide combines with some of the SO_4 ions to form lead sulfate on the positive plate.

The result of this reaction is that the positive plate has a deficiency of electrons and the negative plate has an excess of electrons. **See Figure 12-2.** When the plates are connected together through an external conductor, the electrons from the negative plate flow to the positive plate. This process continues until both plates are converted to lead sulfate, no further chemical action is possible, and the battery is discharged.

During charging, electrical current is forced through the storage battery in the reverse direction as during discharge. Current flows from a higher voltage to a lower voltage. A DC source is connected to the battery with the positive lead clamp connected to the positive plate and the negative clamp connected to the negative plate. If the voltage of the source is greater than the terminal voltage of the battery, electrical current will flow from the source to the battery. The result is that SO_4 ions are driven back into solution where they combine with the H ions of the water and form sulfuric acid. This process continues until both return to their original composition of lead peroxide and spongy lead and the battery is charged.

Sulfuric acid in the electrolyte is consumed when the battery is undergoing discharge and is returned to the electrolyte as the battery is charged. Sulfuric acid has a higher specific gravity than water. Therefore, a measurement of the electrolyte specific gravity gives a good indication of the state of charge in the battery. The *state of charge (SOC)* is a measure of the remaining capacity of a battery relative to the capacity at full charge, usually expressed as a percentage. **See Figure 12-3.**

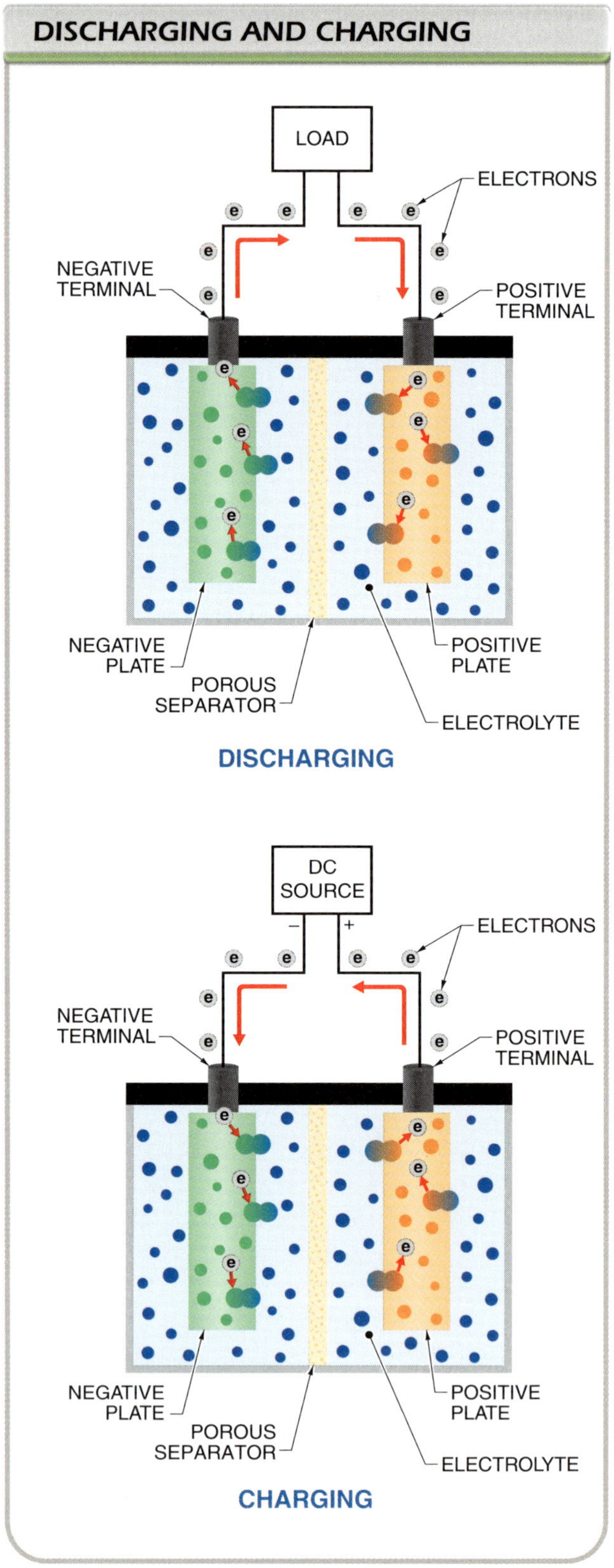

Figure 12-2. During discharge, plates are connected together through an external conductor and electrons from the negative plate flow to the positive plate.

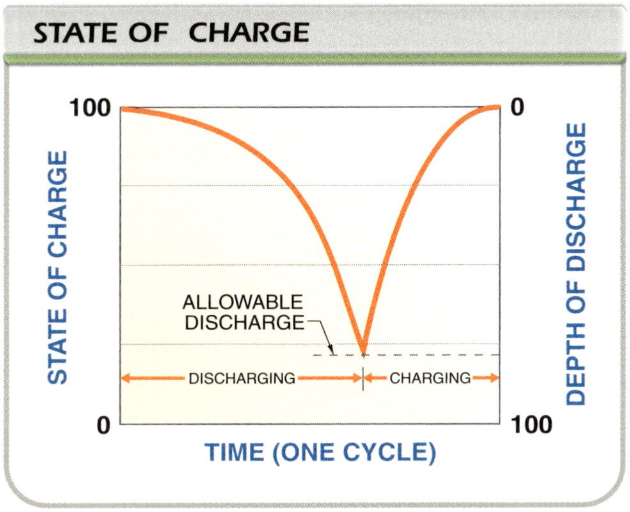

Figure 12-3. The state of charge is a measure of the remaining capacity of a battery relative to the capacity at full charge.

When flooded batteries are charging, oxygen can be generated at the positive electrode from the water in the electrolyte. At the same time, at the negative electrode, hydrogen can be generated from the water. The hydrogen and oxygen escape from the cell. The overall result is water loss from the gassing of the cells. Therefore, flooded battery technology requires periodic water replenishment to maintain optimum operating status. The production of hydrogen requires the battery to be designed with a flame or spark arrestor to ensure that an external spark or flame does not extend into the battery and cause an explosion.

VRLA AGM Battery Technology

Flooded batteries have obvious maintenance and safety concerns. VRLA technology was developed because of manufacturer and customer requests for a low-maintenance battery. VRLA technology is becoming the technology of choice for most manufacturers in hybrid and electric power systems. One disadvantage of a VRLA AGM battery is that the cost is approximately two to three times the cost of a traditional flooded battery of the same capacity.

The plates of a VRLA AGM battery are very compact and rigidly mounted. **See Figure 12-4.** One advantage of VRLA AGM technology is that it does not require periodic maintenance. The battery is completely sealed against hydrogen fumes or leakage, and liquid does not spill out even if the case is broken. It has better shock and vibration resistance than a standard flooded battery and it can survive most freezes. In addition, VRLA AGM batteries have a low self-discharge rate of about 1% to 3% per month. A *self-discharge rate* is the rate at which a battery will discharge without being connected to an external load and is due to the natural chemical reactions in the battery or microshorting in a cell plate.

Figure 12-4. The plates of a VRLA AGM battery are very compact and rigidly mounted.

This means that VRLA AGM batteries can sit in storage for much longer periods without charging than standard flooded batteries. VRLA AGM technology does not have any free internal liquid. It cannot overflow even under severe overcharge. Some manufacturers can offer hydrogen emission rates far below the 4% maximum specified for enclosed spaces or in aircraft.

In a VRLA AGM battery, electrolyte (acid) is contained in the glass mats. This means that there will not be chemical spills even if the battery case is broken. Since a VRLA AGM battery does not contain liquid electrolyte, the battery is considered nonhazardous and shipping costs are lower. In addition, since there is no liquid to freeze and expand, this technology is practically immune to freeze damage.

Discharging and Charging. The charging voltage for a VRLA AGM battery is similar to that of a traditional flooded battery. Therefore, there is usually no need for any special charge algorithms, and there are no compatibility issues with field chargers or charge controls. Internal electrical resistance is extremely low, so internal heating of the battery is very low even when high charge rates are used or if the battery is subjected to very high electrical current discharge loads. Most VRLA AGM manufacturers specify maximum charge or discharge current limits. Personnel who charge these batteries need to adhere strictly to these specifications.

In the event of abnormal operation or excessive gas pressure during the charge cycle, AGM designs include a flame arrestor/safety valve. This valve suppresses explosions that can be generated from internal gases. It also prevents arcs and sparks from entering the cell by permitting the venting of gases during abnormal operation. The valve also maintains pressure in the battery to prevent the release of hydrogen and oxygen during normal operation. The pressure relief valve is typically designed with relief thresholds of 2 psi to 5 psi, depending on battery design. Any gas is vented out the ports through a venting tube. **See Figure 12-5.**

Gas Recombination. One way to minimize gas pressure and the risk of an explosion in a battery is to recombine the hydrogen and oxygen before they are released into the atmosphere. To achieve satisfactory gas recombination efficiency, a way to move the gases between the positive and negative plates is needed. VRLA AGM technology uses gas-phase transfer of oxygen to the negative plates. **See Figure 12-6.** This occurs during charging and prevents the loss of water through electrolysis.

Gas recombination is typically more than 95% efficient. Extremely low amounts of water are consumed during the charging process. Oxygen can pass from the positive plate, where it is generated, to the negative plate, where it reacts with the spongy-lead negative active material. The AGM plate separator is a special, highly porous microfiber glass separator. A continuous path for oxygen transport is created by carefully controlling the saturation level in the mat and the exact balance between electrolyte quantity and porosity. With flooded battery technology, gas recombination is typically less than 30% efficient.

When a flooded battery is tipped, the normal routes for gas exit can become blocked by battery fluid and pressure can build up.

VRLA AGM PRESSURE RELIEF VALVE

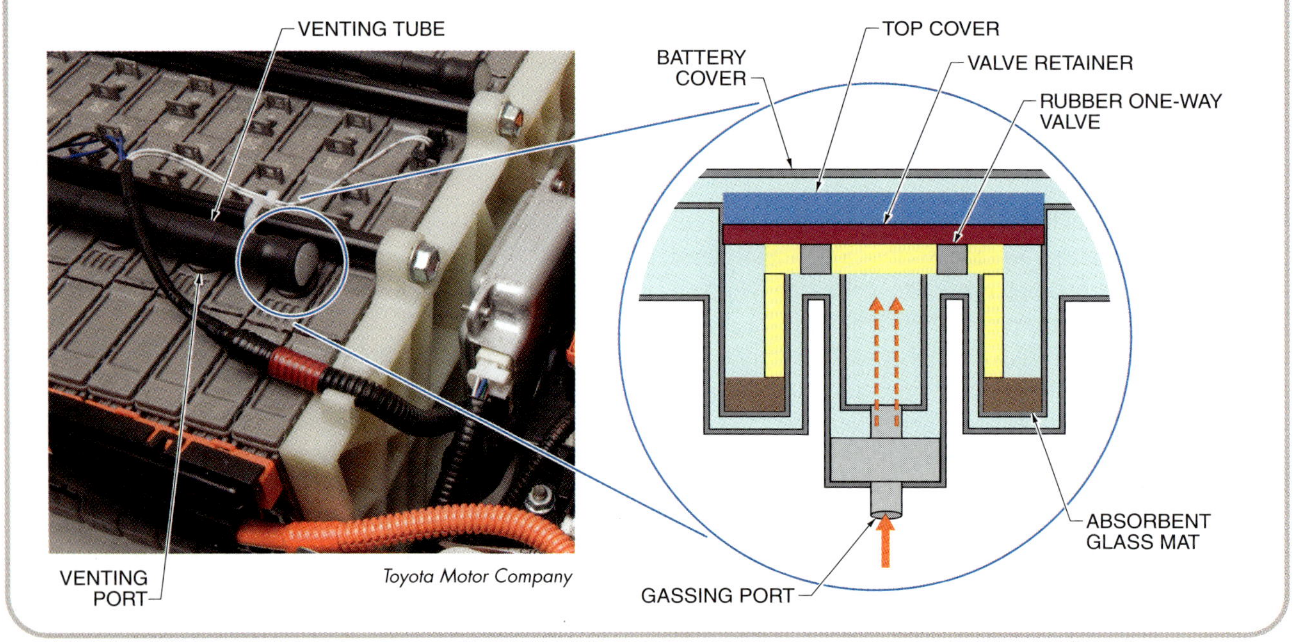

Figure 12-5. In the event of excessive gas pressure in a battery, gas is vented out the ports of a pressure relief valve through a venting tube.

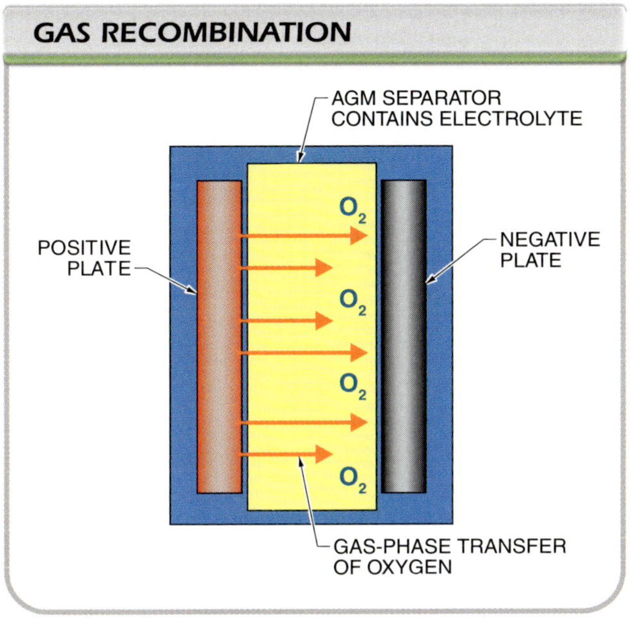

Figure 12-6. VRLA AGM technology uses gas-phase transfer of oxygen to the negative plates.

BATTERY RATINGS

Battery ratings are an important component in the design of hybrid vehicles and their electrical system. In the traditional automotive arena, two main ratings are used for batteries. These are the cold cranking amp (CCA) ratings and reserve capacity (RC) ratings. However, to analyze the battery requirements for a hybrid and electric battery system, amp-hour, energy density, power density, specific energy, and specific power are the ratings used to identify battery performance.

Capacity Rate and Amp-Hours

The period of time and the amount of energy a battery can supply to operate a vehicle is important. The length of time is proportional to the capacity of the battery and discharge rate. The capacity of a battery is determined by discharging the battery at a known constant current until the specified termination voltage is reached. The amount of time it takes to discharge a battery to the termination voltage multiplied by the rate of current at which the battery is discharged determines the capacity rating of the battery.

Capacity rate (C-rate) is the rate at which the capacity of a battery is discharged until its terminal voltage reaches a specified level. **See Figure 12-7.** The C-rate is the amount of electrical current that can be supplied. An *amp-hour (Ah)* is a unit of measure calculated by multiplying the amount of electrical current a battery will deliver (in amps) by the time in which it will deliver the current (in hours) until the battery terminal voltage reaches a specified level. For a 12 V battery, the traditional test procedure is to discharge the battery until the terminal voltage is 10.5 V.

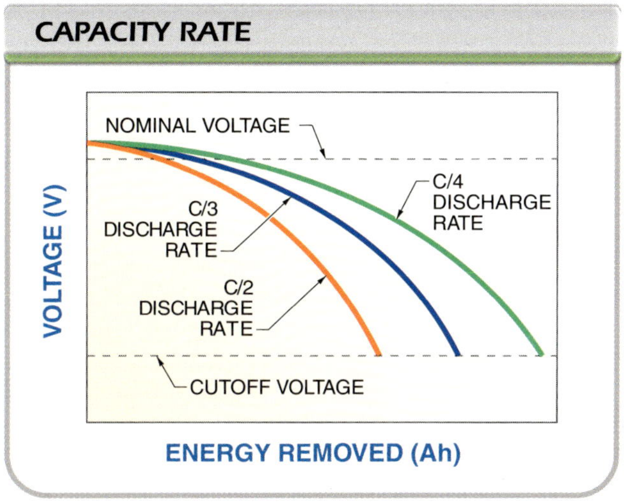

Figure 12-7. Capacity rate (C-rate) is the rate at which the capacity of a battery is discharged until its terminal voltage reaches a specified level.

For example, a 60 Ah battery delivers 60 A for 1 hr, 30 A for 2 hr, 20 A for 3 hr, or any other combination of current and time where the product is 60 Ah. Therefore, its C-rate is 60 A. There are other intervals in which to rate battery capacity. A C/2 rate means that the battery is discharged at its rate divided by 2. A 60 Ah battery can deliver 30 A for 2 hours, so its C/2 rate is 30 A. The same 60 Ah battery has a C/3 rate of 20 A. Other rates can be calculated for the battery.

This does not mean that the battery will consistently (or actually) deliver the rate that is specified by the manufacturer. One of the largest variables to determine the actual amp-hour capacity that a battery can deliver is the condition of the battery at the beginning of discharge. Therefore, a user should not assume that if a battery has a C/2 rate of 30 A that it can actually deliver 30 A for 2 hours.

The common approach to determining C-rate discharge levels is the use of the 1C, 2C, 3C, etc. measurement method. For example, a battery would be discharged at the 1C capacity to ensure that it can deliver that current for one hour. If a battery can deliver 60 A for one hour down to the specified terminal voltage, its 1C rating would be 60 A. This battery can also be tested and calculated to determine C-rates to derive the capacity during different discharge time periods. If the battery has a 1C rating of 60 A, it would be rated at 60 A for 1 hour. The 2C rate means that the battery is discharged at its rate times two. Its 2C rating would be 120 A for ½ hr and its 4C rating would be 240 A for ¼ hr.

Energy Density and Power Density

Energy density is the amount of energy a battery stores per unit volume at a specified discharge rate, usually expressed in watt-hours per liter (Wh/L). *Power density* is the amount of power a battery can deliver per unit volume at a specified state of charge, usually expressed in watts per liter (W/L).

The energy density rating is similar to the amp-hour rating. They are ways of expressing the driving range of a vehicle operating in electric propulsion mode. The amp-hour rating for a traditional PbA battery is typically 50 Ah to 70 Ah. An amp-hour rating that is 65% to 80% less than that of a traditional PbA battery may be all that is necessary to supply energy to crank an engine or for electric launch assist. Therefore, the amp-hour rating of a battery system that is mechanized with a hybrid and electric system could be as low as 6 Ah.

Specific Energy and Specific Power

Specific energy is the amount of energy a battery stores per unit mass at a specified discharge rate, usually expressed in watt-hours per kilogram (Wh/kg). This is a measure of the number of watts the battery can deliver per hour based on its mass. Specific energy is analogous to the reserve capacity (RC) rating. *Specific power* is the amount of power a battery can deliver per unit mass at a specified state of charge, usually expressed in watts per kilogram (W/kg). This is a measure of the instantaneous power a battery can deliver based on its mass. Specific power is analogous to the cold cranking amp (CCA) rating. **See Figure 12-8.**

The choice of battery design and ratings depend on the vehicle requirements. A hybrid and electric power system uses the electric propulsion system for stop/start and to provide a burst of electrical power to assist the ICE during acceleration. The batteries in hybrid and electric systems are typically sized for specific power output rather than an amp-hour rating. This is because the vehicle needs power to meet the stop/start and electric launch requirements of the vehicle. However, it does not need power for extended driving on battery power alone. Therefore, a hybrid and electric system needs a high specific power rating but it does not require a high amp-hour rating.

SPECIFIC ENERGY AND SPECIFIC POWER

Battery Technology	Amp-Hours	Specific Energy Wh/kg	Specific Power W/kg
Lead-Acid (PbA) EV	60	35	200*
NiMH EV	80	60	800†
NiMH HEV	6	40	345‡

* 80% DOD
† 80% DOD
‡ 50% DOD

EV = ELECTRIC VEHICLE
HEV = HYBRID ELECTRIC VEHICLE
NiMH = NICKEL-METAL HYDRIDE BATTERY
DOD = DEPTH OF DISCHARGE

Figure 12-8. Specific energy is the amount of energy a battery stores per unit mass, while specific power is the amount of power a battery can deliver per unit mass.

The design of HEV power systems always involves compromises and tradeoffs between power availability, mass, component dimensions, cost, and the power needs of the system. Each vehicle design also requires different energy and battery control strategies. Vehicle weight, acceleration rate, top speed, emissions level, and miles per gallon targets determine the control strategy that will be designed and used for a particular vehicle. Advanced battery technologies, such as NiMH and lithium ion (Li-ion), can provide amp-hour ratings that are three or more times the amp-hour capacity of PbA battery technology. **See Figure 12-9.** However, advanced battery technologies can cost several times more than PbA technology.

The Panasonic NiMH battery pack has one venting port per module.

BATTERY TECHNOLOGY COMPARISONS

Characteristics	PbA	NiMH	Li-ION
Specific energy (Wh/kg)	30 to 50	60 to 110	150 to 190
Energy density (Wh/L)	75 to 120	220 to 330	220 to 330
Maximum power (W/kg)	Up to 700	Up to 900	Up to 1500
Number of cycles (Charge/Discharge)	400 to 600*	1500	500 to 1000
Technology advantages	Low cost	Very good energy density	Excellent energy and power
Technology disadvantages	Low energy and sudden death	Cost of base materials; performance/temp	Safety of large elements; cost

* sealed technology

Figure 12-9. The design of HEV power systems always involves compromises and tradeoffs between power availability, mass, component dimensions, cost, and the power needs of the system.

Temperature Effects

The capacity of batteries changes with temperature. For example, traditional flooded PbA batteries have lower capacity in cold weather. Therefore, ICE vehicles do not crank as fast or as long in cold weather as in warm/hot weather. As temperature increases, battery capacity increases until a maximum capacity point is reached. PbA battery capacity increases with temperature. The maximum temperature a battery can operate without significantly damaging cell plates is 140°F. **See Figure 12-10.**

Manufacturers must design thermal systems that can either warm or cool batteries to maintain optimum capacity regardless of ambient temperatures. All battery systems need a thermal system that can maintain battery module temperatures for optimum capacity.

TEMPERATURE EFFECTS

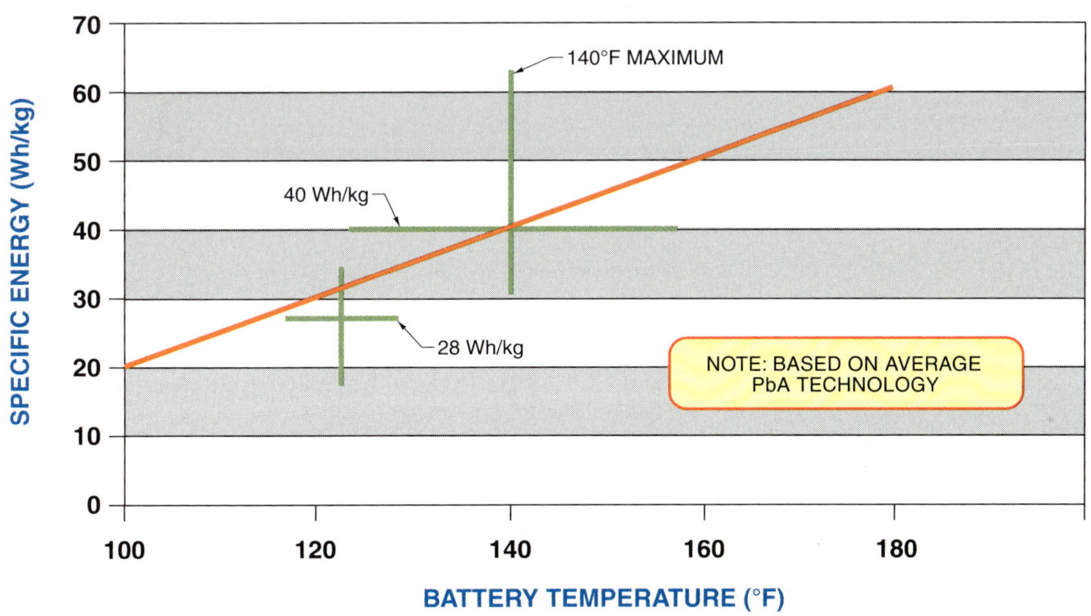

Figure 12-10. As temperature increases, battery capacity increases until a maximum capacity point is reached.

Refer to the *Hybrid Electric Vehicle Technology* CD-ROM for a Quick Quiz® on the chapter contents.

POINTS TO REMEMBER

- Battery-pack module technologies include lead-acid (PbA), nickel-metal hydride (NiMH), and lithium-ion (Li-ion) technologies.
- Two common technologies used to make lead-acid batteries are flooded and absorbent glass mat (AGM) technologies.
- Lead-acid technology has lower specific energy, specific power, and amp-hour ratings than NiMH and Li-ion technologies.
- Valve regulated lead-acid (VRLA) battery technology controls the internal battery pressure with an internal valving system that assists in maintaining correct battery internal pressure during charging.
- State of charge is a measure of the remaining capacity of a battery relative to the capacity at full charge, usually given as a percentage.
- Self-discharge is an internal battery process in which the chemical reactions in the battery cause it to discharge even if there is no load connected to the battery.
- Capacity rate (C-rate) is the rate at which the capacity of a battery is discharged until its terminal voltage reaches a specified level.
- Amp-hour capacity is a measure of how much current a battery can deliver over a specified time and within a given terminal voltage level.
- Energy density is the amount of energy a battery can store based on its size.
- Power density is the amount of power a battery can deliver based on its size.
- Specific energy is how much energy a battery can store based on its mass (weight).
- Specific power is the amount of electrical power a battery can deliver based on its mass (weight).

Refer to the *Hybrid Electric Vehicle Technology* CD-ROM for the Review Questions in Microsoft® Word format.

REVIEW QUESTIONS

1. What would a battery state of charge (SOC) of 30% indicate about the capacity of the battery?
2. What is the C/2 rate for a battery that is specified at 90 Ah?
3. What is the 2C rate for a battery that is specified at 80 Ah?
4. Why can VRLA AMG batteries sit in storage for longer periods than standard flooded batteries?
5. What is the difference between power density and specific power?

12 Battery Construction and Technologies

HYBRID Electric Vehicle TECHNOLOGY

Trade Test

Name ______________________ Date ______________

______ 1. A battery with a rating of 90 Ah could deliver about ___ A for 2 hr.
A. 30
B. 45
C. 90
D. 180

______ 2. If a scan tool provided a data parameter of 65% battery-pack state of charge, the battery pack has ___.
A. 35% of its power rating
B. 35% total resistance
C. 65% total resistance
D. 65% of its amp-hour rating

______ 3. ___ is the electrical current that a battery will deliver in a given amount of time until its terminal voltage reaches 10.5 V.
A. State of charge
B. Amp-hour capacity
C. Specific power
D. Cold cranking amperage

______ 4. The capacity of a PbA battery will increase up to the maximum capacity point when ambient temperature ___.
A. decreases
B. remains the same
C. increases
D. triples

______ 5. ___ battery technology does not fill each battery cell with liquid electrolyte.
A. Microspore
B. VRLA AGM
C. Flooded
D. Lead-peroxide

______ 6. Energy density is the amount of energy a battery can deliver based on its ___.
A. weight
B. size
C. height
D. plate variable

_______________ **7.** ___ is defined as the amount of energy a battery can store based on its weight.

A. Specific energy
B. Specific power
C. Energy density
D. Power density

_______________ **8.** If the amp-hour rating of a battery is 60 Ah, the C/3 rate is ___ A.

A. 10
B. 20
C. 30
D. 180

Nickel-Metal Hydride Technology

13

Nickel-metal hydride battery technology has moved into the forefront of electric, hybrid, and fuel-cell technology since its introduction in the mid-1990s. This technology has become one of the primary energy storage sources for battery-pack systems. It has emerged as the preferred technology for contemporary EVs and HEVs because of the broad specific energy and specific power capabilities.

CHAPTER OBJECTIVES

- Describe nickel-metal hydride self-discharge characteristics.
- Name the electrolyte mixture contained in nickel-metal hydride cells.
- Describe the differences between positive and negative electrodes in a nickel-metal hydride battery.
- Compare the specific energy ratings of lead-acid and nickel-metal hydride technologies.
- Describe nickel-metal hydride charge and discharge operation and characteristics.

KEY TERMS

potassium hydroxide (KOH) electrolyte: An electrolytic solution of potassium hydroxide (KOH) and water.

starved electrolyte battery: A battery with only a small amount of electrolyte. The minimal amount of electrolyte facilitates a diffusion of oxygen to the negative electrode at the end of the charge cycle for an oxygen-recombination reaction.

Peukert effect: The characteristic of a battery in which the battery capacity depends on the discharge current. A higher discharge rate reduces the total power that can be removed from the battery.

NICKEL-METAL HYDRIDE CONSTRUCTION

Rechargeable NiMH batteries are dominant in the market for several reasons. For example, the high electrolyte conductivity permits high power applications, the battery system can be sealed to minimize maintenance and leakage issues, and operation is possible over a wide temperature range. Although NiMH batteries cost more than comparable battery technologies, their long cycle life offsets the higher initial cost. Also, a NiMH battery can deliver higher energy density and lower cost per watt or watt-hour than other batteries. This means that the battery-pack module can be small and compact. **See Figure 13-1.**

Electrolyte

NiMH batteries are called alkaline storage batteries because of the use of alkaline potassium hydroxide (KOH) electrolyte. *Potassium hydroxide (KOH) electrolyte* is an electrolytic solution of potassium hydroxide (KOH) and water. The electrolyte concentration is a major component of cell resistance and remains fairly constant throughout the entire range of charging or discharging. Electrolyte has very high conductivity and usually does not enter into the cell reaction to any significant extent. These characteristics produce a battery with high power performance and a long cycle life.

A minimum amount of electrolyte is used in a sealed cell design with most of the liquid absorbed by the separator and the electrodes. The absorption of electrolyte into a separator is analogous to a baby diaper absorbing and holding liquid. A *starved electrolyte battery* is a battery with only a small amount of electrolyte. The minimal amount of electrolyte facilitates a diffusion of oxygen to the negative electrode at the end of the charge cycle for an oxygen-recombination reaction.

Electrodes

The active materials in NiMH batteries are composed of metal compounds or metallic oxides that are relatively good conductors in a charged state. **See Figure 13-2.** One electrode consists of the compounds nickel oxide and hydroxide.

The electrode only exchanges a proton in the charge/discharge reaction and electron transfer is very rapid. This contributes to a high power capacity. The small change in the size of the electrode between charge and discharge also results in greater mechanical stability and longer cycle life.

Negative Electrode. The active material for a negative electrode is hydrogen. The metal hydride holds hydrogen. As a hydrogen storage material, metal hydride is very efficient in achieving higher volumetric efficiency than liquid hydrogen. Practical materials for NiMH batteries can hold between 1% and 2% hydrogen by weight. A negative electrode has excess capacity compared to a positive electrode to facilitate both overcharge and overdischarge. Therefore, the positive electrode determines the useful capacity of the battery.

NIMH BATTERY MODULE

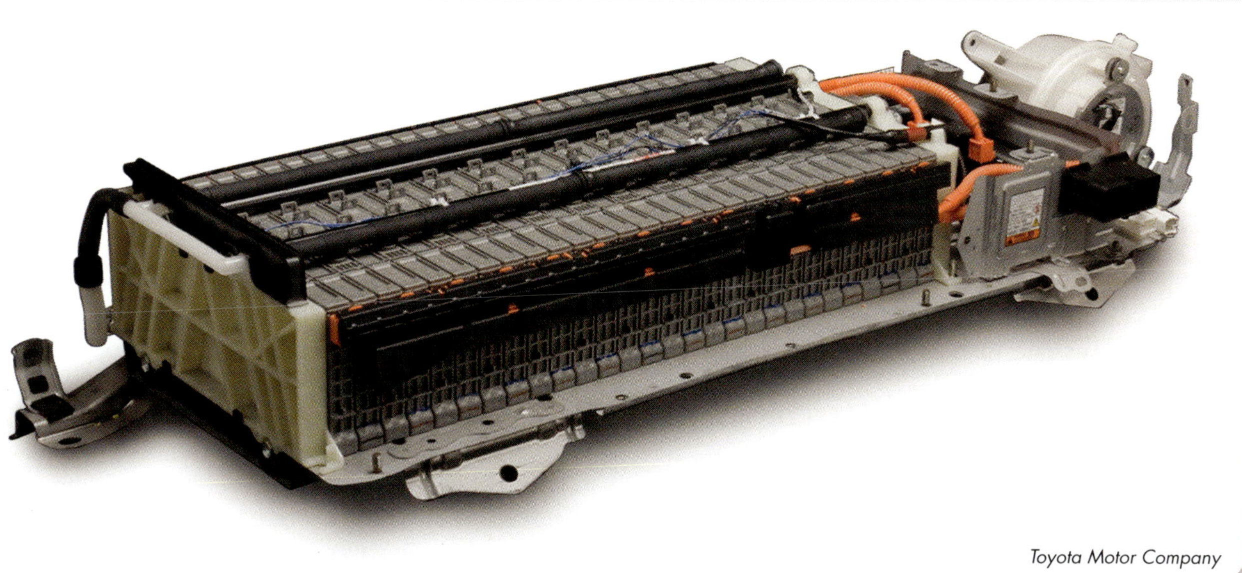

Figure 13-1. A NiMH battery can deliver higher energy density and lower cost per watt or watt-hour than other battery technologies.

ELECTRODES

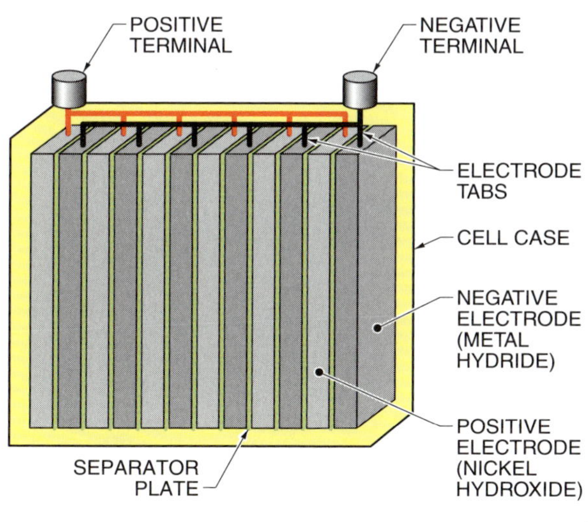

Figure 13-2. The active materials in NiMH batteries are composed of metal compounds or metallic oxides that are relatively good conductors in a charged state.

Positive Electrode. The positive electrode of a NiMH battery contains nickel hydroxide. This is a well-developed electrode material with almost 100 years of history and development. A positive electrode has the same composition as a common nickel cadmium (NiCd) battery. Nickel-based alkaline batteries are attractive because the nickel electrode can be fabricated with very large surface areas. An electrode with a higher surface area has high capacities and high current densities.

Electrolyte does not enter into the electrode reaction. Therefore, conductivity remains at a high level throughout the usable capacity of the battery. In addition, the active material in nickel is insoluble in the KOH electrolyte leading to longer life and greater tolerance to abuse.

NICKEL-METAL HYDRIDE OPERATION

NiMH battery technology has several advantages over PbA or NiCd technology. It has good energy density with the ability to store about three times the energy of a PbA battery. It has a high rate capability and good thermal performance. This technology is made of recyclable nontoxic materials. It also has minimal Peukert effect and there is no hydrogen evolution on overcharge.

The *Peukert effect* is the characteristic of a battery in which the battery capacity depends on the discharge current. A battery with a high Peukert effect has a higher discharge rate that reduces the total power that can be removed from the battery.

NiMH battery technology has several advantages over PbA or NiCd battery technology. It does not contain any toxins and is, therefore, more environmentally friendly than NiCd. NiMH battery technology also has several disadvantages when compared to PbA or NiCd technology. For example, it costs more than PbA. In addition, it has a higher self-discharge rate and requires a more robust energy management system.

In general, NiMH technology is sensitive to overcharging. Deep discharging and constant overcharging contribute to a decline in the life of the battery. Therefore, a constant rate of charge is the recommended charging process.

During a typical drive cycle, a vehicle may have an SOC operating range of about 55% to 60%. **See Figure 13-3.** This operating range facilitates long cycle life. It also allows a low enough SOC range to permit the capture of energy during regenerative braking modes. The battery-pack SOC is a good indicator of static and dynamic battery-pack health.

A NiMH battery is capable of supplying a large amount of power to a load. Batteries with specific power are commercially available with a range of 200 W/kg to greater than 1000 W/kg.

Charging

When a NiMH battery is charging, the positive electrode reaches full charge before the negative electrode and begins to evolve oxygen. The oxygen gas diffuses through the separator to the negative electrode. Diffusion is aided by the starved electrolyte design and the selection of an appropriate separator system. At the negative electrode, oxygen reacts with the hydrogen electrode to produce water.

The negative electrode does not become fully charged, which prevents the generation of hydrogen gas. The charge current must be controlled at the end of the charge and during overcharge to limit oxygen generation to a level below the rate of recombination. This prevents gas and pressure buildup.

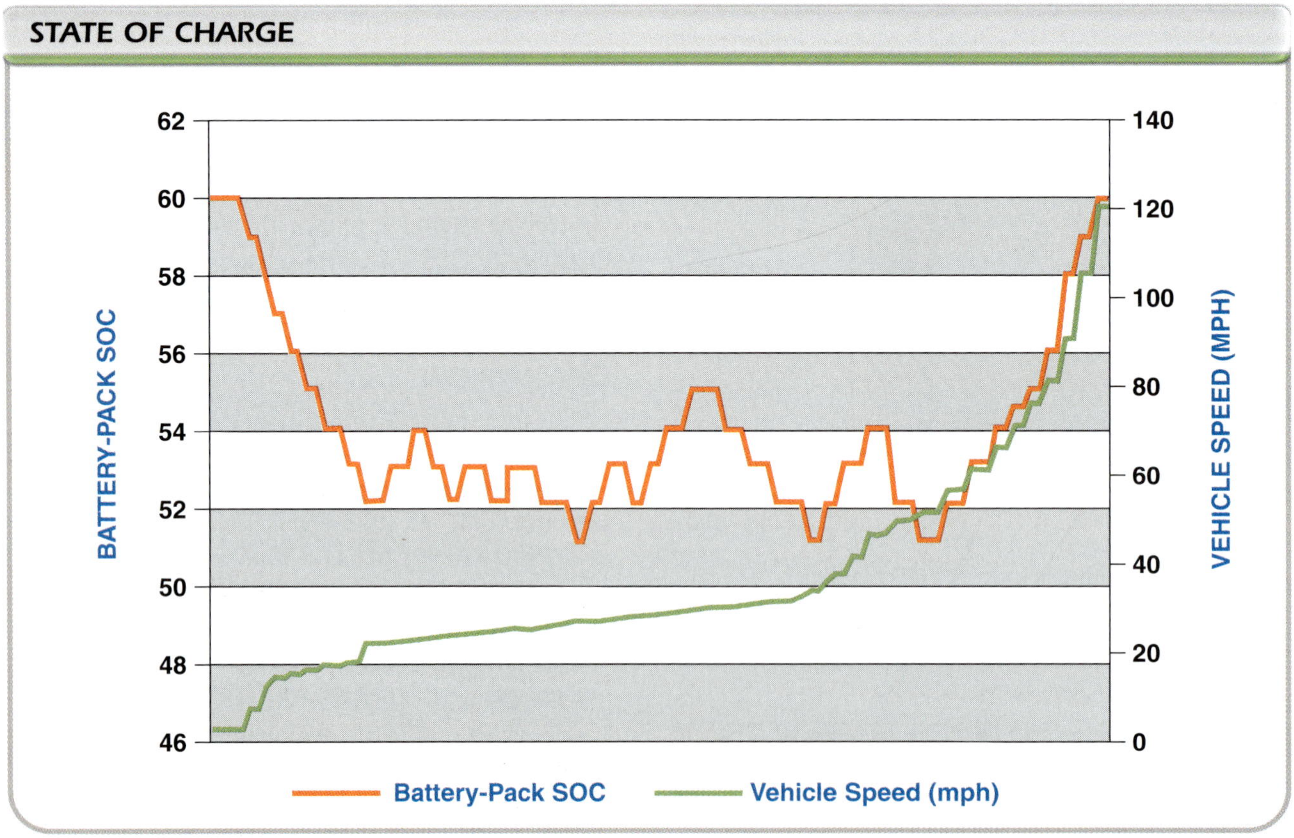

Figure 13-3. During a typical drive cycle, a vehicle may have an SOC operating range of about 55% to 60%.

The electrolyte of a NiMH battery has high conductivity and typically does not react with the anode and cathode. The electrolyte concentration remains fairly constant over the entire range of SOC.

The NiMH cell is designed with a discharge reserve in the negative electrode to minimize gassing and degradation of the cell in the event of overdischarge. The positive electrode determines the useful capacity of the battery.

Overcharge. A NiMH cell uses oxygen recombination to prevent pressure buildup that may result from gases generated after the end of the charge process. Oxygen recombination is based on the use of a negative electrode that has a higher effective capacity than the positive electrode. Overcharge of a battery occurs after the positive electrode no longer has nickel hydroxide to react with the hydroxide from the electrolyte and begins to evolve oxygen.

Oxygen diffuses from the positive electrode through the separator to the negative electrode. At the negative electrode, oxygen recombines with stored hydrogen to form water in the electrolyte. If this oxygen recombination occurs faster than the oxygen that moves away from the positive electrode, the battery maintains its normal pressure. If oxygen recombination occurs slower than the oxygen that moves away from the positive electrode, the battery builds pressure.

NiMH battery cells are constructed with a greater volume of active material on the negative electrode than on the positive electrode to help prevent overpressure. Final pressures in a battery module can range from about 43 psi in a Toyota Prius to about 120 psi in a GM EV1.

Overcharging a NiMH battery can result in permanent loss of capacity and cycle life. If a cell is overcharged and cell pressure increases, there is a corresponding increase in cell temperature that can cause the separator to lose electrolyte. This loss of separator electrolyte causes the separator to dry out and inhibits the proper transport of hydrogen in and out of the electrodes.

If a battery is severely overcharged and an excessive volume of oxygen gas has evolved, the pressure may be released through the safety vent. This removes elements within the cell needed for proper function. To protect against the damaging effects of overcharging, charging software algorithms must be developed.

Discharging

When a NiMH battery is discharged, the chemical reactions are opposite of when it is charged. Hydrogen stored in the metal alloy of the negative electrode is released into the electrolyte to form water. The water then releases a hydrogen ion that is absorbed into the positive electrode to form nickel hydroxide. In NiMH battery cells, the process of moving or transporting hydrogen from the negative to positive electrode absorbs heat. Heat continues to be absorbed until the cell reaches a state of overdischarge. After a battery becomes overdischarged, a secondary reaction occurs within the battery cells that results in a temperature rise.

Overdischarge. There are two phases that occur within battery cells when a NiMH battery is in an overdischarge condition. During phase one, the active material at the positive electrode is depleted and hydrogen gas generation stops. Because the negative electrode has more active material than the positive electrode, it can absorb some of the hydrogen gas evolved by the positive electrode. Any hydrogen not absorbed by the negative electrode is fully depleted of active material. The negative electrode absorbs oxygen, which contributes to the loss of usable capacity.

During the second phase, when both electrodes are fully depleted, extreme overdischarge of NiMH battery cells can result in excessive gassing that results in permanent damage. The damage has two different effects. First, the storage capacity of the negative electrode is reduced when oxygen permanently occupies a hydrogen storage site. Second, excess hydrogen is released through the safety vent. This reduces the amount of hydrogen inside the cell.

NiMH battery modules are grouped together in a battery pack.

Self-discharge. All battery technologies have some self-discharge that occurs when the battery is at rest. Some of the factors contributing to self-discharge include the energy used by the oxygen cycle at high states of charge and particle shorting (grid growth). The contribution to self-discharge from the oxygen cycle is negligible below about 70% SOC. Longer term contributions to self-discharge are caused by chemical ion shuttles that continuously discharge the cell over longer periods.

NiMH batteries have a high self-discharge rate of approximately 5% to 30% per month depending on the technology, SOC, and ambient temperature. This is higher than PbA and NiCd battery technology.

NiMH batteries should be charged to maintain shelf life and should not be stored at full charge. The self-discharge rate is highest for full batteries and drops off somewhat for lower charges. The recommended long-term storage charge is around 40%.

The self-discharge rate is strongly affected by the temperature at which batteries are stored. The self-discharge rate is higher for batteries stored at higher temperatures than for batteries stored at lower temperatures. The rate of self-discharge also depends on the condition of the cell. Batteries stored at higher temperatures and older batteries have higher self-discharge rates. **See Figure 13-4.**

SELF-DISCHARGE

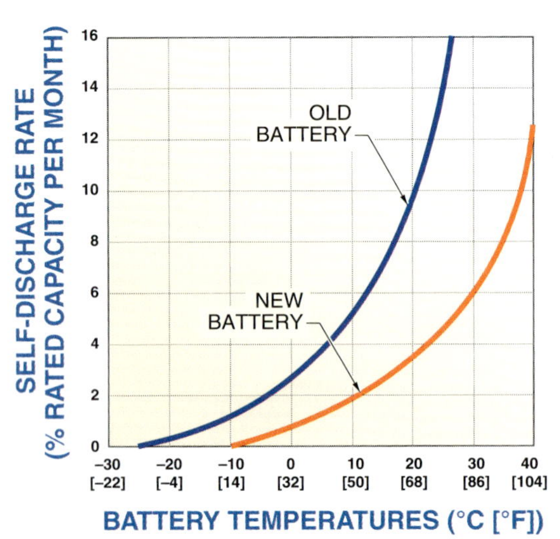

Figure 13-4. Batteries stored at higher temperatures and older batteries have higher self-discharge rates.

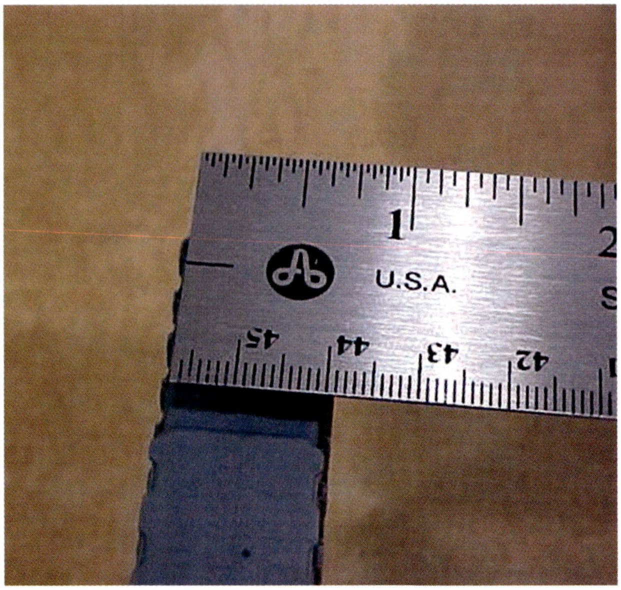

NiMH battery modules are very thin to save space.

Charge and Discharge Electrical Characteristics

The discharge curves of NiMH are very different than those of PbA technology. For NiMH, the charge and discharge curves are very flat throughout the majority of the charge and discharge cycles. Flat voltage output means that dynamic terminal voltage is maintained throughout the battery discharge until the end point. At the end point, the terminal voltage drops severely. In other words, the terminal voltage does not change much during discharge and it cannot be used as a reliable indicator of SOC.

NiMH provides consistent capacity without significant change in terminal voltage. NiMH is favored for its ability to provide a flat voltage output that results in constant terminal voltage levels. **See Figure 13-5.** NiMH exhibits a very linear resistance characteristic for discharge and charge at any given SOC. This allows the battery behavior model to be simplified for EV, HEV, or fuel-cell vehicle applications.

Electric propulsion systems need constant energy supply levels to maintain consistent performance and low amperage levels. A PbA battery has a linear terminal voltage reduction that is proportional to its capacity. Therefore, this technology does not hold a flat output characteristic. It also has a drop at the end of its capacity capability. Therefore, its specific energy performance significantly lags NiMH technology.

CHARGE AND DISCHARGE

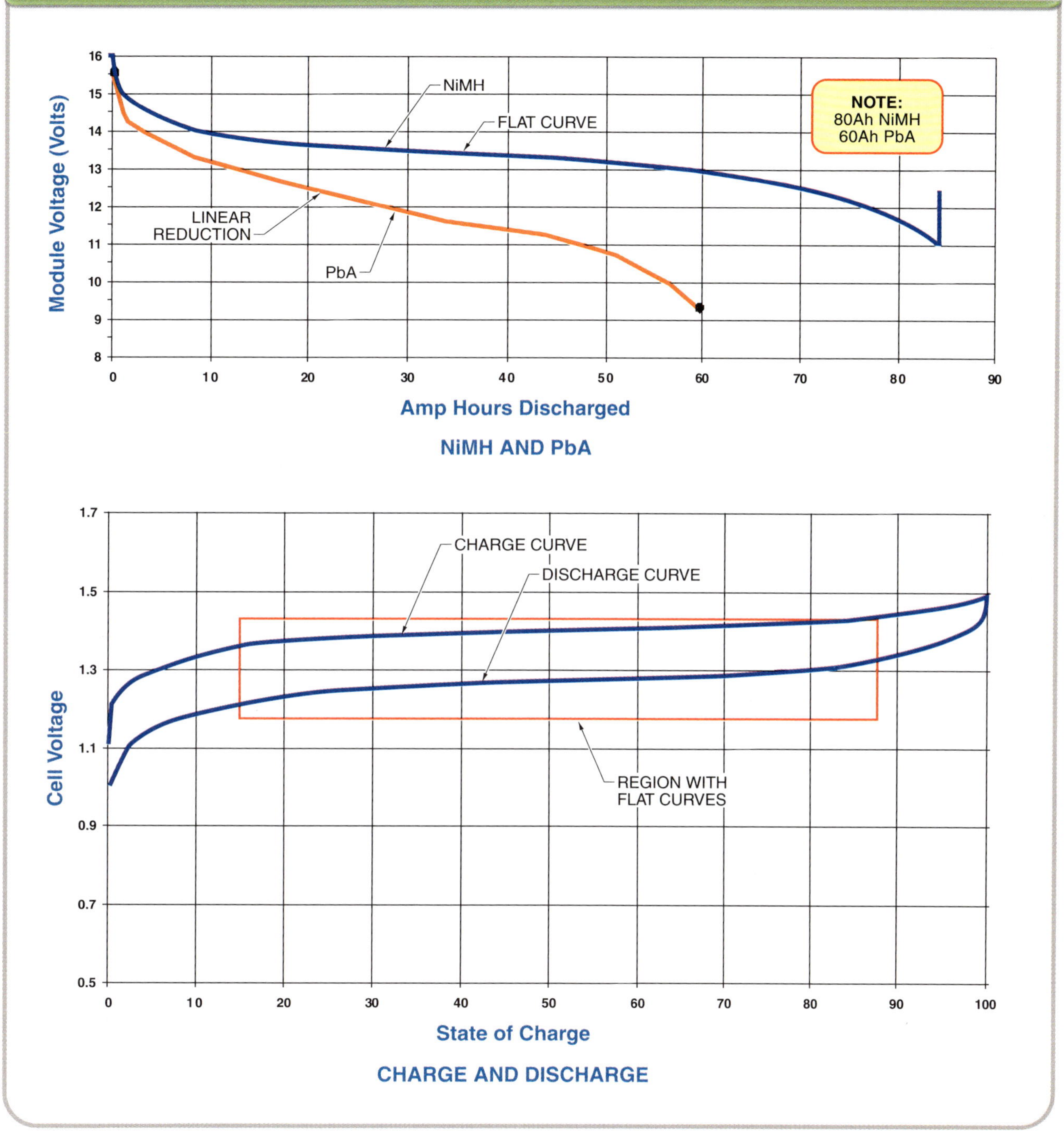

Figure 13-5. NiMH is favored for its ability to provide a flat voltage output that results in constant terminal voltage levels.

Refer to the *Hybrid Electric Vehicle Technology* CD-ROM for a Quick Quiz® on the chapter contents.

POINTS TO REMEMBER

- NiMH battery modules, as with AGM battery technology, use a design in which the electrolyte is stored in the plates and separators. NiMH does not use a flooded electrolyte design.
- NiMH technology uses potassium hydroxide (KOH) electrolyte.
- NiMH battery technology has higher specific energy, higher specific power, and a higher amp-hour rating than lead-acid technology.
- The discharge curve of NiMH technology has a distinct advantage over PbA battery technology by providing a very flat, consistent terminal voltage, which results in a stable, steady voltage during discharge. Lead-acid technology has a linear voltage drop during discharge that results in terminal voltage dropping during discharge.
- Lead acid batteries have a lower rate of self-discharge when compared with NiMH technology.
- When storing NiMH batteries it is recommended that the state of charge be approximately 40% in order to reduce the effects of self-discharge.

Refer to the *Hybrid Electric Vehicle Technology* CD-ROM for the Review Questions in Microsoft® Word format.

REVIEW QUESTIONS

1. If a battery with a high Peukert effect is discharged very quickly, how will the power output be different than a battery with a low Peukert effect?
2. What two operating parameters contribute to a decline in the life of a NiMH battery?
3. Why is it that the terminal voltage of NiMH batteries cannot be used to determine SOC?
4. Is the self-discharge rate for a NiMH battery higher or lower when it is stored at full charge?
5. Is the self-discharge rate for a NiMH battery higher or lower when it is stored at a low temperature?

13 Nickel-Metal Hydride Technology

HYBRID Electric Vehicle TECHNOLOGY

Trade Test

Name ______________________ Date ______________

______ **1.** The terminal voltage of NiMH batteries cannot be used to determine SOC because ___.
A. terminal voltage stays constant throughout a discharge
B. resistance constantly changes
C. ambient temperature causes erratic behavior
D. terminal voltage has no relationship with SOC

______ **2.** A nickel-metal hydride battery can store about ___ times as much energy as a lead-acid battery.
A. two
B. three
C. four
D. five

______ **3.** The electrolyte solution used in a nickel-metal hydride battery module is a combination of ___ and water.
A. sulfuric acid
B. bromide
C. boron nitrate
D. potassium hydroxide

______ **4.** A NiMH battery does not include ___ in its design.
A. flooded electrolyte
B. positive plates
C. negative plates
D. KOH electrolyte

______ **5.** The cells of a nickel-metal hydride battery can lose significant capacity if the battery is ___.
A. stored at less than 50% capacity
B. severely undercharged
C. connected in series
D. severely overcharged

______ **6.** A NiMH cell uses ___ to prevent pressure buildup that may result from gases generated after the end of the charge process.
A. hydrogen storage
B. self-discharge
C. overcharging
D. oxygen recombination

_______________ **7.** The NiMH battery is called an alkaline storage battery because of its use of alkaline ___ electrolyte.

A. potassium hydroxide (KOH)
B. sodium hydroxide (NaOH)
C. sulfuric acid (H2SO4)
D. metal hydroxide (MOH)

Battery-Pack and Energy Management Systems

14

Electrical power is essential to a hybrid propulsion system. The energy management system controls the flow of electrical power in a hybrid vehicle. Electrical energy is stored in the battery pack. Battery packs are placed in many various locations in different vehicles. The characteristics of series-aided and parallel-aided battery connections are important parts of battery-pack operation. Higher voltage systems supply smaller electric current to deliver the same power. This allows the use of lighter and less expensive cables, wires, transistors, and cooling systems.

CHAPTER OBJECTIVES

- Explain Battery-Pack technology.
- Demonstrate how to calculate electrical power (watts) from voltage and current.
- Compare the electrical characteristics of connecting batteries in a series configuration and in a parallel configuration.

KEY TERMS

electrical power: The amount of work done by electric current. This is equivalent to the rate at which electrical energy is transmitted. The SI unit of power is the watt.

cylindrical battery: A battery module that is cylindrical in shape.

prismatic battery: A battery module that is rectangular in shape.

series circuit: A circuit with only one current path for electric current flow.

parallel circuit: A circuit with two or more electrical circuit branches for electric current flow.

BATTERY-PACK AND ENERGY MANAGEMENT SYSTEMS

Although the electric drive motor and power inverter are central to a hybrid propulsion system, propulsion cannot be completed without a form of electrical power. *Electrical power* is the amount of work done by electric current. This is equivalent to the rate at which electrical energy is transmitted. The SI unit of power is the watt. The energy management system (EMS) used to store and control electrical power is the battery pack. **See Figure 14-1.**

BATTERY PACKS

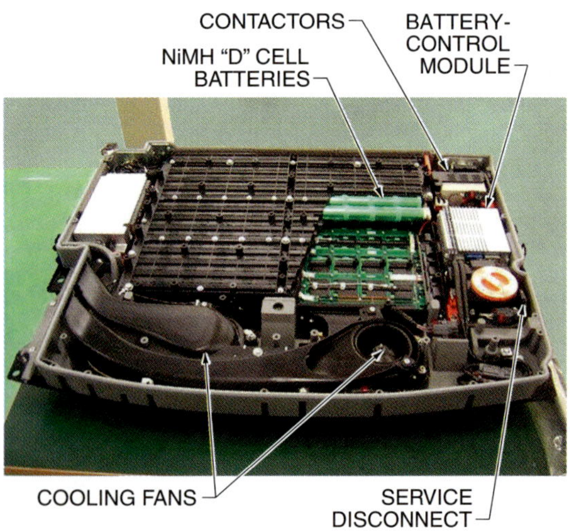

Figure 14-1. The energy management system (EMS) used to store and control electric power is the battery pack.

Battery packs are found in EVs and HEVs. Historically, the battery pack has been the factor limiting the range of an EV. It does not limit the range of an HEV vehicle because a battery pack also uses an ICE for power. However, battery pack performance remains central to HEV and fuel-cell vehicle operation.

One difference between pure EV and HEV battery packs is the amp-hour (Ah) capacity. An EV needs to store enough energy for its entire operating range. HEV technology needs a smaller amp-hour rating than a pure EV. HEV technology needs to store enough energy for starting and assist. It can generate electricity with its on-board power generating system.

The battery pack (single or multibattery) is a source of power for a hybrid and electric system. The technology used in EV and HEV battery packs for decades has been lead-acid (PbA) technology, but this technology is phasing out. Most manufacturers are using advanced battery designs such as nickel-metal hydride (NiMH) or lithium-ion (Li-ion) technologies.

Regardless of the battery technology, the battery is required to deliver energy to the power inverter for it to deliver controlled power to the electric drive motor and ancillary vehicle system. The battery controller needs to deliver infinitely controlled power to these systems to satisfy the needs of the vehicle and customer.

For example, both the Honda Insight and Civic hybrids use cylindrical, D-size NiMH batteries. A *cylindrical battery* is a battery module that is cylindrical in shape. Other battery modules may use a prismatic design, instead of a cylindrical design. A *prismatic battery* is a battery module that is rectangular in shape. Prismatic batteries are used in the Toyota Prius.

In contemporary HEV designs, the battery pack is located in areas of the vehicle not typically thought of for power electronics. Battery-pack energy management systems may be located in the trunk area, behind the rear seat, in the main cabin (under seats or rear load floor), or under the vehicle.

The battery pack connects to other controllers through the vehicle data communications network. **See Figure 14-2.** The data communications network consists of batteries, a fuse, interconnection cables, cabling and connections, contactors (relays), a battery-pack thermal system, and shunt or Hall-effect sensors. Energy management thermal systems use small fans to force air through the battery pack to maintain proper temperature.

BATTERY-PACK COMPONENTS

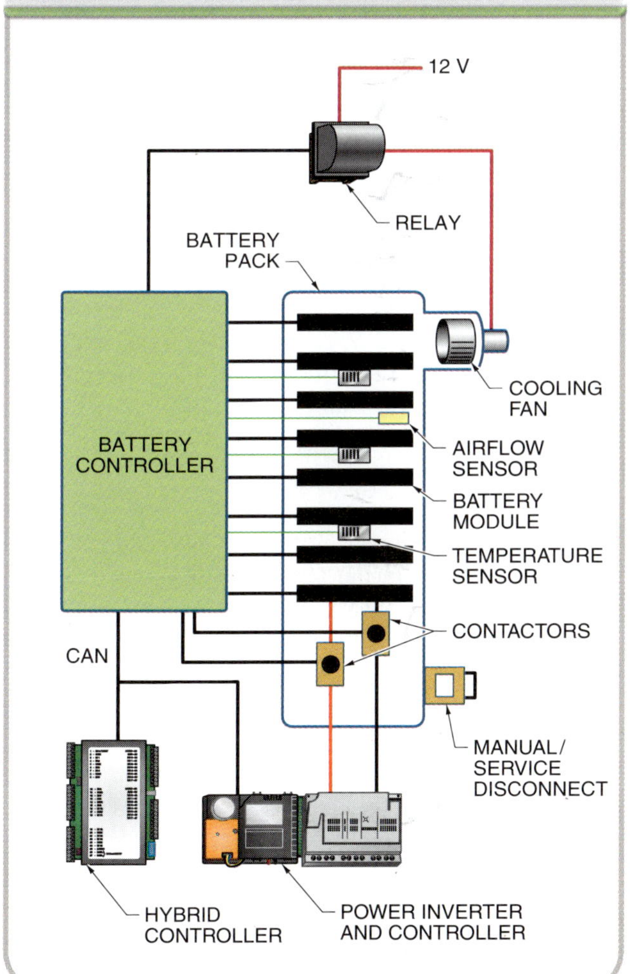

Figure 14-2. The battery pack connects to other controllers through the vehicle data communications network.

Power Law

The power law can be used when analyzing any electrical circuit, especially a circuit that demands a high amount of electrical power. The power law states the relationship between power, current, and voltage. This provides a mathematical understanding of why higher voltage systems have many advantages over lower voltage systems. The power law can be used to analyze any electrical circuit, whether in an ICE cranking motor or in any electrical circuit in a vehicle. **See Figure 14-3.** The power law formula is as follows:

$$P = E \times I \text{ or } E = \frac{P}{I} \text{ or } I = \frac{P}{E}$$

where

P = power (in W)
E = voltage (in V)
I = current (in A)

OHM'S LAW AND POWER LAW FORMULA

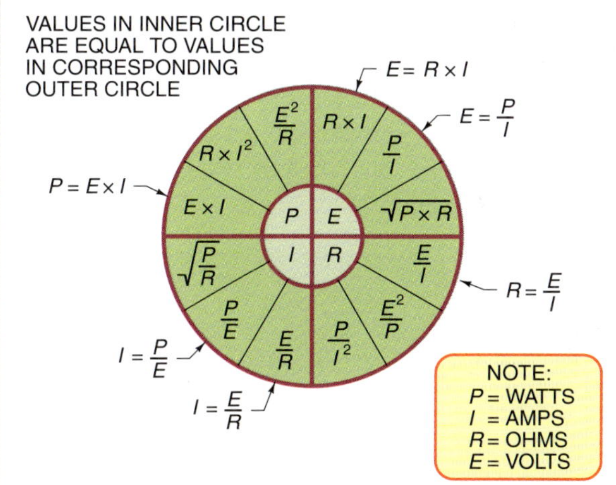

Basic Formulas		
$E = R \times I$	$I = \frac{E}{R}$	$R = \frac{E}{I}$
$E = \frac{P}{I}$	$I = \frac{P}{E}$	$P = E \times I$

Figure 14-3. The power law states the relationship between power, current, and voltage.

Ohm's law states that the current (in A) in a circuit is equal to the voltage (in V) divided by the resistance (in Ω).

For example, a vehicle with a 12 V system and a heater blower motor could draw 4 A when the motor is running. The amount of power drawn by the blower motor is calculated as follows:

$$P = E \times I$$
$$P = 12 \times 4$$
$$P = \mathbf{48\ W}$$

This shows that 48 W are required to power the motor. In an electrical circuit, the power used by a load is the same if the power is supplied at a different voltage. A device that has a specific power rating always uses its rated power to do the same amount of work. For example, if the same motor is powered with a 24 V source, it still uses 48 W. The amount of current is calculated as follows:

$$I = \frac{P}{E}$$
$$I = \frac{48}{24}$$
$$I = \mathbf{2\ A}$$

The power law can be thought of as a teeter-totter. If a load is placed into a circuit and the voltage is increased, the power law shows that the current decreases. If the voltage is decreased, the power law shows that the current increases. The power law concepts help explain circuit behavior and can be helpful when troubleshooting any electrical circuit.

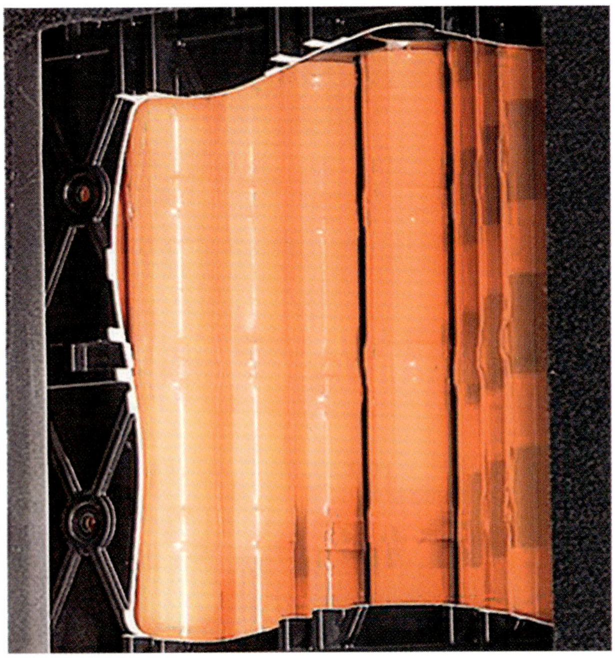

Some hybrid vehicles use cylindrical, D-size NiMH batteries in the battery pack.

The power law can also be applied to compare the current drawn from a motor by a 12 V power system with the current drawn by a 24 V power system. The higher voltage system draws less current, which means it can use smaller conductors. This reduces the amount of expensive copper wire and reduces the weight of the electrical system.

The most common occurrence of power loss in an electrical circuit is when heat is produced from electric current that flows through resistance, whether it is a wire, motor, or bulb. The electric current is converted to waste heat. Formulas can be used to show the relationship between power, current, and resistance. Power loss due to resistance is calculated as follows:

$$P = I^2R$$

where

P = power (in W)

I = current (in A)

R = resistance (in Ω)

A typical automotive 12 V battery actually delivers about 14 V. Components are specified for use in a 12 V circuit, but the battery could actually deliver about 14 V, depending on the SOC. A hybrid vehicle may have components specified for use in a 36 V circuit, but the battery could actually deliver about 42 V.

The power law shows that when voltage is tripled from 12 V to 36 V, current is ⅓ of the current for the lower voltage system. For example, if a 14 V circuit draws 2 A, then a 42 V circuit draws 0.667 A. If the 14 V circuit has 7 Ω of resistance, the power lost is calculated as follows:

$$P = I^2R$$
$$P = 2^2 \times 7$$
$$P = 4 \times 7$$
$$P = \mathbf{28\ W}$$

The power lost in the 42 V circuit is calculated as follows:

$$P = I^2R$$
$$P = 0.667^2 \times 7$$
$$P = 0.444 \times 7$$
$$P = \mathbf{3.1\ W}$$

The power lost in the 42 V circuit is ⅑ of the power lost in the 14 V circuit. This shows that power loss is much less in a higher voltage system.

A disadvantage of lower voltages is the increased current needed to deliver the same amount of power. If the current increases, the wire and cable, stator wire gauge, and power transistor size must increase. Cooling requirements for the power transistors must also increase. By increasing the voltage, the sizes of these components can be reduced while supplying the same amount of power.

There are limits to the voltage available for vehicle systems. Even if a 1000 V battery pack were available, there may not be a power transistor that could operate at the same level and still provide high levels of electric current.

Power loss is a major issue with electrical and electronic systems because good thermal conditions and a normally operating cooling system are required to maintain the safe operating temperature of all devices, cables, and wires. Heat as a result of power loss is one of the major contributors of electric component failure.

Series-Aiding Connections

Series aiding, or connecting batteries in a continuous series circuit, is important in understanding hybrid and electric systems. A *series circuit* is a circuit with only one current path for electric current flow. The power law can be used to calculate power, current, and voltage for more than one battery. By connecting two batteries in a series string, or series aiding, the voltages can be added together. The currents of the batteries are not added together.

If two 12 V batteries are connected in series, the sum of the voltages is 24 V. If each battery can supply 500 A, then the two batteries in series also supply 500 A. A circuit with two 12 V batteries connected in series would be rated at 24 V and 500 A. **See Figure 14-4.** The negative and positive terminals are connected to each other.

Parallel-Aiding Connections

Batteries can be connected in a parallel circuit. A *parallel circuit* is a circuit with two or more electrical circuit branches for electric current flow. When connecting batteries in parallel, the positive terminal of one battery is connected to the positive terminal of the other. The negative terminal of one battery is connected to the negative terminal of the other. **See Figure 14-5.** The supply voltage remains at 12 V. Because there are two batteries in parallel, electric current sources add together. The total electric current for the circuit is 1000 A.

There are numerous historical examples of vehicle systems that have used parallel battery connections. Heavy-duty truck (diesel) engines and recreational vehicle electrical systems are two primary examples of applications with parallel battery connections due to their unusually high electric current requirements. The batteries were typically connected in parallel to provide the cranking motor with a higher cold cranking amp (CCA) and amp-hour capacity, especially for cold weather engine cranking operation.

> Series-aiding battery connections are used to increase the voltage in a battery pack. The voltages add when batteries are in series. Parallel-aiding battery connections are used to increase the current in a battery pack.

SERIES-AIDING CONNECTIONS

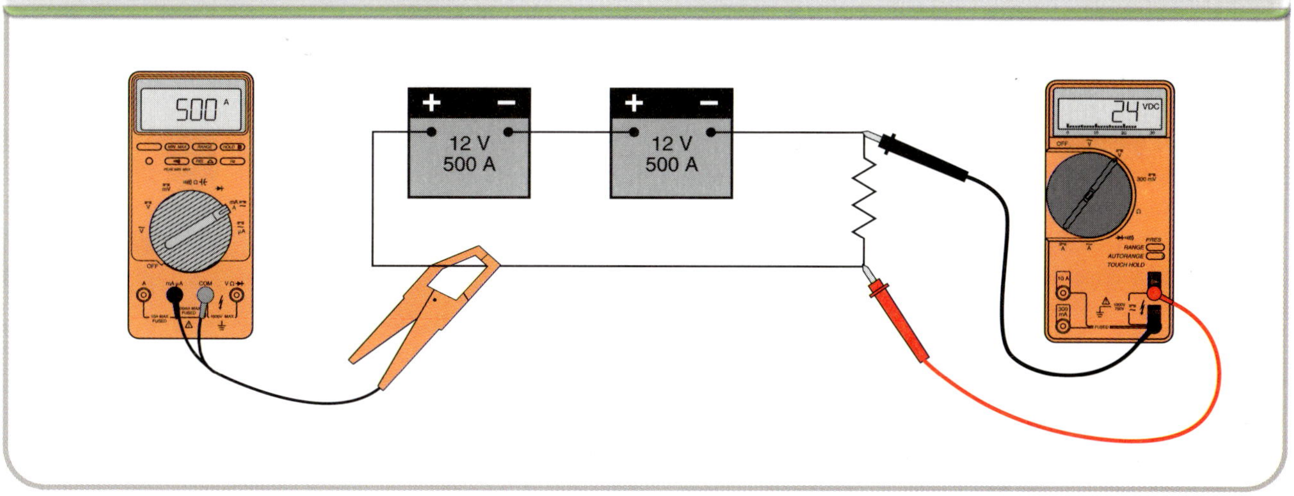

Figure 14-4. A series circuit is a circuit with only one electric current path for current flow.

PARALLEL-AIDING CONNECTIONS

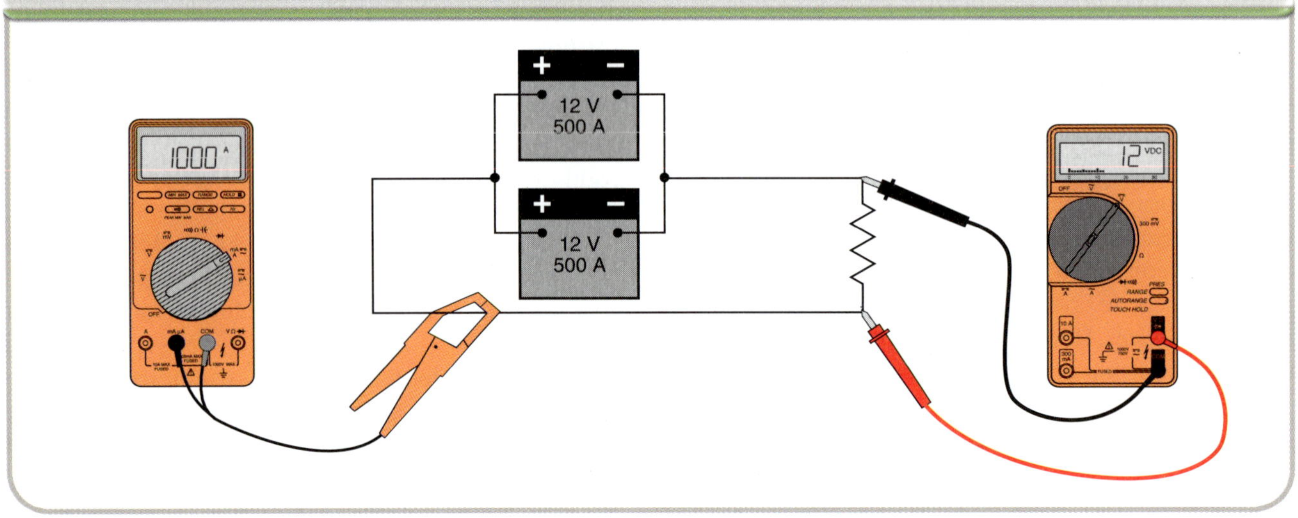

Figure 14-5. A parallel circuit is a circuit with two or more electrical circuit branches for current flow.

REGENERATIVE BRAKING

One attribute of regenerative braking (regen) is that when batteries are near depletion, the electric machine is commanded by the controller to transfer large amounts of electrical power to the battery system to replenish energy. During regen or coast down, the drive motor becomes a generator and directs energy back to the battery pack. **See Figure 14-6.**

Extreme levels of regenerative braking can be dangerous to occupants of the vehicle. Kinetic energy drives the axles, which in turn drives the electric machine. If motor power output is not controlled, the dynamic electric braking effect could cause the vehicle to decelerate rapidly. This raises concern regarding vehicle stability, braking, and occupant safety.

The electric machine can produce hundreds of volts and amps that could overcharge the batteries, especially if the battery pack was near a full SOC. The battery controller and the electric machine controller must communicate to limit the maximum regenerative braking power to a safe level.

The battery controller must smoothly transfer power to prevent sudden deceleration. At the same time, it must transfer enough energy to the battery to recover the energy that was lost during vehicle acceleration. In addition, the battery controller must completely inhibit regen if the battery is at a full SOC or until the system voltage is at a safe level.

REGENERATIVE BRAKING

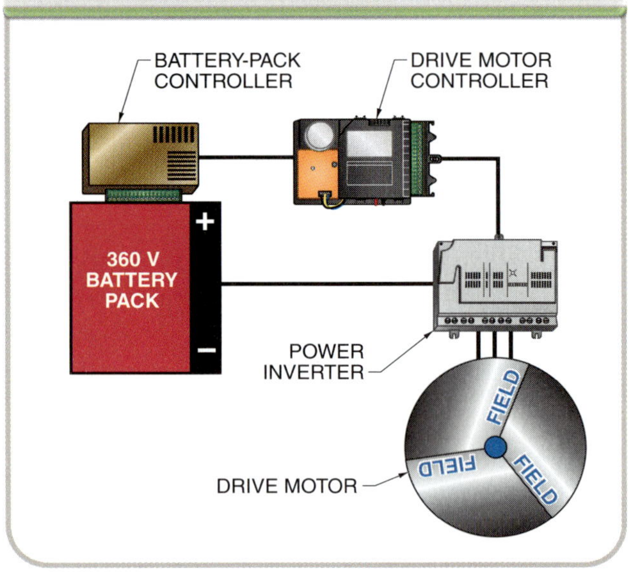

Figure 14-6. During regenerative braking or coast down, the drive motor becomes a generator and directs energy back to the battery pack.

Refer to the *Hybrid Electric Vehicle Technology* CD-ROM for a Quick Quiz® on the chapter contents.

POINTS TO REMEMBER

- A battery-pack is a type of energy management system (EMS).
- A battery-pack hardware system includes batteries, fuses, interconnection cables, power inverter cabling and connections, current sensors, contactors (relays), a thermal (cooling) system, and a control system.
- Battery modules can be of a cylindrical or prismatic design.
- The battery-pack used with energy management systems may be located in the trunk area, behind the rear seat upright, in the main cabin (under seats or rear load floor), or under the vehicle.
- Energy management thermal systems use small fans to force air through the battery pack to maintain proper temperature.
- When connecting two 12 V batteries in series, the total voltage doubles and the ability to supply electric current remains the same. The electric current capability will be the output that could be delivered from the rating of one battery.
- When connecting two 12 V batteries in parallel, the total voltage remains at 12 V and the ability to supply electric current doubles. The electric current capability will be the output that could be delivered from the rating of both batteries.
- Electrical power is calculated by multiplying voltage and electric current.
- A battery-pack controller controls the energy exiting and entering the battery modules by monitoring electric current, battery-pack voltage, module voltages, module temperatures, needs of the power inverter, and needs of the accessory systems.

Refer to the *Hybrid Electric Vehicle Technology* CD-ROM for the Review Questions in Microsoft® Word format.

REVIEW QUESTIONS

1. What is the physical shape of a prismatic battery? RECTANGULAR
2. Define the power law. $P = E \times I$
3. Explain how to calculate power loss caused by resistance in a circuit. $P = I^2 \times R$
4. List two advantages of using a higher voltage system instead of a lower voltage system.
5. Describe the difference between a series-aiding connection and a parallel-aiding connection.

14 Battery-Pack and Energy Management Systems

HYBRID Electric Vehicle TECHNOLOGY

Trade Test

Name ______________________ Date ______________

______ **1.** The battery pack supplies ___ W when the battery-pack voltage is 300 V and current is 100 A.
A. 3
B. 30
C. 3000
D. 30,000

______ **2.** If two 12 V, 500 A batteries are placed in series, the output is ___ V and ___ A.
A. 12; 500
B. 12; 1000
C. 24; 500
D. 24; 1000

______ **3.** If two 12 V, 500 A batteries are placed in parallel, the output is ___ V and ___ A.
A. 12; 500
B. 12; 1000
C. 24; 500
D. 24; 1000

______ **4.** Most manufacturers use ___ batteries in HEVs.
A. NiCd and PbA
B. NiMH and NiCd
C. PbA and Li-ion
D. Li-ion and NiMH

______ **5.** The function of the battery-pack system is to ___.
A. act as a capacitor
B. store and deliver electrical power
C. store voltage
D. convert electrical power from DC to AC

______ **6.** The most common reason for power loss in electrical circuits is due to ___.
A. voltage loss
B. current loss
C. waste heat
D. regenerative braking

_______________ 7. If a battery-pack controller senses a high SOC level during regenerative braking, the battery-pack controller ___.

A. requests more regenerative braking power
B. does not allow regenerative braking power
C. restricts the use of ancillary electrical systems
D. forces the vehicle into a coast mode

Hybrid Vehicle Braking Systems

15

Modern hydraulic braking systems have received a number of changes throughout the past two decades. Newer electric propulsion systems have led to a new level of technology and systems integration. Brake systems must now work with electric machines, power inverters, and energy management systems to produce the correct level of braking energy and provide the operator a safe stopping feel. Advanced technology braking combines traditional friction braking with regenerative braking and uses negative torque from the electric traction system to generate electricity. The braking forces from these two systems must be blended to provide a smooth, safe vehicle braking system.

CHAPTER OBJECTIVES

- Provide examples of components of hybrid braking systems.
- Describe how regenerative and hydraulic braking are affected by battery-pack SOC.
- Describe the purpose of the regenerative braking mode.
- Describe the two different categories of braking systems operation.
- Identify components and describe the operation of the hill-hold control function.
- Explain the difference between the hill-hold and creep-aid systems.

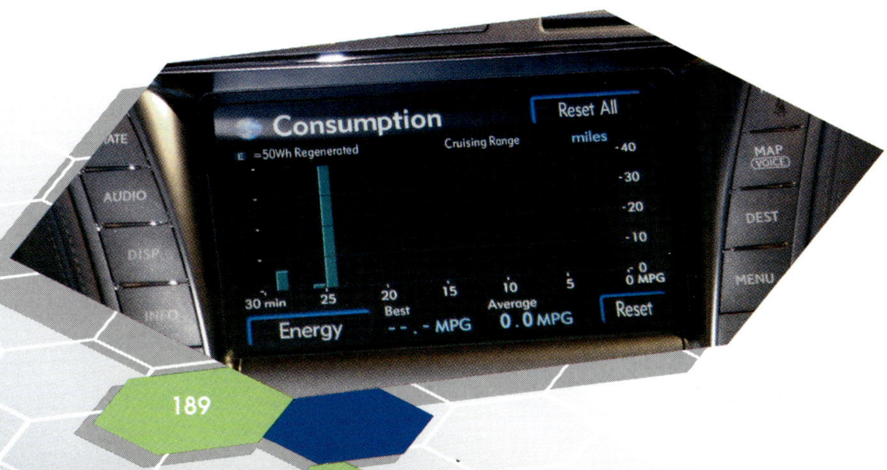

KEY TERMS

brake system power supply: A backup system used to ensure that an electric motor with brake pressure can stop a vehicle if the primary system fails. The brake system power supply usually consists of a capacitor bank.

hill-hold system: A type of vehicle propulsion system that uses an electric motor to prevent vehicle rollback down a hill. The hill-hold feature is used until the current draw becomes too high, at which time the hydraulic system is also engaged to hold the vehicle.

creep-aid system: A type of vehicle braking system that uses a series of solenoid valves and hydraulic pressure on the rear and, in certain modes, the front brakes to prevent a vehicle from rolling back on an incline.

BRAKING SYSTEM COMPONENTS

The braking systems used in hybrid vehicles are significantly different from the braking systems used in traditional automotive vehicles. The most significant differences are the regenerative braking system and the creep-aid and hill-hold features of the hybrid braking system. These braking modes, along with the antilock braking system (ABS) and traction control system (TCS), are integrated into almost every subsystem on the vehicle. The control system must remain stable under all driving conditions while blending the regen, creep-aid, and hill-hold modes.

Hybrid braking system hardware, such as the disc brake, drum brake, and hydraulic system components, is similar to traditional vehicle braking systems. However, the control strategy and the TCS regen mode make hybrid vehicle braking systems different from traditional vehicle braking systems.

Toyota Prius Components

Most of the components of the Toyota Prius vehicle braking system are very similar to the braking system of a traditional vehicle. Two components that are significantly different are the brake actuator assembly and the stroke (brake pedal) simulator. **See Figure 15-1.**

Brake Actuator Assembly. The brake actuator assembly consists of the electric pump, hydraulic power source, and the hydraulic control sections. The electric pump pressurizes the hydraulic brake fluid and sends it to an accumulator that stores this brake fluid under pressure for use during brake applications. There are two master cylinder pressure sensors, four wheel cylinder pressure sensors, and an accumulator pressure sensor located in the brake actuator assembly.

TOYOTA PRIUS COMPONENTS

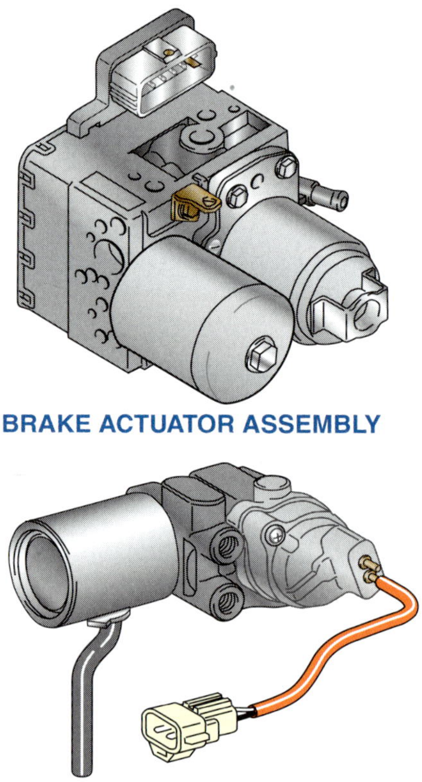

Toyota Motor Company

Figure 15-1. The brake actuator assembly and the stroke (brake pedal) simulator are unique components in a hybrid vehicle braking system.

A hybrid vehicle system stops the engine when the vehicle stops. This eliminates the vacuum used in the braking system. Therefore, an electric motor must be used to assist in maintaining brake pressure when the engine is OFF.

Stroke (Brake Pedal) Simulator. The stroke (brake pedal) simulator provides information on the movement of a brake pedal during braking. The stroke simulator provides a signal to the brake system controller that the operator has depressed the brake pedal. The controller monitors the length and the rate of change of the pedal stroke.

The rate of change indicates the speed that the pedal is being depressed to a specific point. A low rate of change means the operator has gently pressed the brake pedal to slowly brake the vehicle. A high rate of change means the operator has quickly pressed the brake pedal to quickly brake the vehicle.

The first movement of the stroke simulator signals the controller to provide regenerative braking only with no hydraulic braking assist. The first movement of the master cylinder stroke does not permit hydraulic fluid to flow into the brake system. During the initial portion of the stroke, only regen braking is initialized in order to maximize energy capture that can be transferred to the battery pack.

Stroke Sensor. A stroke sensor provides pedal position data to the ABS controller to begin braking. It contains two types of coil springs with different compression rates. The springs provide the operator with the feel of a traditional hydraulic braking system without brake system hydraulics being applied. This brake feel is important because only regen is used when the system initially begins the braking event.

If more braking is required to more aggressively slow the vehicle, the master cylinder stroke will begin transferring hydraulic fluid to the brake system. This results in blending electric and hydraulic braking. The transfer of hydraulic fluid continues to capture energy while also using the hydraulic system to reduce vehicle speed more aggressively. This permits the system to easily recover and to provide complete control of an electrical energy braking event.

Frederick Lanchester patented the first disc brake caliper system in 1902, although it did not come into common use until the 1950s.

The regen and base braking system work together to provide vehicle braking until electrical energy capture is no longer necessary. When energy capture is no longer necessary, the braking system operates very similar to a traditional vehicle.

BRAKING SYSTEM OPERATION

Hybrid braking systems are essential for ensuring that HEV systems are operating at the highest possible efficiencies. The braking systems are designed to capture kinetic energy from a moving vehicle by changing the operation of electric machine drive motor(s) from propulsion to regeneration. They then transfer this electrical energy to the battery pack for storage until energy is needed.

Blending Regenerative and Hydraulic Braking

A hybrid braking system is designed to blend friction braking with the negative torque of electric machine drive motor(s) during regenerative braking. Without regenerative braking, the vehicle hydraulic braking system would provide all of the vehicle braking.

Both hydraulic and regenerative braking contributions are variable because the battery-pack temperature and SOC are major factors in determining how much regenerative braking energy is transferred to the battery pack. Only the energy used for accelerating can be captured. Energy used in maintaining speed cannot be captured. The amount of energy recaptured during regen depends on the battery-pack SOC, battery temperature, operator input (foot pressure on brake pedal), antilock brake operation, and intervention by the traction control system.

In addition, a hybrid braking system is designed to provide hydraulic pressure to braking systems with the engine stopped. It uses a brake system to supply electric power to the hydraulic system in the event of low or malfunctioning vehicle electrical system (12 V).

A *brake system power supply* is a backup system used to ensure that an electric motor with brake pressure can stop a vehicle if the primary system fails. The brake system power supply usually consists of a capacitor bank. For example, a Toyota/Lexus design uses an auxiliary power supply with a capacitor energy bank to supply electrical energy. **See Figure 15-2.**

BRAKE SYSTEM POWER SUPPLY

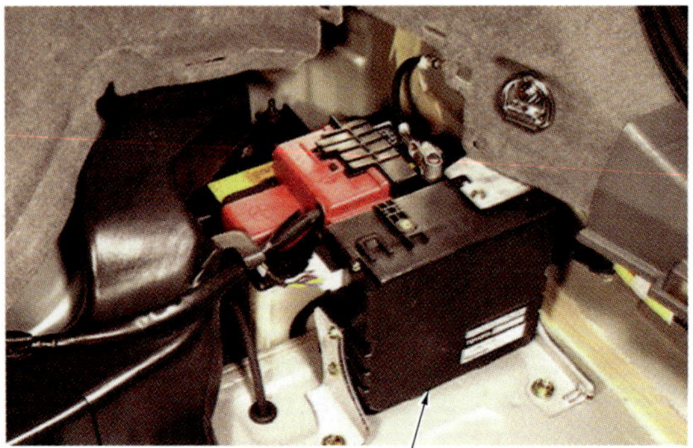

Figure 15-2. A Toyota/Lexus design uses an auxiliary power supply with a capacitor energy bank to supply electrical energy in the event that the primary vehicle electrical system fails.

Regenerative braking systems are categorized by how the operation is integrated with the vehicle braking control system. Parallel operation is where friction braking is always used with regen when the brake pedal is pressed. This is used in 2000–2005 Honda products. Series operation is where regen is an active system that minimizes friction braking to maximize the amount of energy recovered by regen. The regenerative braking system, ABS, and friction braking system are integrated and operate as one total system.

Braking Controls

Hybrid braking systems are highly integrated and involve most of the vehicle controllers to ensure high efficiency while maintaining a high degree of braking safety and vehicle stability. **See Figure 15-3.** Each manufacturer may have different methods to control regen and hydraulic braking power, but the basic control strategies are similar.

In many braking systems, four controllers provide input to the hybrid controller so that it can properly integrate all the information. In a typical situation, the operator depresses the brake pedal. The braking system reports the braking event to the ABS/TCS controller. Then, the ABS/TCS control reports the event data to the hybrid controller. At the same time, the hybrid controller continuously receives messages from the engine, battery pack, and power inverter controllers.

Regen and SOC. If the battery-pack controller sends a signal to the hybrid controller that it has the capacity to absorb energy (low SOC), then the hybrid controller sends a message to the power inverter controller to send regen braking energy to the battery pack. The power inverter controller monitors its own current as well as the feedback from the electric machine rpm sensor (resolver) in order to monitor the electrical current supplied to the battery.

The power inverter controller provides energy until the battery-pack controller signals that the battery has reached its target SOC. The power inverter controller then reduces or disallows regen braking and does not allow energy to be sent to the battery pack. If the battery-pack temperature is higher than a specified threshold, the controller disallows regen braking. When this occurs, the hydraulic braking system provides all the braking power.

If the battery-pack controller sends a signal to the hybrid controller that it is at the target SOC value for a given battery-pack temperature and voltage, the hybrid controller does not send a signal to the power inverter controller. The operator is then required to apply more pressure on the brake pedal because there is no regen. There may or may not be braking assist from engine compression, depending on when the fuel cutoff point is initiated for engine shutdown on deceleration. Engine compression braking requires less aggressive brake pedal pressure from the operator and affects how the regen braking strategy is implemented.

The operator control panel can often be configured to show the amount of energy recaptured during regenerative braking.

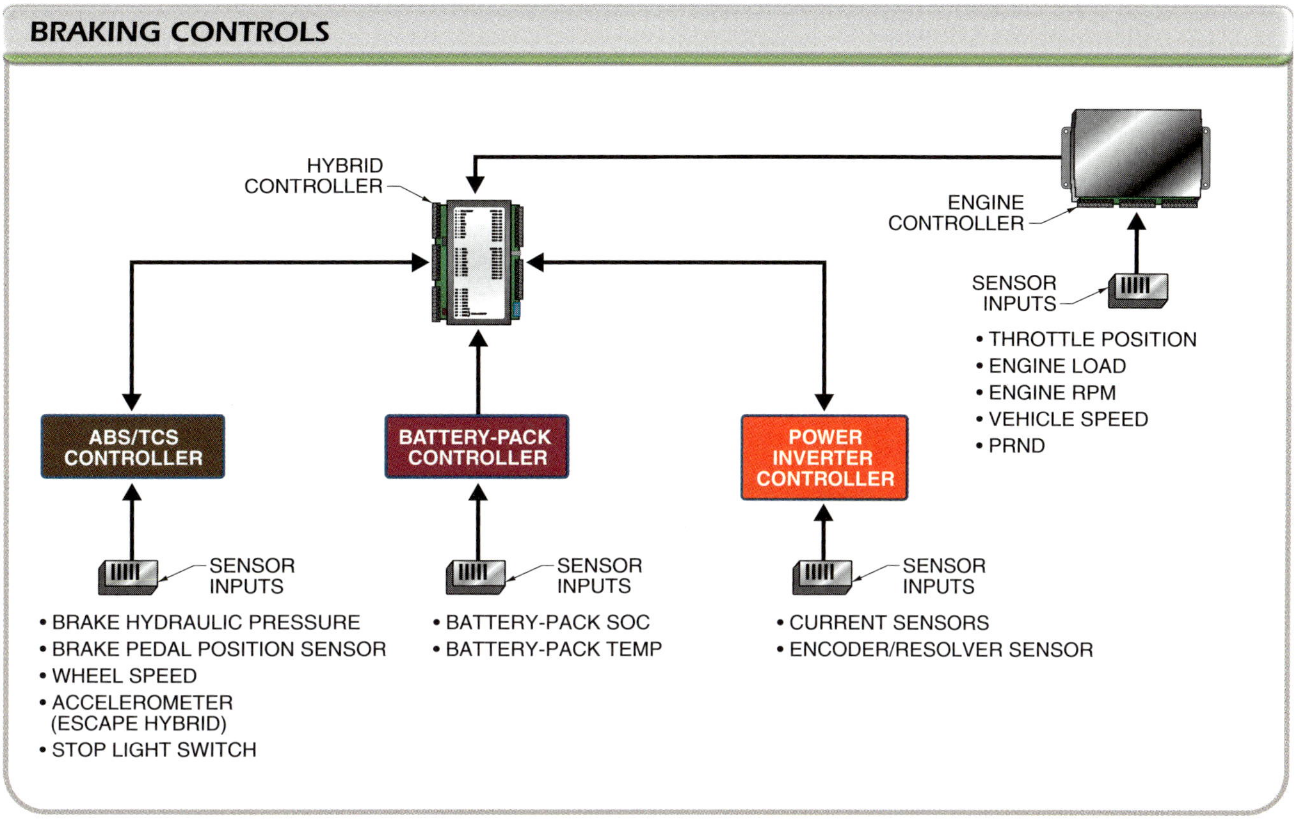

Figure 15-3. Hybrid braking systems are highly integrated and involve most of the vehicle controllers.

Some hybrid models, such as the 2005 Ford Escape Hybrid, may use accelerometers. Accelerometers measure gravitational forces to calculate the rate of acceleration or deceleration. They provide information to the brake controller about the rate at which the vehicle is increasing or reducing speed. The data provided can assist in controlling a regenerative braking energy event.

Hill-Hold Systems

Hybrid vehicles have a number of special operating modes that are not included in traditional vehicle systems. One of these special operating modes is the hill-hold control system, also called an uphill-assist control system. A *hill-hold system* is a type of vehicle propulsion system that uses an electric motor to prevent vehicle rollback down a hill. The hill-hold feature is used until the current draw becomes too high, at which time the hydraulic system is also engaged to hold the vehicle.

When a hybrid vehicle encounters a stop sign or stoplight, the engine enters idle-stop mode. If this happens while the vehicle is on an incline, there must be a method to provide braking with the engine OFF. With the engine OFF, there is nothing to stop the vehicle from rolling backward and there is no vacuum for a brake booster.

Two common designs are used with the Toyota Prius and Honda Civic Hybrid systems. Even though different control strategies are used, neither system allows the vehicle to roll backward when the engine is in an idle-stop state or when the engine transitions from an idle-stop to an auto-start state.

Toyota Prius Hill-Hold. The Toyota Prius has a hill-hold system that uses both the electric propulsion and hybrid braking system. The use of this combination applies the necessary drivetrain torque and/or friction braking to prevent vehicle rollback. **See Figure 15-4.** When a vehicle is on an incline and the engine is in idle-stop mode, the hill-hold control system prevents the vehicle from moving backward. It also prevents the vehicle from moving backward when the brake is released for an engine auto-start.

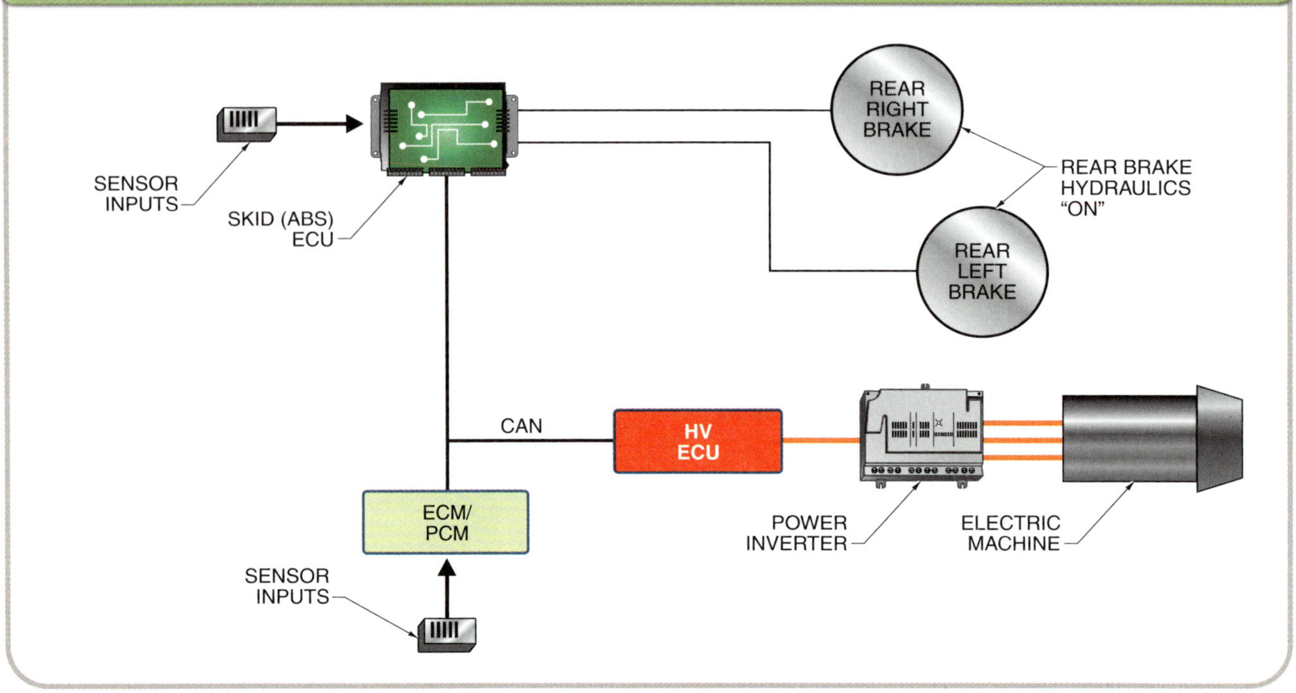

Figure 15-4. The Toyota Prius has a hill-hold system that uses both the electric propulsion and hybrid braking system to prevent vehicle rollback.

The electric traction system, via the drive motor, provides an offsetting forward torque to counter the backward motion of the vehicle. It also increases torque to hold the vehicle position.

If the slope is too steep and the hybrid controller senses that the drive motor torque will not hold the vehicle position, it can command the skid (ABS) ECU to apply the rear brakes. This inhibits the vehicle from receding farther. A software control algorithm compares motor current (torque), wheel speed sensors, motor shaft position, brake system pressure, brake pedal position, battery-pack inputs, and ECM/PCM inputs to determine how much drive motor torque to apply and when to apply the rear brakes.

The Ford Escape Hybrid has a hill-hold mode that is similar to the Toyota Prius. It also uses the electric drive motor to provide torque for holding the vehicle on an incline. If the electric machine cannot hold the vehicle, the operator merely increases the brake pedal force to use hydraulic pressure/friction brakes to hold the vehicle.

Honda Civic Creep-Aid Control. The 2003–2005 Honda Civic Hybrid has a hill-hold control system that is significantly different from the Toyota Prius hill-hold system. Honda uses the term "creep-aid control system" for their system. A *creep-aid system* is a type of vehicle braking system that uses a series of solenoid valves and hydraulic pressure on the rear and, in certain modes, the front brakes to prevent a vehicle from rolling back on an incline. **See Figure 15-5.** The system contains many of the same inputs and outputs as the Toyota Prius system.

When the vehicle stops, the TCS enables the creep-aid system solenoids A and B, which control diagonal hydraulic pressure to the front and rear brakes. The Honda creep-aid system maintains brake application during engine idle-stop when the vehicle is stopped. It prevents the vehicle from rolling as the engine transitions from auto-stop to auto-start, during the lag-time between when the brake pedal is depressed and when the accelerator pedal is pressed. The engine will crank and auto-start instantly when the brake pedal is released.

Humphry Davy developed the first electric light in 1800. Thomas Edison later advanced the development of the light bulb.

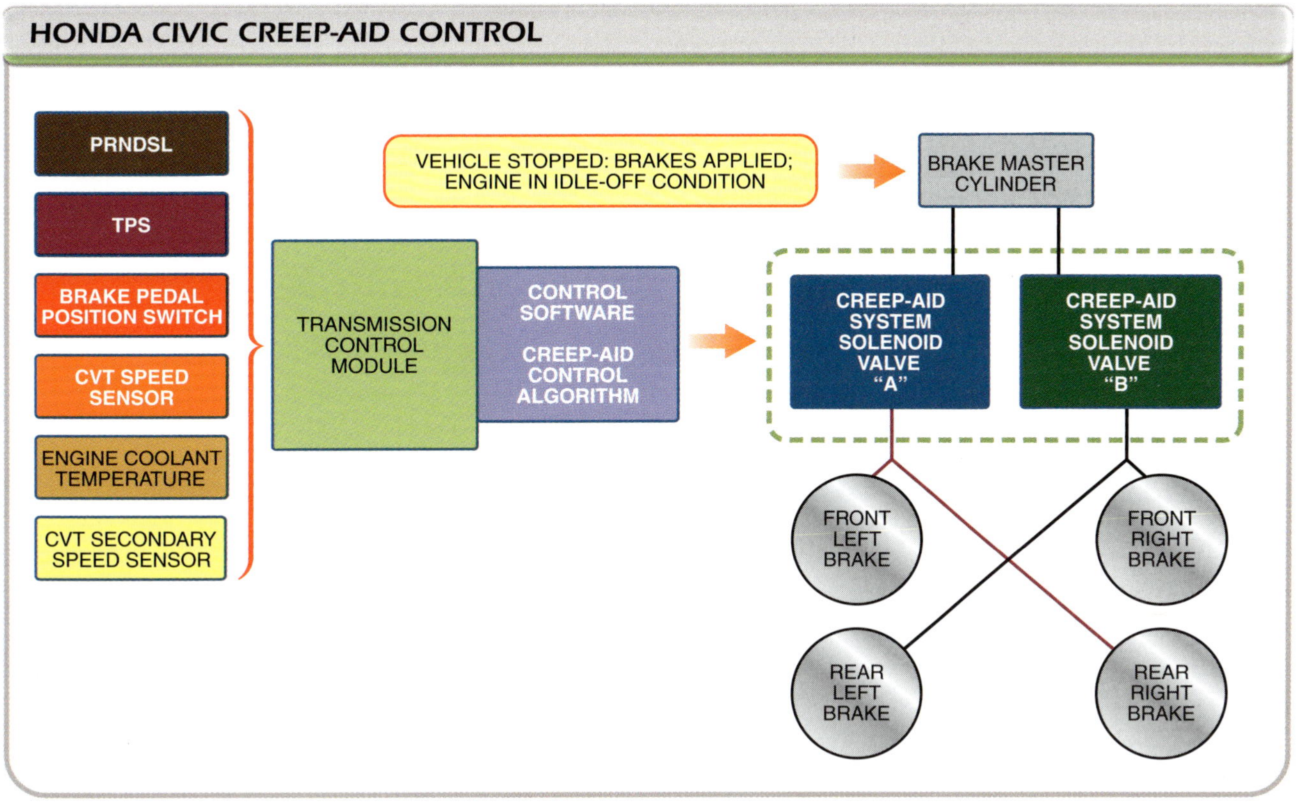

Figure 15-5. A creep-aid system uses a series of solenoid valves in the brake system to prevent the vehicle from rolling back while on an incline.

Refer to the *Hybrid Electric Vehicle Technology* CD-ROM for a Quick Quiz® on the chapter contents.

POINTS TO REMEMBER

- Hybrid braking systems must operate with the electric propulsion system to stop the vehicle.
- Regenerative braking systems combine hydraulic braking and regen braking (negative torque) to capture vehicle kinetic energy and generate electricity to recharge the battery-pack modules.
- The amount of regenerative braking provided to the vehicle depends on the battery-pack state of charge (SOC) and temperature.
- If a battery pack is at a high SOC level, only hydraulic braking is used so that the battery-pack modules are not overcharged.
- Hybrid braking systems typically use a brake stroke simulator and sensor to signal the brake controller when the brake pedal is applied. This is to initialize regen and/or to blend hydraulic and electric machine braking.
- In the event that the vehicle has lost 12 V auxiliary battery power, some vehicles use the brake system power supply to ensure that the electric motors develop brake pressure and have electrical power to stop the vehicle.
- The controllers that are typically used to control hydraulic and regenerative braking are the ABS/TCS, battery pack, power inverter, hybrid, and engine controllers.

- Hybrid vehicles use engine idle-stop to turn off the engine when the vehicle is not moving. If the vehicle is on an incline, a special hill-hold feature is used to prevent vehicle rollback.
- The hill-hold feature is used until the current draw becomes too high, at which time the hydraulic system is also engaged to hold the vehicle.
- The creep-aid system uses a series of solenoid valves and hydraulic pressure on the rear and, in certain modes, the front brakes to prevent the vehicle from rolling back while on an incline.

Refer to the *Hybrid Electric Vehicle Technology* CD-ROM for the Review Questions in Microsoft® Word format.

REVIEW QUESTIONS

1. Describe the hill-hold system of a hybrid braking system.
2. What is the purpose of a stroke simulator?
3. How are regen and the hydraulic/friction braking system affected when the battery-pack SOC is low?
4. Explain the difference between series and parallel hybrid braking systems.
5. What is the purpose of a brake system power supply?

15 Hybrid Vehicle Braking Systems

HYBRID Electric Vehicle TECHNOLOGY

Trade Test

Name ______________________ Date ____________

______________ **1.** A capacitor bank system is used by ___ in the hybrid braking system.
A. Toyota
B. Ford
C. Honda
D. all of the above

______________ **2.** The purpose of the regenerative braking mode is to ___.
A. provide better braking
B. capture/recover vehicle energy
C. increase brake lining operating life
D. provide propulsion assist

______________ **3.** The two categories of hybrid braking systems are ___ systems.
A. series and parallel
B. regenerative and base
C. electric front and hybrid rear
D. electric and solenoid control

______________ **4.** The two systems used with hill-hold to hold a vehicle on a grade or hill are ___ systems.
A. hydraulic and mechanical
B. mechanical and braking
C. hydraulic and electric
D. mechanical and electric

______________ **5.** The primary reason that the hill-hold or creep-aid systems are used in hybrid vehicles is because ___.
A. there is no engine vacuum during stops
B. they have higher braking efficiency
C. fewer components wear out
D. they are simpler than traditional hydraulic systems

______________ **6.** If the battery-pack SOC level is low while braking, the controller directs the braking system to ___.
A. disable electric braking and use 100% hydraulic braking
B. use more hydraulic braking force than electric braking force
C. use both electric braking force and hydraulic braking force
D. use more electric braking force than hydraulic braking force

____________________ **7.** Honda hybrid products are categorized as ___ braking systems.

A. series
B. parallel
C. series-parallel
D. hydraulic-series

____________________ **8.** The Toyota Prius is categorized as a ___ braking system.

A. series
B. parallel
C. series-parallel
D. hydraulic-series

____________________ **9.** In the Toyota Prius braking system, the stroke (brake pedal) simulator and sensor provide ___ to the brake system controller.

A. the length and rate of change of the pedal stroke
B. hydraulic pressure of the braking system
C. brake pedal pressure and feel
D. creep-aid control and solenoid status

Hybrid Climate Control Systems

16

Vehicle climate control systems historically have used belt-driven compressors, and climate control system performance was tied directly to vehicle and engine speed. When electric, hybrid, and fuel-cell vehicles were introduced, it became necessary to add a method to cool the vehicle cabin when the vehicle was not moving and the engine not operating. Therefore, the electric compressor was chosen as the method to provide vehicle cabin climate control regardless of vehicle or engine speed.

CHAPTER OBJECTIVES

- Explain why some hybrid vehicles use electric A/C compressors.
- List the correct type of oil to use in an electric A/C compressor.
- Describe the results of using the incorrect oil in an electric A/C compressor.
- Explain the A/C system power inverter.
- Describe the auxiliary engine coolant pump system in a typical hybrid vehicle.
- Describe the V/Hz method of A/C compressor control.
- Describe the type of compressor design used in a hybrid vehicle.
- Define the acronym COP.
- Describe the operating stages of a Honda Accord A/C compressor.
- Describe the basic parameters of engine idle-stop mode as it relates to the A/C system.

KEY TERMS

polyolester (POE) oil: A compressor oil that provides a high level of electrical isolation to minimize isolation fault conditions.

polyalkylene glycol (PAG) oil: A compressor oil that provides a low level of electrical isolation and has a high water saturation value.

volts per hertz (V/Hz) motor control: An electric motor control strategy in which the frequency and voltage applied to a motor is changed to provide a corresponding change in motor speed and torque levels.

coefficient of performance (COP): The ratio of output power to input power in a compressor. The COP is a measurement that determines the ratio of power required to operate a compressor to the amount of heat transferred. It can be used to compare an electric compressor to a traditional belt-driven compressor.

engine idle-stop mode: An operating mode in which a controller stops or shuts off an engine when a vehicle comes to a stop. This mode saves fuel during city driving and, therefore, increases the city fuel economy rating and efficiency of the vehicle.

engine auto-start mode: An operating mode in which an engine is cranked and started during acceleration or a low battery pack is recharged by a high-voltage system.

CLIMATE CONTROL SYSTEM COMPONENTS

A climate control system includes the cabin (vehicle interior) air conditioning (A/C) and associated heating systems. A traditional vehicle A/C system uses 12 V control and power systems and the accompanying mechanical components such as the condenser, evaporator, thermostatic expansion valve (TXV), orifice tube, accumulator/receiver dryer, and various hoses/tubes. Traditional 12 V control and power systems pose minimal electrical danger to technicians diagnosing or repairing the system.

Climate control systems in hybrid electric vehicles (HEVs) have different components and operate at higher voltages. The power electronics system and associated control system is significantly different from traditional climate control systems. In addition, cabin heating systems have been significantly modified in hybrid vehicles.

Electric Compressors

An advantage of HEVs is the high-voltage system that can be used for power electronics and electric machines. High-voltage electric compressors are used in some HEVs to increase overall vehicle efficiency. Compressors also permit the A/C system to operate during periods that the engine is OFF, such as during the engine idle-off condition to reduce fuel consumption.

There are HEVs that can use belt-driven compressors instead of electric compressors. The climate and engine control strategies are very different for a belt-driven compressor than for an electric compressor. A typical high-voltage electric compressor is a scroll compressor. **See Figure 16-1.** Electric scroll compressors can be about 40% smaller and 50% lighter than conventional belt-driven A/C compressors.

ELECTRIC COMPRESSORS

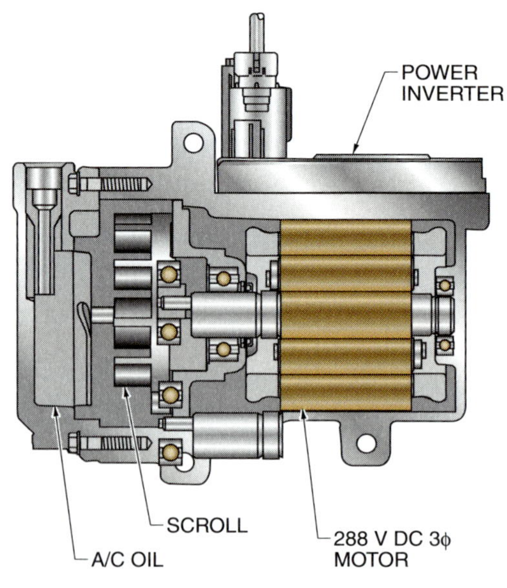

Figure 16-1. A typical high-voltage electric compressor is a scroll compressor.

Compressor Oil

As with traditional belt-driven A/C compressors, electric compressors also circulate lubricating oil through an A/C system. The high-voltage A/C compressor is exposed to the circulating system oil. The machine stator and associated wiring must remain electrically isolated from the vehicle chassis because electric compressors use internal high-voltage electric machines. An electrical leak path through the compressor housing to the vehicle chassis ground could occur if the electrical insulating properties of the oil are too low.

The correct A/C oil must be used in order to ensure that enough resistance is maintained. Using the wrong oil can compromise the electrical integrity of the high-voltage system. Two common compressor oils are polyolester (POE) oil and polyalkylene glycol (PAG) oil. *Polyolester (POE) oil* is a compressor oil that provides a high level of electrical isolation to minimize isolation fault conditions. *Polyalkylene glycol (PAG) oil* is a compressor oil that provides a low level of electrical isolation and has a high water saturation value. The water saturation value is approximately four times higher than that of POE oil.

The electrical isolation properties of PAG oil decrease as it absorbs high concentrations of water. The low resistance provides an electrical leak path from the motor stator to the vehicle chassis. This can result in an isolation fault condition. POE oil has higher electrical resistance than PAG oil. If low-resistance PAG oil were added to an A/C system that requires high-resistance POE oil, the lower resistance of the mixture could reduce the insulating resistance enough to cause an electrical isolation fault condition.

When the A/C system is turned OFF and begins to cool, condensed water can collect in the compressor if it is the lowest point of the A/C system. Water and oil reside within the compressor and can cause the isolation properties of the oil to decrease when the oil absorbs the water. PAG oil has a higher ability to absorb water than POE oil. The water saturation value of PAG is about 10,000 ppm, while the water saturation value for POE is 2500 ppm.

For example, if 8% to 10% PAG oil is mixed with POE, the compressor startup resistance will be significantly below the minimum resistance specification. With this mixture, resistance of the electric circuit will be significantly low for the first few seconds of compressor start. After about 3 sec to 10 sec, the resistance could increase into the normal 20 MΩ to 60 MΩ range. An isolation fault condition could develop during the first few seconds.

Compressor Power Inverter

A common design for an electric A/C compressor uses high-voltage electric power supplied by an A/C power inverter. **See Figure 16-2.** For example, the 201 V DC battery pack in a 2004–2008 Toyota Prius is supplied to the traction (main) A/C power inverter. The electric A/C compressor is a Denso ES18, 3ϕ electric machine. The inverter changes the 201 V DC to 201 V AC.

The power inverter must receive its control commands from a controller and control software. The A/C electronic control unit (ECU) system controls cabin temperature by varying compressor speed based on A/C sensor inputs, occupant demand, or flow of refrigerant through the TXV or orifice tube.

Another design of an electric A/C compressor uses a different mechanical and electrical architecture. The power inverter of the Lexus RX400h is integrated into the A/C compressor, also built by Denso. With this design, two high-voltage DC wires are connected to the compressor. The integrated power inverter inverts the DC power to 3ϕ AC electric power. Although this compressor has an integrated power inverter system, its operation is essentially identical to the Toyota Prius electric compressor that uses an external A/C power inverter assembly.

Auxiliary Coolant Pump

A climate control system includes A/C systems and cabin heating systems. When a vehicle is stopped and the engine is turned OFF, coolant is pumped through an auxiliary coolant pump to transfer heat from the heater core into the vehicle cabin. **See Figure 16-3.**

Honda hybrids use combination electric and belt-driven compressors.

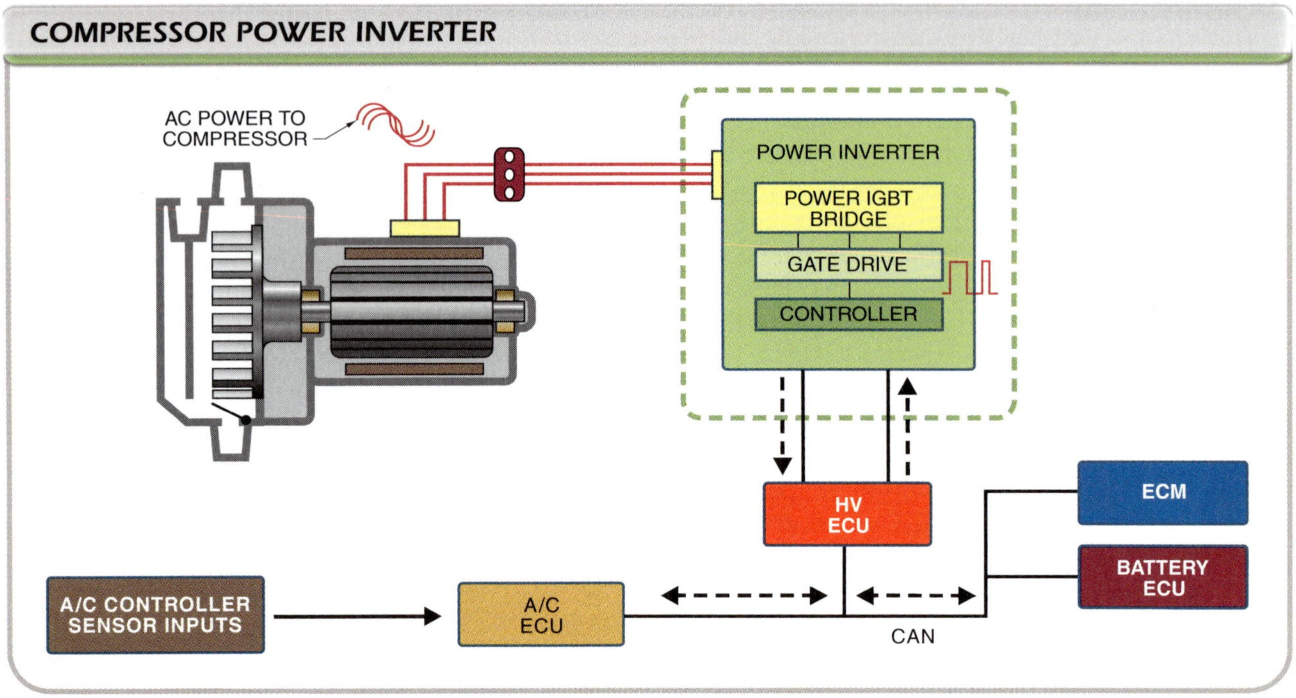

Figure 16-2. A common design for an electric A/C compressor uses high-voltage electric power supplied by an A/C power inverter.

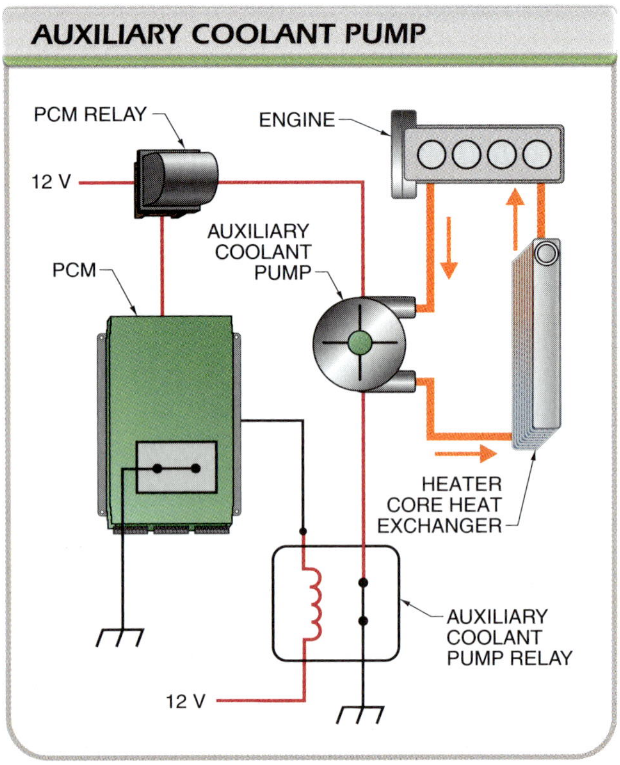

Figure 16-3. When a vehicle is stopped and the engine turned OFF, coolant is pumped through an auxiliary coolant pump to transfer heat from the heater core into the vehicle cabin.

The Ford Escape has a powertrain control module (PCM) that provides a ground path through a driver transistor for the auxiliary coolant pump relay. The relay either provides 12 V power or a ground for the coolant pump. When the engine transitions from idle-off to auto-start, the 12 V auxiliary coolant pump is commanded OFF by the PCM. This removes the ground from the relay. After the engine starts, the engine-driven coolant pump provides coolant circulation.

Other vehicle models, such as the Toyota Highlander, Lexus RX400h, and Honda Accord, have auxiliary coolant pump systems that are similar to the Ford Escape Hybrid circuit. The auxiliary coolant pump motor is typically a 12 V brush DC motor. A single-speed pump is activated by a relay, which is controlled by one of the vehicle controllers.

CLIMATE CONTROL SYSTEM OPERATION

The climate control system provides operator comfort during vehicle operation. In order to optimize operator comfort and vehicle efficiency, the system must control the compressor speed. The compressor design significantly affects vehicle efficiency. In addition, many hybrid vehicles include an engine idle-stop mode where the engine shuts off while the vehicle is stopped. The climate control system must continue to operate when the engine is OFF.

Speed Control

Accessory motors are typically speed controlled instead of torque controlled. A speed control strategy tracks only the speed of the motor and does not calculate a torque output level. The electric A/C compressor can be easily controlled to a desired target speed with a few simple sensor values and control software. Other motors controlled with speed control are radiator cooling fan motors and climate control blower motors.

The compressor speed determines the rate that refrigerant is pumped through the A/C system. A faster compressor speed means that more refrigerant is pumped through the system to increase cooling. A slower compressor speed means that less refrigerant is pumped through the system to decrease cooling. In other words, heat can be removed from the vehicle cabin at a higher rate when the refrigerant can be pumped through the system at a higher rate.

Optimum efficiency can be attained for any desired condition by varying compressor speeds or stopping it during specific cooling load demands. The control software contains a "map" of the most efficient operating points for any given heat load. The compressor speed is mapped for a given heat load condition and this mapping provides infinitely variable compressor speeds. This optimizes the efficiency of an electric compressor. A traditional belt-driven A/C compressor cannot be controlled as well as an electric compressor.

Compressor speed is a major element in determining the cooling effectiveness of the A/C system. However, it must work with refrigerant control valves to provide overall cooling control of the vehicle cabin.

Charles Kettering and Thomas Midgley are credited with discovering the chemical Freon that was used in air conditioning systems.

Voltage-Frequency Speed Control

A common method of electric motor control is speed control in which the controller sets the actual speed of the motor. This is not the same as the torque control used with the propulsion motor. Speed control adjusts the voltage and frequency of the AC power at the motor. *Volts per hertz (V/Hz) motor control* is an electric motor control strategy in which the frequency and voltage applied to a motor is changed to provide a corresponding change in motor speed and torque levels. This method is much simpler than torque control, and the software control algorithms are much simpler and less costly to develop.

A V/Hz motor controller changes the power output to maintain a desired ratio of the voltage to frequency. The ratio is calculated to ensure that motor speed stability is maintained throughout the entire speed range of the compressor motor. When the proper ratio is not maintained, the motor speed is not easily controlled because it becomes very erratic and unstable.

When discussing the control of an A/C compressor motor, motor speed determines how much refrigerant gets circulated through the A/C system. The amount of refrigerant determines the cooling effectiveness of the system. Therefore, vehicle cabin cooling can be controlled by accurately controlling motor speed.

When the control system increases the frequency to increase the motor speed, the voltage also needs to increase. When the control system decreases the frequency to decrease the motor speed, the voltage also needs to decrease. Therefore, as frequency increases there is a necessary requirement to increase motor torque to continue an incremental increase in speed.

By increasing voltage, while increasing the frequency, the motor receives the necessary electrical current to continue creating torque and to acquire the target speed. The result is that when the frequency is increased, motor torque increases. However, it is the frequency command that determines compressor operation.

If the frequency were increased or decreased in an effort to control motor speed, the motor would become erratic. This is because a commercial 1ϕ electric motor is designed to operate on a constant 60 Hz at 120 VAC or 240 VAC. For example, a 240 VAC electric motor can be designed to operate at 60 Hz. The ratio of voltage to frequency is 4 ($240 \div 60 = 4$). The motor in this example operates at a ratio of 4:1.

The correct voltage level must be selected for each change in frequency to achieve optimum motor speed control and overall motor speed stability. **See Figure 16-4.** The ratio is based on motor type, number of poles or phase, rotor design, motor power (HP), desired torque at given speed, etc. The ratios will not always be an exact linear step change to control motor speed. However, a typical motor has a broad range where the changes to frequency and voltage are linear. In this range, whenever the frequency is changed by 1 Hz, the voltage needs to be changed by 4 V.

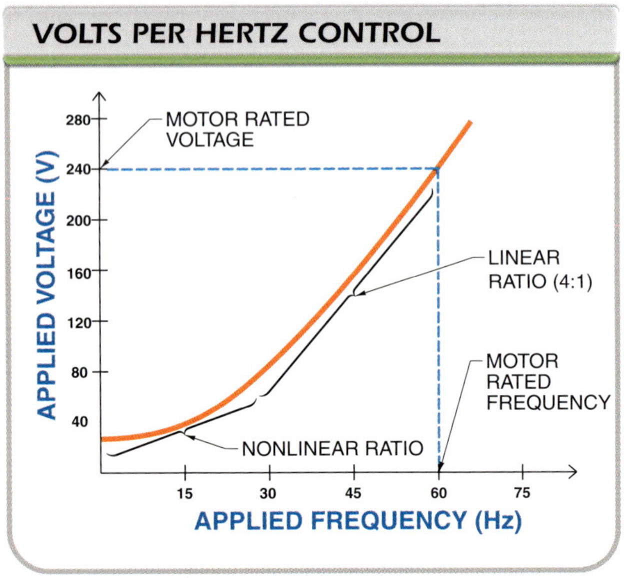

Figure 16-4. The correct voltage level must be selected for each change in frequency to achieve optimum motor speed control and overall motor speed stability.

Compressor Performance

The performance of the compressor used in a climate control system depends on the design and intended purpose. The performance specifications of the vehicle drive the choice of compressor for the vehicle. If the vehicle were to be marketed as an efficiency vehicle design, then a compressor with lower power consumption would probably be specified.

There are important differences between electric and belt-driven compressors. An electric compressor may have comparable performance to a belt-driven compressor. **See Figure 16-5.** However, electric compressors are not necessarily used for an increase in thermal performance. They are used to increase overall vehicle efficiency and to reduce vehicle weight.

Efficiency, cooling capacity, power consumption, and COP depend on the design of the compressor. With hybrid vehicles, efficiency and power consumption drive the choice of a compressor. For example, the Toyota Prius electric compressor cannot match the cooling capacity of a traditional belt-driven compressor, but the electric compressor consumes much less power and is more efficient.

Coefficient of Performance. When A/C compressors are compared, one of the most important measures that is used to compare compressor performance is the coefficient of performance (COP). The *coefficient of performance (COP)* is the ratio of output power to input power in a compressor. The COP is a measurement that determines the ratio of power required to operate a compressor to the amount of heat transferred. It can be used to compare an electric compressor to a traditional belt-driven compressor. A belt-driven model can remove more heat for optimal A/C, but an electric compressor is more energy efficient.

A higher COP means that the compressor is more efficient than a compressor with a lower COP. In addition, the COP changes throughout the drive cycle and depends on ambient temperatures. **See Figure 16-6.** The COP is higher at cooler ambient temperatures. The COP is calculated by using the following formula:

$$COP = \frac{P_{out}}{P_{in}}$$

where

COP = coefficient of performance

P_{out} = output power (in kW)

P_{in} = input power (in kW)

A dual compressor is actually two compressors in one comprised of a belt-driven compressor and an electric compressor.

POWER CONSUMPTION COMPARISON

Speed	Compressor Type	Power Consumption*	Cooling Capacity*	Cooling COP
Idle	Electric	1.8	3.9	2.2
	Belt-driven	2.2	4.0	1.8
25 mph	Electric	3.0	5.8	1.8
	Belt-driven	4.0	5.8	1.3
60 mph	Electric	4.5	5.9	1.3
	Belt-driven	4.5	5.3	1.1

* in kW

Figure 16-5. The power consumed by a compressor depends on vehicle speed and ambient temperature.

COMPRESSOR PERFORMANCE

Speed	Ambient Temperature	Compressor Type	Approximate COP
Idle	82°F (28°C)	Belt	2.50
		Electric	2.56
Idle	100°F (38°C)	Belt	2.30
		Electric	1.70
1500 rpm–50 mph	82°F (28°C)	Belt	2.30
		Electric	2.56
1500 rpm–50 mph	100°F (38°C)	Belt	1.50
		Electric	1.80

Figure 16-6. An electric compressor may have comparable performance to a belt-driven compressor.

For example, an A/C system may use 1.0 kW of input power to remove 3.0 kW of heat. The COP is calculated as follows:

$$COP = \frac{P_{out}}{P_{in}}$$

$$COP = \frac{3.0}{1.0}$$

$$COP = \mathbf{3}$$

Power Consumption. Another comparison between compressors is the amount of power required to operate the compressor (power consumption) to remove cabin heat. A belt-driven compressor system is measured by how much mechanical (shaft) power is required to operate the compressor to remove heat. The power, in W or kW, is calculated from the mechanical power required to operate the compressor so that an accurate comparison with the electric compressor can be made.

The power consumed by a compressor depends on vehicle speed and ambient temperature. **See Figure 16-7.** For example, the compressor of the Toyota Prius Denso ES18 uses significantly more electrical power at low vehicle speeds and higher ambient temperatures. More heat can be removed from the A/C condenser at higher vehicle speeds. This lowers the condenser pressure.

> Both Toyota and Honda state that if the wrong oil is introduced into the A/C system, all of the main A/C parts, such as the evaporator, condenser, receiver/drier, and compressor, must be replaced. They also recommend that a manifold gauge set be dedicated to hybrid use when charging the system.

The electrical power demand is reduced because the compressor pumps against a reduced pressure. Fuel economy comparisons of the belt-driven and electric compressor indicate that the electric compressor is superior when all efficiencies and fuel consumption calculations are considered.

Combination Electric and Belt-Driven Compressors

There are A/C system designs that integrate both electric and belt-driven technologies. These dual compressors are used in the 2005–2007 Honda Accord Hybrid A/C system. They are designed as a dual scroll compressor system. **See Figure 16-8.** The belt-driven compressor has a displacement of 70 cc per revolution. The electric compressor has a displacement of 15 cc per revolution. The magnetic clutch works with the belt pulley to engage and disengage the belt-driven scroll compressor.

The A/C compressor inverter can deliver up to 1 kW (1.35 HP) of power. The compressor motor is a 3ϕ, brushless, 6-pole permanent magnet electric machine. The compressor draws 6 A at full load in the sun at 86°F (30°C) and 3.5 A at 25 mph out of the sun at the same temperature.

The Honda power inverter and control system is similar to the Toyota Prius system and many of the control methods are similar. As with the Toyota and Lexus electric A/C compressor systems, the electric compressor system of the Honda uses a V/Hz control system.

Honda Accord Control Logic. The Honda Accord Hybrid has a dual compressor system with a unique A/C control logic system. The basic control logic system uses only the belt-driven or electric compressor to optimize energy use. The control can also use both compressors to get maximum cooling capacity. By design, the total capacity of the two compressors operating at the same time is comparable to the lower boundaries of a traditional belt-driven system.

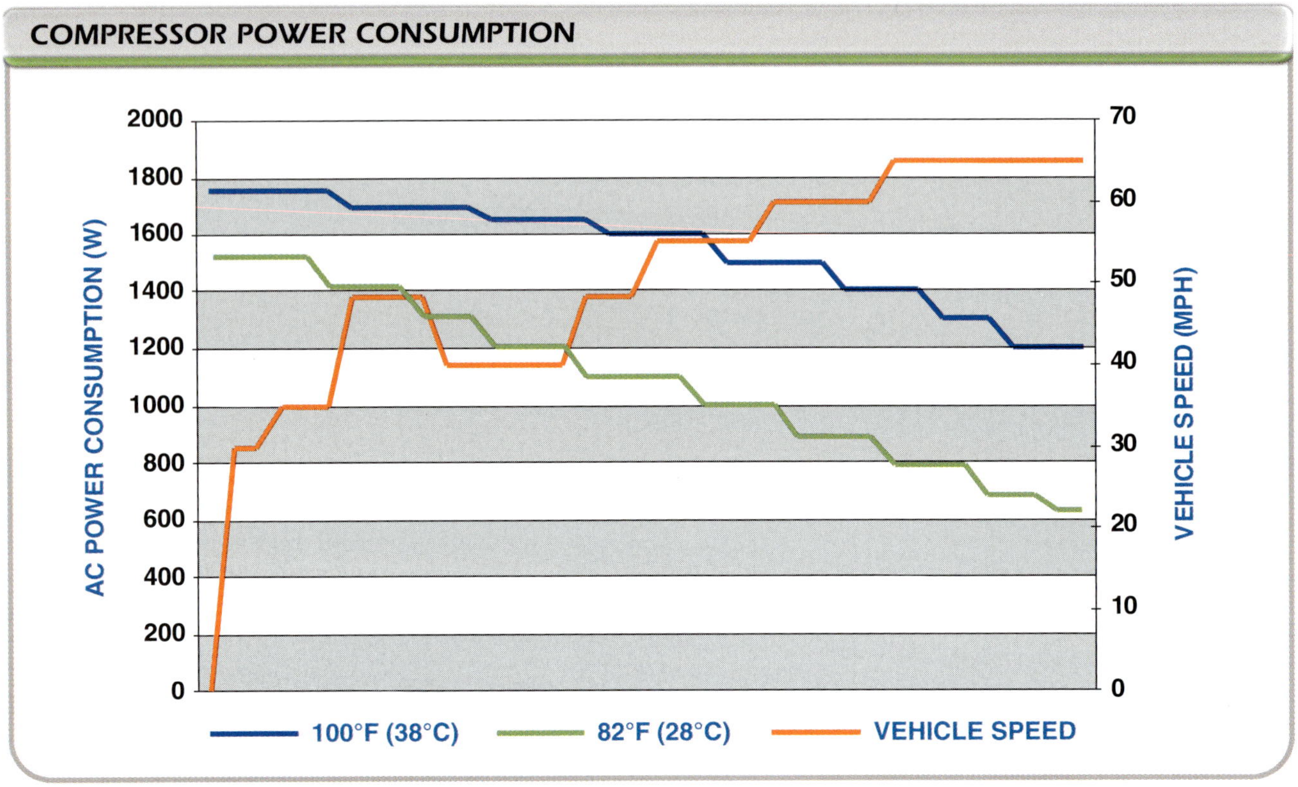

Figure 16-7. The COP changes throughout the drive cycle and depends on ambient temperatures.

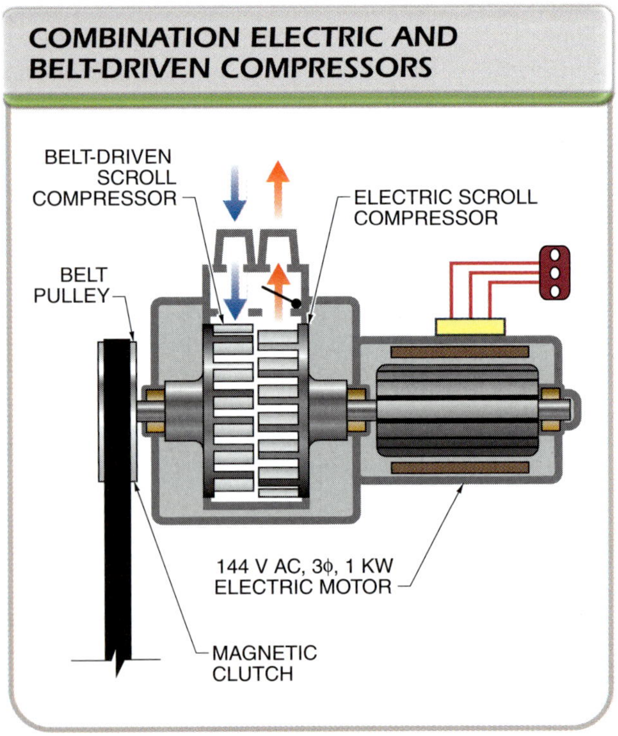

Figure 16-8. A Honda A/C system integrates both electric and belt-driven technologies.

The logic of the dual compressor system is applied in three stages. Stage 1 is used at startup. The control system uses both compressors to get maximum cooling capacity. **See Figure 16-9.** For example, if the capacity of the belt-driven compressor is 5.8 kW at 3000 rpm and the capacity of the electric compressor is 1 kW at 3000 rpm, the controller can operate both compressors and use 6.8 kW of power for cooling. The motor speed decreases as the evaporator temperature difference decreases.

In Stage 2, the control system uses the mechanical compressor only. This is used when a stable cabin temperature is reached. It is also used when the fan blower speed is set on high and the electric compressor cannot maintain the targeted temperature at the selected blower speed.

In Stage 3, the control system uses the electric motor operation only. This is used when a stable cabin temperature is reached. The PCM determines the speed of the compressor based on standard inputs. Stage 3 is used when the electric compressor is able to maintain a desired temperature. However, if the cabin temperature increases because the load changes, the electric compressor may no longer be able to keep up and the system switches back to Stage 2.

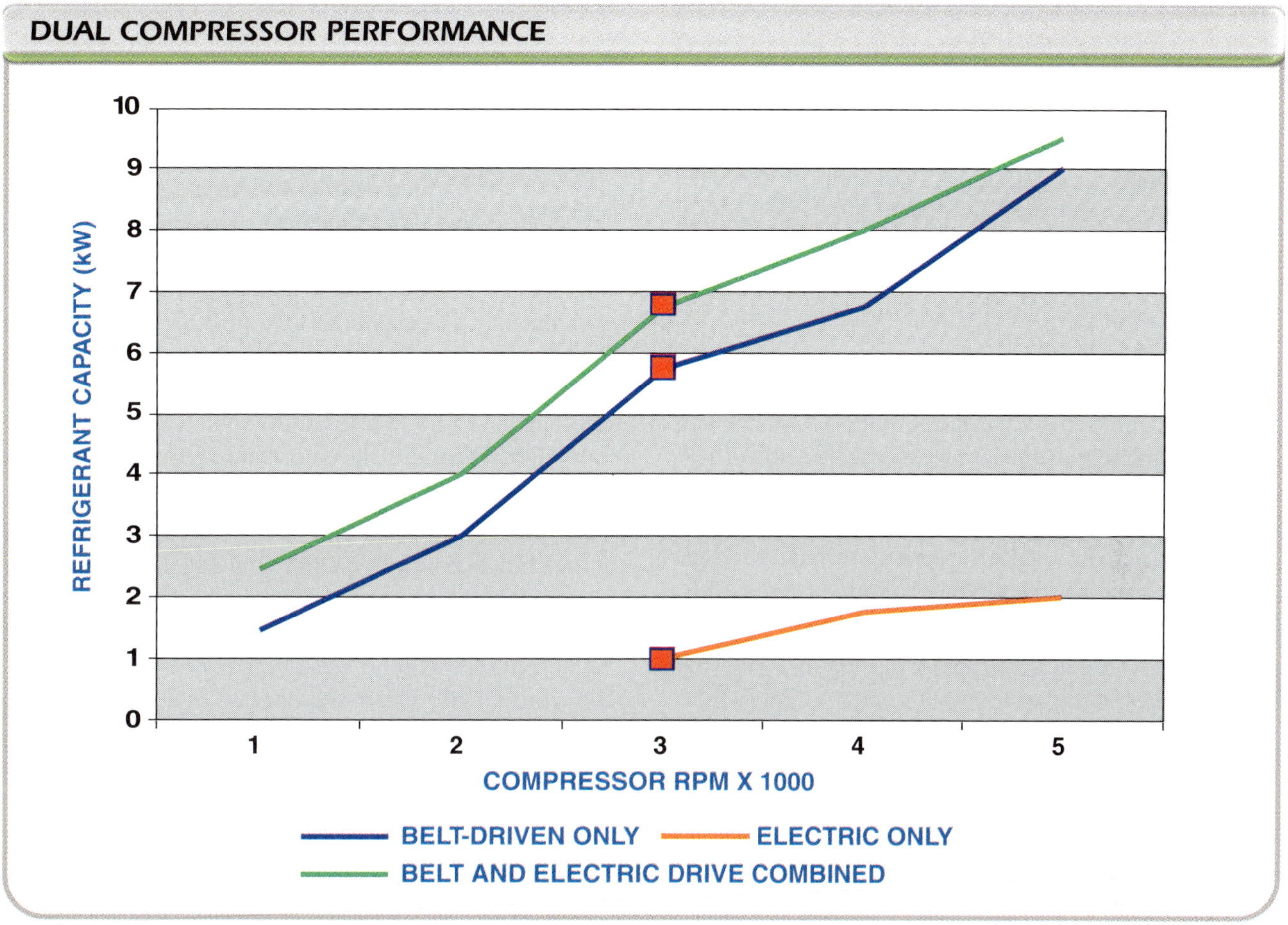

Figure 16-9. The Honda control system uses both compressors to get maximum cooling capacity.

This system cannot transition directly from Stage 3 to Stage 1, or from Stage 1 to Stage 3. It must transition through all stages sequentially.

Engine Idle-Stop Mode

Engine idle-stop mode is an operating mode in which a controller stops or shuts off an engine when a vehicle comes to a stop. This mode saves fuel during city driving and, therefore, increases the city fuel economy rating and efficiency of the vehicle. When the controller stops the engine, the climate control systems must continue to provide cooling (or heating). Different manufacturers use different strategies for operating climate control systems and idle-stop mode.

Extra precautions must be taken when climate control systems are functioning or there is high electrical demand and the hood is open. Hybrid vehicles that are equipped with engine idle-stop operation are also equipped with engine auto-start operation.

Engine auto-start mode is an operating mode in which an engine is cranked and started during acceleration or a low battery pack is recharged by a high-voltage system. The safety of anyone engaging the vehicle is compromised any time a system can automatically start the engine without an operator in the vehicle to control it. This is especially a problem when the hood is open and moving parts are accessible.

There are a few procedures to follow if vehicle service requires the hood to be open or undercar maintenance/service to be performed. First, the vehicle should not be left ON when it is unattended. If the vehicle is unattended, the keys should be removed from the ignition or the vehicle depowered to inhibit idle-off or auto-start operation. Repairs should not be performed that place a technician near rotating parts, fluids under pressure, or high-voltage power when the engine can auto-start from an idle-off condition. If there is any doubt about the safety of the system, the key should be removed or the system depowered.

2001–2003 Toyota Prius. The engine can restart if the climate control system is ON and the battery SOC is low. The engine can also restart if there is a change to a higher electrical load due to increased accessory load. If the A/C is set to maximum or to defrost, the engine operates continuously because it has a belt-driven compressor. This model does not have a hood-open safety switch to disengage the auto-start. Therefore, the engine can restart if the climate control system is ON, even if the hood is open.

2004–2008 Toyota Prius. The engine can restart if the climate control system is ON and the battery SOC is low. The engine can also restart if there is a change to a higher electrical load due to increased accessory load. If the A/C is set to maximum or to defrost, the engine can start at any time. This model does not have a hood-open safety switch to disengage the auto-start. Therefore, the engine can restart if the climate control system is ON, even if the hood is open.

2003–2005 Honda Civic Hybrid. The engine can restart if the climate control system is ON and the battery SOC is low. The engine can also restart if there is a change to a higher electrical load due to increased accessory load. If the A/C is set to maximum or to defrost, the engine operates continuously because it has a belt-driven compressor. The engine runs continuously if the ambient air temperature is less than 41°F (5°C). This model does not have a hood-open safety switch to disengage the auto-start. Therefore, the engine can restart if the climate control system is ON, even if the hood is open.

2005–2007 Honda Accord Hybrid. The engine can restart if the climate control system is ON and the battery SOC is low. The engine can also restart if there is a change to a higher electrical load due to increased accessory load. If the A/C is set to defrost, the engine operates continuously. The engine runs continuously if the ambient air temperature is less than or equal to 56°F (13°C), or if it is equal to or more than 100°F (38°C). This model does not have a hood-open safety switch to disengage the auto-start. Therefore, the engine can restart if the climate control system is ON, even if the hood is open.

2005–2007 Ford Escape Hybrid. The engine can restart if the climate control system is ON and the battery SOC is low. If the A/C is set to defrost, the engine restarts because it has a belt-driven compressor. If the A/C is requested ON with no battery pack cooling requested, the engine will restart and operate as a cycling clutch system. This model does not have a hood-open safety switch to disengage the auto-start. Therefore, the engine can restart if the climate control system is ON. It does have a hood switch as a theft deterrent system, but the hood switch does not inhibit an engine restart.

Refer to the *Hybrid Electric Vehicle Technology* CD-ROM for a Quick Quiz® on the chapter contents.

POINTS TO REMEMBER

- The two major A/C compressor technologies used in hybrid vehicles are belt-driven and electric compressors.
- Coefficient of performance is the ratio of output power to input power.
- It is important to use the correct oil in an electric compressor to ensure that the compressor does not develop an isolation fault condition.
- POE oil has been typically recommended by manufacturers for most A/C systems because it has a lower water saturation value than PAG oil.
- Electric A/C compressors are powered by 3ϕ electric machines and require the use of a power inverter to operate.
- Some compressors use an internal 3ϕ power inverter, while other compressors may have an integrated power inverter.
- Electric compressor speed is controlled with V/Hz speed control.
- A belt-driven compressor can remove more heat than an electric compressor, but the electric compressor is more energy efficient.

- Some Honda A/C compressors combine belt-driven and electric technologies into one unit to gain the highest efficiency possible.
- Vehicle climate control systems work with the hybrid engine idle-stop mode to control the A/C compressor operation and performance.
- Because the engine can auto-stop for long periods, a hybrid vehicle uses a low-voltage heating pump to circulate coolant through the heater core to provide cabin heating.
- Precautions must be taken when performing any analysis or repair on a hybrid vehicle engine compartment. If the key is in the ignition, the engine can crank and run anytime there is a change in any of the vehicle systems. The key or keyless device must be removed from the vehicle so that the engine cannot be cranked by the system auto-start or by another person.

Refer to the *Hybrid Electric Vehicle Technology* CD-ROM for the Review Questions in Microsoft® Word format.

REVIEW QUESTIONS

1. How does the size of an electric A/C compressor compare to a belt-driven A/C compressor?
2. Why should an electric A/C compressor system use only POE oil instead of a PAG-based oil?
3. Explain the operating characteristics of the volts per hertz (V/Hz) electric A/C compressor control.
4. What does a higher COP number indicate about an A/C system?
5. Explain the three stages of the Honda Accord Hybrid A/C system.

16 Hybrid Climate Control Systems

HYBRID Electric Vehicle TECHNOLOGY

Trade Test

Name ______________________ Date ______________

______________ **1.** Electric A/C compressors are used in some hybrid electric vehicle systems because they ___.
A. are more efficient
B. weigh less
C. come in small sizes
D. all of the above

______________ **2.** A ___ compressor is the type of compressor typically used in hybrid electric vehicle A/C systems.
A. swash plate
B. scroll
C. variable pitch
D. fixed displacement

______________ **3.** An electric A/C compressor uses more power at ___ regardless of vehicle speed.
A. higher temperatures
B. lower temperatures
C. temperatures less than 41°F
D. none of the above

______________ **4.** The Honda Accord A/C control system uses ___ stage(s) of control.
A. one
B. two
C. three
D. four

______________ **5.** When a Honda Accord Hybrid control system is in a mechanical-only operation, the compressor is operating at stage ___.
A. one
B. two
C. three
D. four

______________ **6.** The Honda A/C dual compressor provides the most cooling when operating in ___ mode.
A. mechanical-only
B. electrical-only
C. mechanical and electrical
D. either mechanical or electrical

_______________ **7.** A hybrid vehicle can auto-start when the ___.
A. battery pack has a low SOC
B. battery pack has a high SOC
C. key is removed from the vehicle
D. system is depowered

_______________ **8.** The 12 V auxiliary coolant pump system operation is used only when the ___.
A. A/C system is operating
B. vehicle is in engine idle-stop mode
C. engine is turned OFF
D. vehicle is braking

_______________ **9.** The compressor oil specified for electric A/C compressors is ___ oil.
A. polyalkylene glycol (PAG)
B. mineral
C. polyolester (POE)
D. standard engine

_______________ **10.** When comparing A/C compressors, COP is used to compare ___.
A. vehicle speed and refrigerant flow
B. output and input power
C. refrigerant and vent outlet temperatures
D. inlet and outlet ambient air temperatures

_______________ **11.** The Lexus RX400h Hybrid electric compressor power inverter is located in the ___.
A. hybrid ECU
B. A/C controller
C. traction (main) system power inverter
D. compressor

_______________ **12.** The 2001–2006 Toyota Prius electric compressor power inverter is located in the ___.
A. hybrid ECU
B. A/C controller
C. traction (main) system power inverter
D. compressor

_______________ **13.** One method that controls the operating speed of an electric compressor is the use of ___ speed control.
A. V/Hz
B. torque/angle
C. V/torque
D. HP/amplitude

_______________ **14.** In a Honda Hybrid A/C system, the ___ compressor system provides the cooling in Stage 2.
A. electric
B. belt-driven
C. electric/belt-driven
D. none of the above

HYBRID Electric Vehicle TECHNOLOGY

Hybrid Vehicle Jump Start and Jump Start Assist Programs

17

Jump starting a vehicle with a depleted battery has been a standard procedure in the automotive arena for many years. However, depending on the vehicle, jump starting a hybrid vehicle may require a different procedure for safety. To prevent possible damage, it is important to know the exact jump start procedure.

CHAPTER OBJECTIVES

- Explain why traditional jump start connections will not jump start all hybrid vehicle types.
- Describe the Ford Escape Hybrid jump start assist procedure and identify the start control switch.
- Describe methods for jump starting, or jump start assisting, different hybrid vehicles.

KEY TERMS

jump start assist: A method by which a 12 V auxiliary battery and an onboard vehicle DC-DC boost converter are used to recover (charge) a high-voltage battery pack enough to crank an engine using a high-voltage electric machine.

JUMP STARTING PROCEDURES

A hybrid electric vehicle has many procedures and systems that are very similar to a traditional vehicle. Jump starting is an area that may or may not be the same as a traditional vehicle, depending on the HEV system and manufacturer. It is important for a technician to understand the jump starting concepts that are specific to HEV products.

Toyota Prius

Toyota provides procedures to jump start only the auxiliary 12 V battery. The Prius does not use the auxiliary battery to crank the engine. If the auxiliary battery is depleted and the high-voltage battery is adequately charged, the vehicle will not crank because the controllers cannot operate at low system voltages. With the engine OFF, the auxiliary battery provides electrical energy to power all of the traditional vehicle systems. If it is depleted, the electronic controllers and ancillary systems are unable to provide control and electrical power to the vehicle systems. Therefore, the vehicle will not crank or operate until the 12 V battery is provided energy via a jump start. **See Figure 17-1.**

The high-voltage battery pack is used to provide electrical energy to the generator (MG1) inside the transmission so that it can crank the engine for starting. Toyota does not provide procedures to charge or jump start the high-voltage battery. If the 12 V battery is adequately charged, but the high-voltage battery pack is depleted, the engine will not start. The vehicle must be towed to a Toyota dealership for service.

Some models of the Prius have a jump start terminal in the underhood fuse box. In addition, the operator could connect directly to the battery terminal if the interior compartment closeout panels are removed. The auxiliary battery for the 2001–2008 Prius is located in the trunk area. The battery is smaller than a traditional vehicle cranking battery and provides less electrical capacity.

The Ford Escape uses jump start assist in which the 12 V battery charges the high-voltage battery pack.

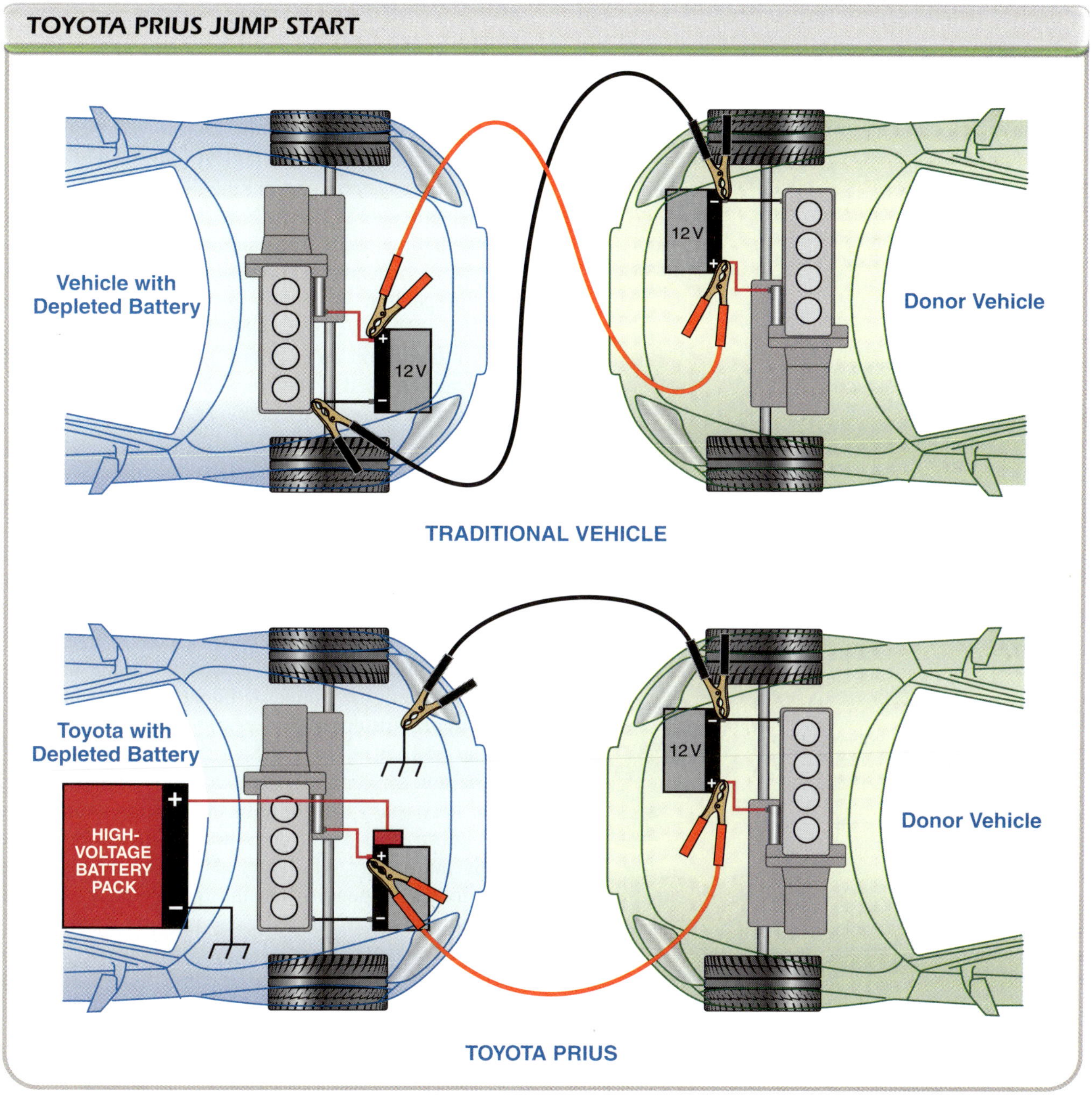

Figure 17-1. A Toyota Prius will not crank or operate until the 12 V battery is provided energy via a jump start.

Honda Hybrid Products

Honda hybrid systems are very similar to traditional vehicles. As with Toyota systems, the 12 V battery provides electrical power to base vehicle functions. Also, the high-voltage battery pack provides electrical energy for cranking the engine within most typical ambient temperature ranges.

> Charles Kettering patented an electrical ignition system for automobiles that eliminated the need for hand cranking. Kettering and Clyde Coleman installed the electric cranking motor in a Cadillac on February 17, 1911.

Special procedures are not required to jump start Honda systems and products even though they are mechanized with high-voltage battery systems. A Honda Hybrid vehicle uses a 12 V cranking motor to provide engine cranking in the event of a malfunction of the high-voltage battery pack or when ambient temperatures are outside the safe operating envelope. If the high-voltage battery service switch is moved to the OFF position, the hybrid controller automatically commands the system to crank on the 12 V cranking motor. **See Figure 17-2.**

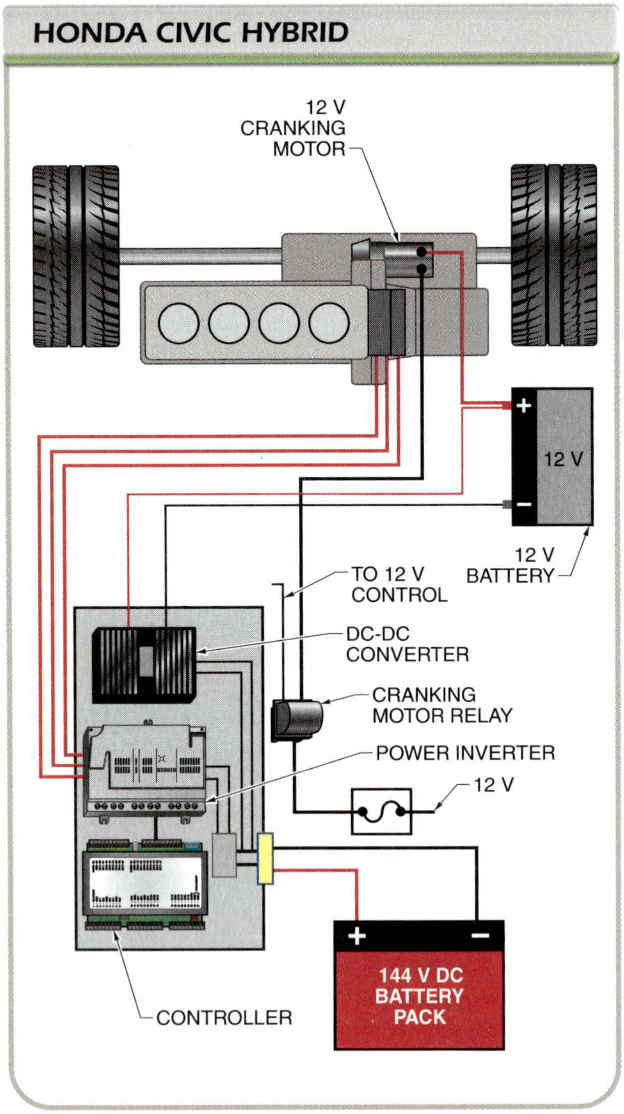

Figure 17-2. If a high-voltage battery service switch is OFF, the hybrid controller automatically commands the system to crank on the 12 V cranking motor.

2005–2008 Ford Escape Hybrid Jump Start Assist

The Ford Escape Hybrid uses a high-voltage battery pack and one of the two high-voltage electric machines inside its transaxle to crank the engine. It does not use a 12 V cranking motor system. It relies solely on the high-voltage system for cranking the engine. The jump starting procedure for the Ford Escape Hybrid is so unique that the term "jump starting" does not apply to the system. Instead, the Ford Escape uses the jump start assist procedure.

Jump start assist is a method by which a 12 V auxiliary battery and an onboard vehicle DC-DC boost converter are used to recover (charge) a high-voltage battery pack enough to crank an engine using a high-voltage electric machine. The difference between the jump start and jump start assist procedures has to do with the electrical energy requirements. A low-voltage battery system can supply enough electrical energy to the high-voltage battery pack if a boost converter system is used to boost the voltage from a 12 V battery.

The size and cost of the boost converter module would be significant if the desired outcome were an immediate engine crank. A boost converter system sized to provide enough energy to attain an immediate engine crank in a jump start assist would be expensive and hard to implement. Because of this limitation, the jump start assist procedure must be implemented.

A jump start assist does not permit an instantaneous jump start. The elapsed time could be up to 24 minutes, depending on the battery-pack state of charge. Jump start assist does provide a means to recover the battery-pack capacity to a level that can crank the engine. The core of the procedure is to place the ignition in the OFF position and then access the jump start button located in the driver side foot well area. **See Figure 17-3.**

It is very important to read the owner's manual or a technical manual before attempting to jump start a hybrid vehicle. Different manufacturers and models have different jump-start procedures. Some vehicles use a 12 V battery for starting. Other vehicles use a high-voltage battery.

FORD ESCAPE HYBRID

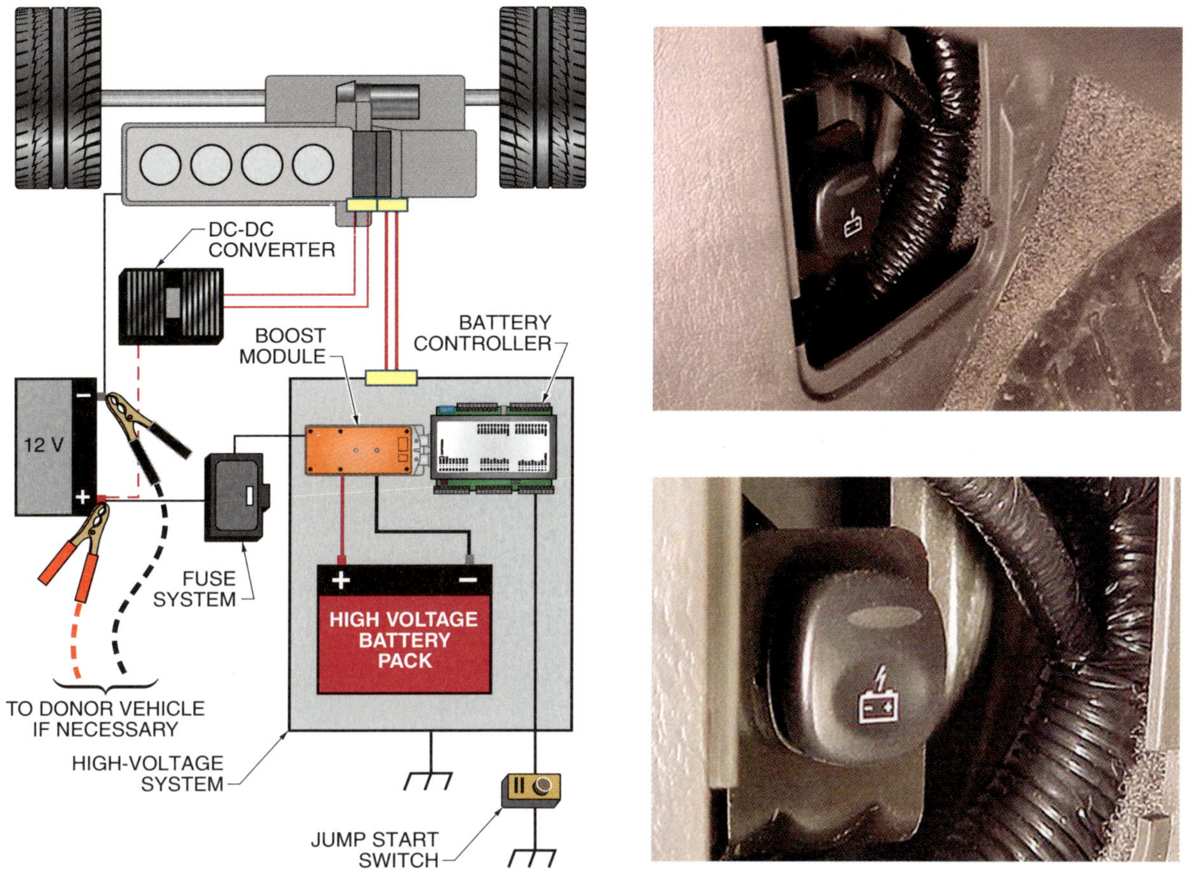

Figure 17-3. Jump start assist is a method by which a 12 V auxiliary battery and an onboard vehicle DC-DC boost converter are used to charge a high-voltage battery pack enough to crank an engine with a high-voltage electric machine.

GM Parallel Hybrid Truck

As with other vehicle manufactures, GM provides specific jump starting procedures that need to be followed by the operator or technician during a jump start event. The GM parallel hybrid truck (PHT) system uses a 12 V auxiliary battery to operate the traditional low-voltage vehicle electrical systems. It also uses a 36 V battery-pack system to provide power to the hybrid power train. **See Figure 17-4.** The PHT jump starting procedure provides the option of jump starting the vehicle through the 12 V auxiliary battery, the 36 V battery pack, or both.

There are two methods for jump starting the PHT through the 12 V auxiliary battery. The first method is the traditional jump starting method, which requires jump start cables and a donor vehicle to jump start the 12 V battery. After connecting the jumper cables, the key for the PHT needs to be held in the START position for at least 3 seconds.

If the vehicle does not crank, the key needs to be placed in the LOCK position for 5 minutes. This procedure can be repeated several times. If the engine still does not crank, it needs a technician to service it.

The second method requires an operator or technician equipped with a Tech 2 serial data tool, or scan tool. Within the Tech 2 menus, a special function command called "42 V to 12 V Command" permits the operator or technician to command the system to use the 36 V battery system to charge the 12 V battery.

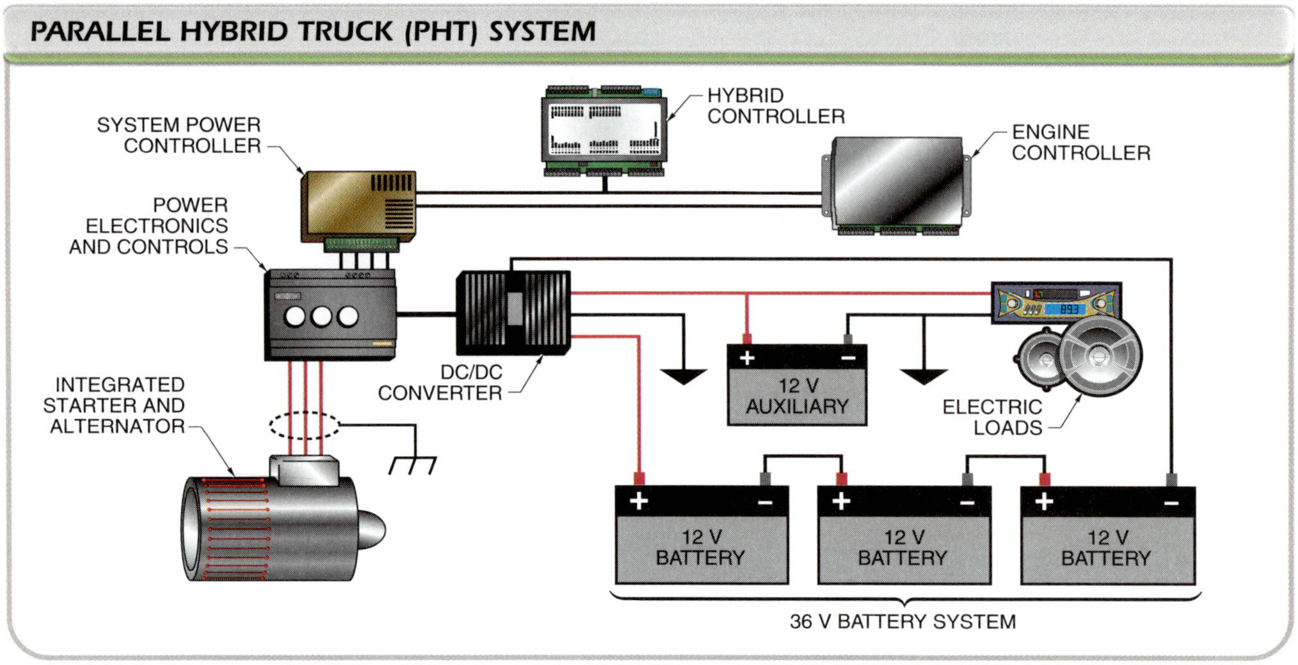

Figure 17-4. The GM parallel hybrid truck (PHT) system uses a 12 V auxiliary battery to operate the traditional low-voltage vehicle electrical systems with a 36 V battery-pack system to provide power to the hybrid power train.

The PHT system also has control commands to allow the 12 V battery to jump start, or boost, assist the 36 V battery system. If the 36 V battery has been depleted to the point that it will not crank the engine, the 12 V auxiliary battery can be commanded to transfer energy to the 36 V battery pack. However, the owner's manual recommends that a donor vehicle be used to accomplish this procedure so that the PHT 12 V auxiliary battery energy does not become depleted. It may take 10 to 20 minutes to complete the jump start assist procedure. If the 36 V battery pack still does not crank the engine, the vehicle must be analyzed and repaired by a qualified technician.

GM Belted Alternator Starter System

As with the GM PHT system, the GM belted alternator starter (BAS) system uses 12 V and 36 V battery systems. The 12 V battery provides electrical energy to the low-voltage vehicle systems. The 12 V battery also provides energy for the first cranking event of a drive cycle or when there is a failure of the 36 V battery system. The 36 V battery pack provides energy for the hybrid propulsion modes. **See Figure 17-5.** These modes include auto-starting the engine during a drive cycle.

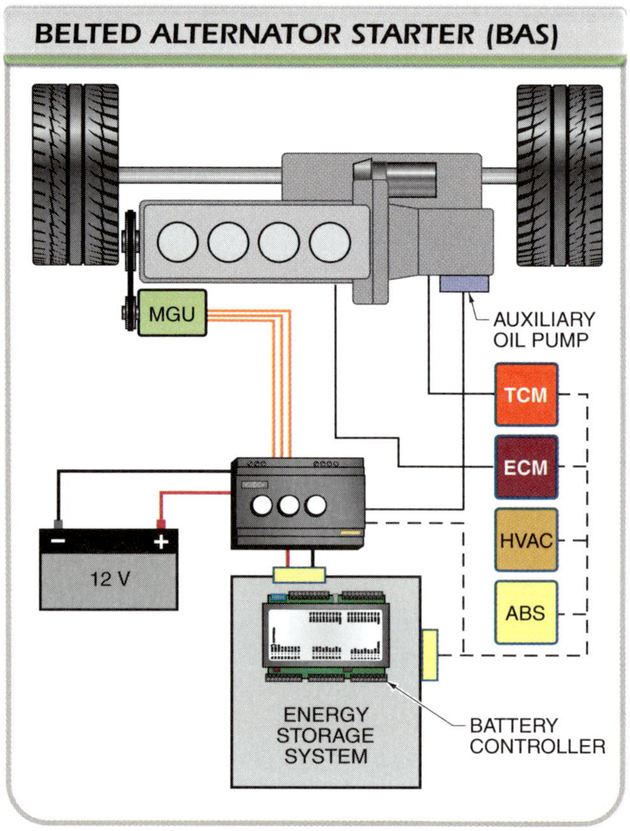

Figure 17-5. A 36 V battery pack provides energy for hybrid propulsion modes.

With a BAS system, only traditional jump starting methods on the 12 V auxiliary battery can be used. There is no method of transferring energy from the 12 V battery to the 36 V battery (boost charging) in the event of failure of the 36 V system. Therefore, jump starting for a BAS vehicle is confined to the 12 V auxiliary battery. Traditional jump starting procedures are used to jump start the 12 V battery. If there are any operational problems with the 36 V battery system, the vehicle must be serviced by a qualified technician.

Refer to the *Hybrid Electric Vehicle Technology* CD-ROM for a Quick Quiz® on the chapter contents.

POINTS TO REMEMBER

- Not all hybrid engines are cranked and started with a 12 V battery and a 12 V cranking motor system.
- The hybrid engines that do retain a 12 V battery and a 12 V cranking motor can be jump started using traditional methods.
- Most hybrid systems do not use a 12 V battery system and cranking motor to crank the engine. These systems use the transaxle electric machine with energy from the high-voltage battery pack to crank the engine.
- Most manufacturers do not have an onboard vehicle method of recovering (charging) the high-voltage battery pack if it is depleted.
- Some vehicles have a jump start assist system that permits the 12 V auxiliary battery to be used with an onboard vehicle DC-DC boost converter to recover the high-voltage battery pack enough to crank the engine with the high-voltage electric machine.

Refer to the *Hybrid Electric Vehicle Technology* CD-ROM for the Review Questions in Microsoft® Word format.

REVIEW QUESTIONS

1. What procedure should be followed if the high-voltage battery in a Toyota Prius is depleted?
2. Define the term "jump start assist" and name the hybrid vehicle product(s) that use this method.
3. What procedure should be followed if a Honda Hybrid needs a jump start?
4. Which battery power system provides power for the first cranking event of a GM BAS vehicle?
5. What procedure should be followed if a GM PHT vehicle needs a jump start?

17 Hybrid Vehicle Jump Start and Jump Start Assist Programs

HYBRID Electric Vehicle TECHNOLOGY

Trade Test

Name ____________________ Date ____________________

______ **1.** The Toyota Prius permits a jump start or jump start assist of the ___ system.
- A. low-voltage battery-pack
- B. low-voltage auxiliary battery
- C. high-voltage battery-pack
- D. high-voltage auxiliary battery

______ **2.** The Ford Escape Hybrid permits a jump start or jump start assist of the ___ system.
- A. low-voltage auxiliary battery
- B. high-voltage battery-pack
- C. low-voltage battery-pack
- D. both A and B

______ **3.** A Honda Hybrid can be jump started like a conventional vehicle because ___.
- A. it uses a high-voltage relay
- B. it has a 12 V cranking motor system
- C. the engine has an air starter
- D. the vehicle has a backup 12 V battery

______ **4.** The 12 V battery of the 2001–2008 Toyota Prius is located ___.
- A. behind the rear seat upright
- B. in the engine compartment
- C. within the battery pack
- D. in the left well of the trunk

______ **5.** The GM belted alternator starter (BAS) system permits a jump start of the ___ system.
- A. high-voltage battery-pack
- B. low-voltage auxiliary battery
- C. high-voltage auxiliary battery
- D. both A and B

______ **6.** The GM parallel hybrid truck (PHT) system permits a jump start of the ___ system.
- A. high-voltage battery-pack
- B. low-voltage auxiliary battery
- C. high-voltage auxiliary battery
- D. both A and B

_______________ **7.** Some models of the Toyota Prius have a jump start terminal located ___.

A. in the underhood fusebox

B. in the fusebox in the cabin

C. at the high-voltage battery pack

D. in the trunk

HYBRID
Electric Vehicle
TECHNOLOGY

GM FlexPower PHT

18

High-voltage hybrid systems are the most common systems available in the hybrid automotive market. Low-voltage hybrid systems are much less expensive and can provide many of the same benefits. General Motors (GM) was the first manufacturer to release a production 42 V hybrid system with a parallel hybrid truck (PHT) system in the Silverado/Sierra truck family. The concept of the 42 V system is now being applied to passenger cars.

CHAPTER OBJECTIVES

- Describe the design and function of the heating/ventilation and air conditioning (HVAC) control module.
- Describe the design and function of the starter-generator control module (SGCM).
- Describe the design and function of the energy storage control module (ESCM).
- Describe the design and function of the electrohydraulic power steering module (EHPSM).
- Explain how using the HVAC system affects engine idle-stop.
- Explain how engine idle-stop, fuel cutoff, and auto-start are used to reduce fuel consumption.
- Explain the operation of the auxiliary power outlet (APO) system.

KEY TERMS

hybrid control module (HCM): A control module that functions as the master of all vehicle controllers on the data bus in a hybrid vehicle.

powertrain control module (PCM): A control module that controls vehicle engine functions and reports data to other controllers on a data bus.

starter-generator control module (SGCM): A control module that controls electrical power to a starter-generator and includes a DC-DC buck-boost converter system to charge a 12 V auxiliary battery. The converter can also charge a 36 V battery pack from the 12 V system if necessary.

energy storage control module (ESCM): A control module that controls the functions within an energy storage module (ESM) and reports data to other controllers on a data bus.

electrohydraulic power steering module (EHPSM): A control module that controls the electrically driven hydraulic motor to provide hydraulic pressure for steering and braking assist.

heating, ventilating, and air conditioning (HVAC) control module: A control module that controls air conditioning and heating functions and reports data to other controllers on a data bus.

auxiliary power outlet (APO) system: A system that provides 120 V AC standby power to outlets located in a vehicle cargo bed and a cabin under the passenger seat.

GM FLEXPOWER PHT CONTROL MODULES

General Motors (GM) released its FlexPower Silverado and Sierra 42 V parallel hybrid truck (PHT) in 2004. These vehicles feature significant enhancement in fuel economy, emissions reduction, and power generation. The electric machine assembly is located within the transmission bell housing area. The stator is attached to the engine block and is designed with high efficiency copper bar electrical conductors in place of traditional copper wire windings to increase machine efficiency. The rotor bolts directly to the engine crankshaft and rotates inside the stator.

The electric machine unit and 258 mm diameter torque converter are mounted between the engine and transmission. An auxiliary transmission oil pump is located within the transmission and is used to enable the auto-start feature by providing the necessary hydraulic line pressure to permit immediate torque transfer when the engine is started.

Hybrid Control Module

A *hybrid control module (HCM)* is a control module that functions as the master of all vehicle controllers on the data bus in a hybrid vehicle. **See Figure 18-1.** It is mounted to the left of the engine radiator. The HCM controls automatic engine idle-stop and auto-start as well as the energy that flows into and out of the 14 V and 42 V battery pack. This module enables and disables power for both the auxiliary power outlets and transmission/driveline active damping.

HYBRID CONTROL MODULE (HCM)

General Motors

Figure 18-1. The hybrid control module (HCM) functions as the master of all vehicle controllers on the data bus in a hybrid vehicle.

The HCM also controls electric machine speed during transmission downshift assists. The downshift assists support the engine during unfueled downshifts from fourth gear to third and from third gear to second. The HCM calculates a target engine speed and uses the electric machine unit to maintain that speed. The electric machine unit can control engine speed by using a torque command that resides within the upper and lower torque limits specified by HCM calibrations.

As a matter of safety in the event of an accident, the HCM disables all energy generation after airbag deployment has been triggered. Additionally, the HCM controls the auxiliary transmission pump, power maintenance relay, and auxiliary heater pump.

Powertrain Control Module

A *powertrain control module (PCM)* is a control module that controls vehicle engine functions and reports data to other controllers on a data bus. The PCM initiates and controls engine cranking and starting. It also stores fault codes for the energy storage control module (ESCM), electrohydraulic power steering module (EHPSM), starter-generator control module (SGCM), and the HCM.

Starter-Generator Control Module

A *starter-generator control module (SGCM)* is a control module that controls electrical power to a starter-generator and includes a DC-DC buck-boost converter system to charge a 12 V auxiliary battery. **See Figure 18-2.** The converter can also charge a 36 V battery pack from the 12 V system if necessary.

The SGCM functions to control electrical energy generated by the electric machine. The energy is used to generate torque for cranking the engine and generating electrical power. Specifically, the SGCM provides electrical power to the electric machine to control overall engine cranking, torque, and speed; permits torque smoothing and active damping; and controls the generation of electrical power. The SGCM also controls the accessory power module, including 14 V DC for vehicle nonhybrid system components, 42 V DC for hybrid components, 120 V AC for auxiliary power outlets, and 28 V AC for engine cranking.

The SGCM controls the 3ϕ inversion and conversion processes necessary to convert 42 V DC to 28 V AC for starting, to 14 V DC for the traditional alternator function, and to 120 V AC to power electrical outlets. The SGCM also controls AC conversion for recharging the 36 V battery system, and it controls 12 V DC power to jump start the 36 V battery pack.

SGCM Cooling System. In addition to the traditional engine cooling system, there is a separate dedicated cooling system for the SGCM. **See Figure 18-3.** Its purpose is to protect the SGCM internal electronic components from thermal damage. The system includes an auxiliary radiator coupled to an electric radiator fan and a secondary surge tank with a pressure cap. The coolant pump is activated when the SGCM is functioning. The infinitely variable radiator fan activates as needed.

> The hybrid control module functions as the master of all vehicle controllers on the data bus. It also controls engine idle-stop and auto-start.

STARTER-GENERATOR CONTROL MODULE (SGCM)

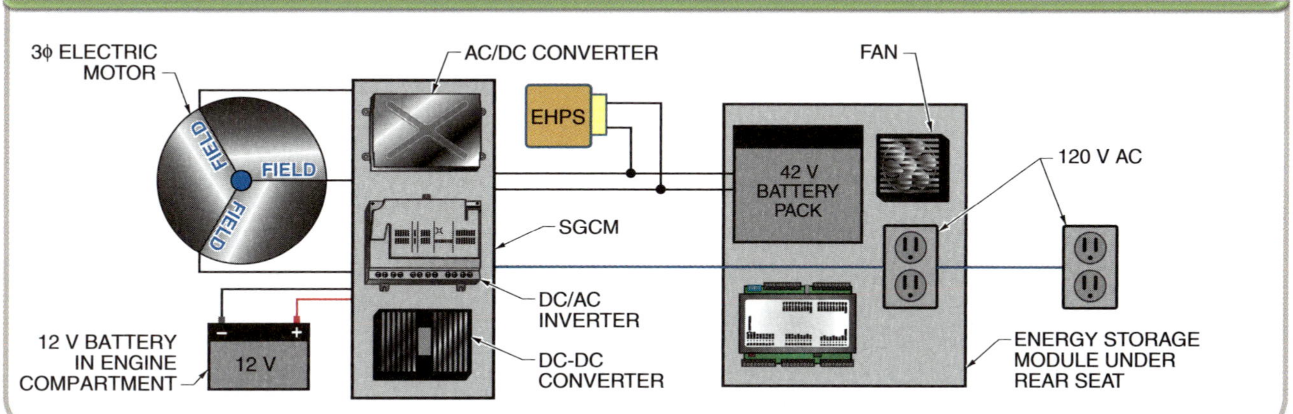

Figure 18-2. The starter-generator control module (SGCM) controls electrical power to the starter-generator.

SGCM COOLING SYSTEM

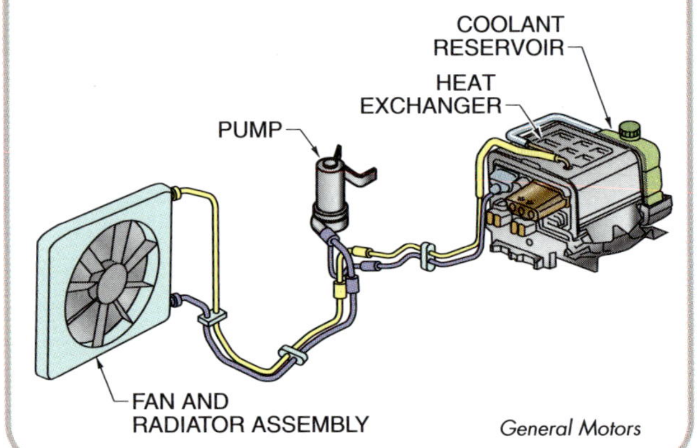

Figure 18-3. There is a separate dedicated cooling system for the SGCM.

At higher temperatures, the SGCM generates hot vapor internally. This hot vapor is cooled with coolant from the secondary cooling system through a heat transfer surface at the top of the SGCM. The heat is dissipated through the auxiliary radiator coupled to the electric fan.

The secondary cooling system is located near the right rear of the engine compartment on the passenger side. The radiator fan is located behind the auxiliary radiator. The auxiliary radiator is located directly in front of the condenser. The auxiliary radiator and the fan module are located directly in front of the engine radiator and condenser.

Energy Storage Control Module

The *energy storage control module (ESCM)* is a control module that controls the functions within an energy storage module (ESM) and reports data to other controllers on a data bus. The ESCM is located in the ESM and controls the energy storage module fan and battery cutoff switch. **See Figure 18-4.**

The primary function of the ESM is to provide electrical power for the electrohydraulic power steering and the electric machine. It also communicates the battery-pack status, including each battery module state of charge (SOC), current, and engine idle-stop available time. In addition, it communicates battery-pack constraints, such as the minimum SOC levels and the available charge and discharge power. A traditional 12 V battery is located under the hood and supplies electrical power to all other electrical loads.

ENERGY STORAGE CONTROL MODULE (ESCM)

Figure 18-4. The ESCM is located in the ESM and controls the energy storage module fan and battery cutoff switch.

The ESM is a component of the energy management system containing elements that both provide and store electrical energy. The energy management system is made up of the ESM component and other controllers, sensors, switches, software, and interconnections that are engineered to control the 42 V energy system. The ESM provides the 42 V electrical power to the SGCM via two cables.

The ESM package assembly contains three VRLA batteries for the 42 V system. These deep-cycle batteries are designed with absorbent glass mat (AGM) technology. Each of the three 12 V AGM batteries has about 50 Ah capacity. The batteries are stored in a single ESM module that is mounted under the rear seat. A manual/service disconnect switch provides a method to disconnect power from the remainder of the vehicle. The switch is a green lever located on the outside of the ESM.

Electrohydraulic Power Steering Module

An *electrohydraulic power steering module (EHPSM)* is a control module that controls the electrically driven hydraulic motor to provide hydraulic pressure for steering and braking assist. The EHPSM communicates the status of electrohydraulic power steering (EHPS) parameters, including brake pedal position and its rate of change, power consumed from the 42 V bus, and the EHPS system operating status. **See Figure 18-5.**

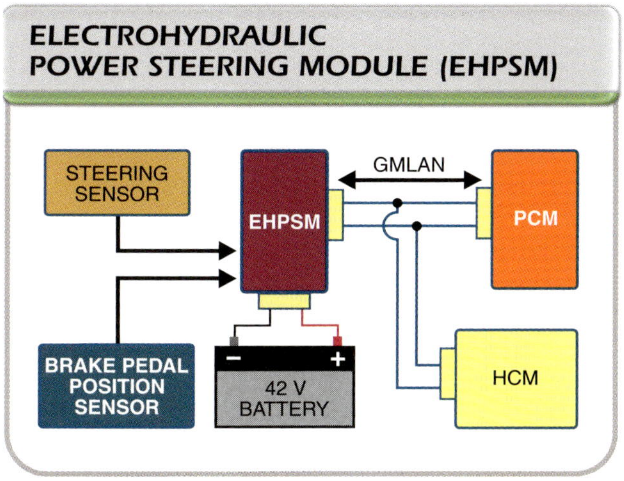

Figure 18-5. A steering sensor and a brake pedal position sensor provide input to the electrohydraulic power steering module (EHPSM).

The hybrid EHPS system is different from the power steering system in traditional vehicles because of the hybrid idle-stop mode. By using an electric motor to drive the pump, the engine-independent EHPS operates even when the engine is shut off to conserve fuel.

The 42 V variable-effort EHPS system is designed with an electric motor to drive the power steering pump to develop power steering pressure. A sensor on the steering column assembly provides steering wheel rate of turn data to the controller enabling the controller to increase output flow as a function of increasing steering wheel rate to prevent momentary high steering effort. The electric motor, pump, and control module are integrated into a singular engine-mounted unit. The 42 V power level is used to minimize electrical current.

The EHPS steering system has two operating modes. The first is the ON mode that uses the lowest possible EHPS output for a given operating condition. The ON mode allows hydraulic output to be varied between the preset idle flow and the maximum flow based on vehicle speed, steering wheel turn rate, brake pedal apply rate, or any combination of the three.

The second mode is the continuous mode for the auxiliary power outlets, which eliminates current consumption by the EHPS when the vehicle is in park. In this mode, the EHPS system shuts down if it does not receive input signal information in a specified amount of time.

If the vehicle speed, steering wheel rate, or brake apply rate changes, or if the driver shifts from park, the EHPS reenergizes to provide hydraulic pressure to the power steering rack or steering box system. The EHPS control strategy enables the system to provide speed-dependent, variable-effort, steering assist. This reduces the effort required to turn the steering wheel for improved stability as vehicle speed increases.

Vehicle Braking with EHPS. Hydroboost hydraulic brake boosters are used for braking assistance. The EHPS system provides the hydraulic power for the brake booster in addition to the hydraulic power for the steering gear. These two systems are hydraulically connected in series, with the brake booster closer to EHPS pump outlet pressure.

The system includes a unique 1800 psi pressure valve, a long stroke 34 mm diameter master cylinder, and a pedal position sensor. The brake pedal assembly sensor provides pedal position data to the EHPS controller enabling it to adjust hydraulic output as required. This brake system provides brake feel and performance that is identical to other GM trucks. In response to quick brake applications, the EHPS system can increase the maximum flow of brake fluid to the booster more quickly than an engine-driven pump system.

HVAC Control Module

The *heating, ventilating, and air conditioning (HVAC) control module* is a control module that controls air conditioning and heating functions and reports data to other controllers on a data bus. **See Figure 18-6.** The vehicle heating system is configured to allow engine coolant to circulate when the engine is stopped. An auxiliary coolant pump has been added to the traditional heater system to provide coolant flow to the HVAC system when the engine cycles during auto-start and idle-stop. The HVAC control module works with the PCM to control the pump.

The hybrid vehicle has other changes to the air conditioning system. An algorithm change in the HVAC control module software extends air conditioning performance capability when the engine is OFF. Traditional GM truck air conditioning systems use outside air and delay the use of recirculated air to prevent excessive blower noise.

The air conditioning system in the hybrid truck operates differently than traditional systems. The HVAC system recirculates cabin air when the engine goes into engine idle-stop mode because the compressor does not operate in this mode. The HVAC blower motor speed is reduced during the engine idle-stop mode to maintain a quiet cabin.

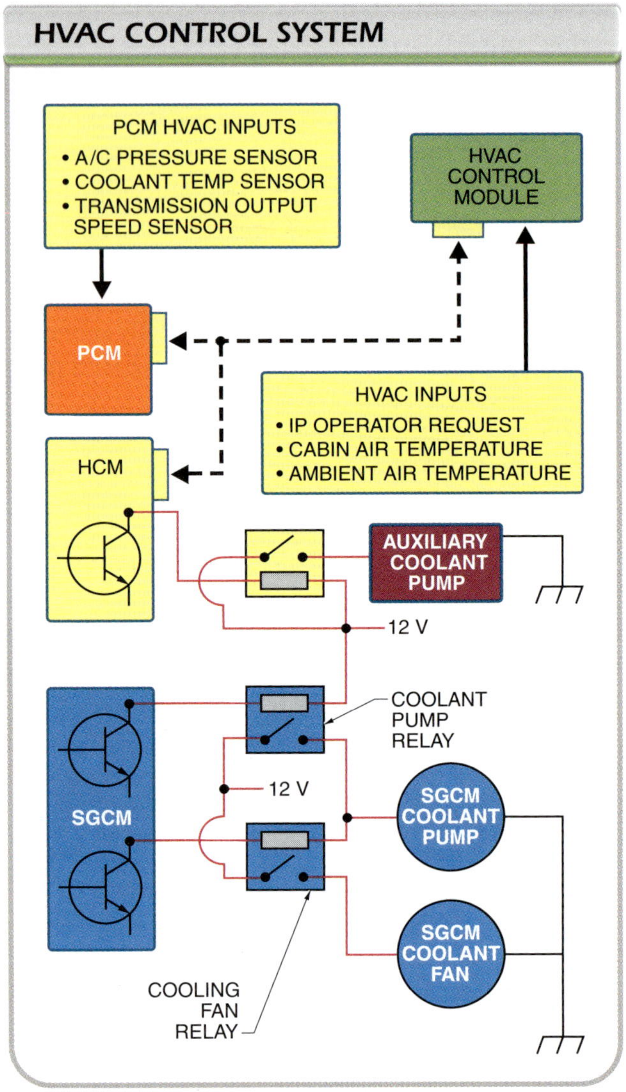

Figure 18-6. The heating/ventilation and air conditioning (HVAC) control module controls air conditioning and heating functions.

GM FLEXPOWER PHT OPERATION

GM trucks are designed to provide substantial trailering capabilities. However, this causes high fuel consumption and vehicle emissions. The GM Flex-Power PHT addresses these problems with improved fuel economy and more efficient vehicle operations. The 42 V hybrid system is targeted to improve fuel economy by 10% in composite city/highway miles and 15% city only miles over traditional GM trucks with an equivalent powertrain.

Hybrid vehicle engine cranking and starting is designed to be quicker and quieter than traditional vehicles. At the first cranking event of a drive cycle, the engine is cranked and started as with any traditional engine. Further improvements include the torque converter lockup in second, third, or fourth gear. This mechanically couples the transmission into direct drive modes, similar to that of a manual transmission.

Fuel Economy

The PHT design uses a compact 14 kW (19 HP) AC induction electric machine that functions as the cranking motor (starter) and generator. **See Figure 18-7.** This electric machine is located between the engine and transmission. The electric machine provides fast and quiet starting power and permits automatic engine idle-stop, fuel cutoff, and auto-start to conserve fuel. The electric machine also smoothes out any driveline surges, generates electrical current to charge batteries, runs auxiliary power outlets, and provides coast-down regenerative braking as an aid to fuel economy.

Early torque converter clutch lockup and surge smoothing is a function in which the transmission shift schedules have been modified to provide earlier shifts. This permits the engine to operate at a lower speed. Using the torque converter application in second, third, and fourth gears permits the transmission to direct drive the vehicle in a manner similar to a manual transmission. Driveline forces from the inertia of the vehicle maintain engine rotation when the fuel is cutoff.

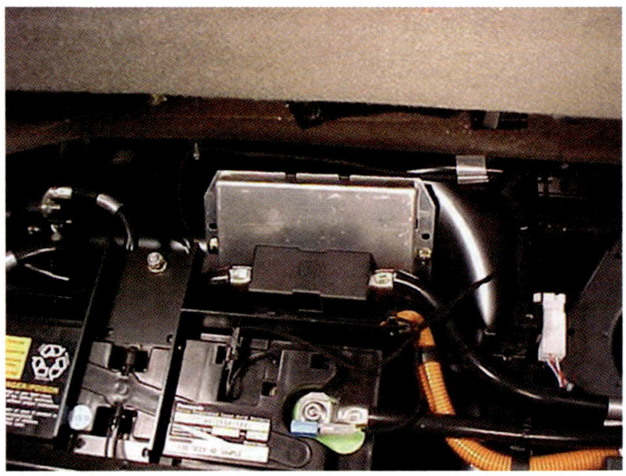

The energy storage control module is located in the energy storage module near the battery.

PHT FEATURES

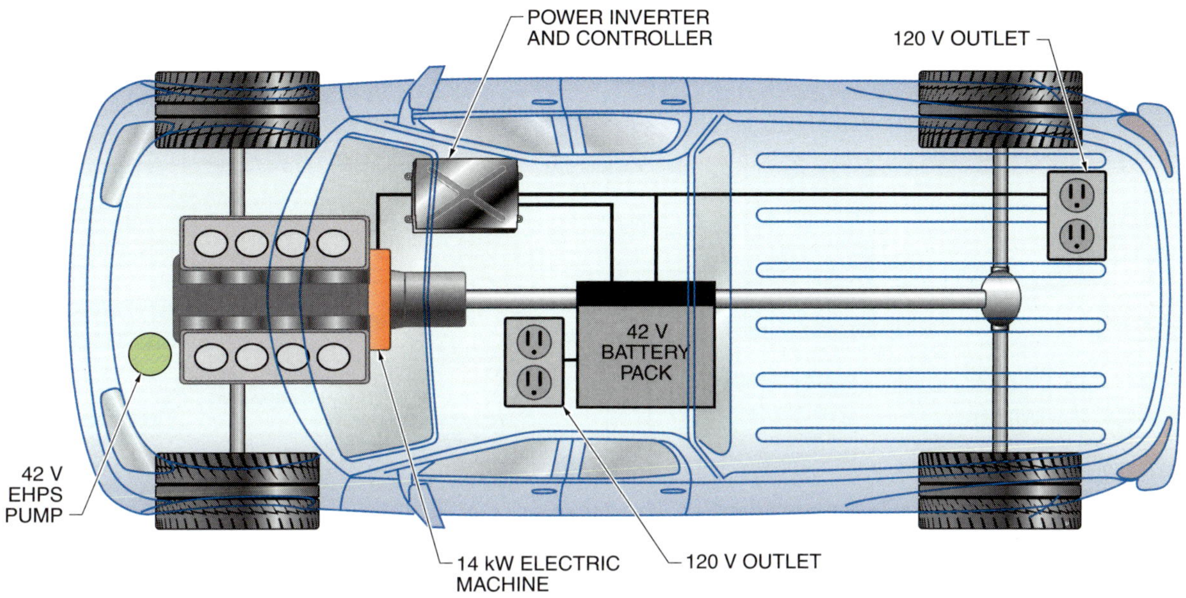

Figure 18-7. The PHT design uses a compact 14 kW AC induction electric machine that functions as the cranking motor (starter) and generator.

Regenerative braking captures vehicle kinetic energy during braking. Coast-down regenerative braking captures energy from the vehicle when the accelerator pedal is released while coasting. To enhance this benefit, fourth gear has also been modified to provide more aggressive braking. The EHPS system provides power assistance as a function of driver demand.

Engine Idle-Stop Mode

Engine idle-stop mode is one of several design features used to improve fuel economy. During engine idle-stop, the controller automatically stops the engine to conserve fuel when the vehicle is idling. **See Figure 18-8.** This mode occurs only after the engine and transmission reach operating temperature.

If the operator depresses the brake pedal with the gear selector in D3 or D4, or releases the accelerator pedal at speeds lower than 13 mph, the engine transitions into engine idle-stop mode. Normal functions, such as power steering and recirculating conditioned air in the cabin, continue to operate.

When the vehicle is at a complete stop, the engine will continue to run if the driver releases the brake pedal, even slightly, or has the shifter in neutral, reverse, second gear, or first gear. The engine continues to run, if necessary, and does not transition into engine idle-stop mode on extremely hot or cold days to provide optimum interior comfort or windshield defrost. It may also continue to run to charge a low battery pack.

To maintain full vehicle operation and safety, the engine will only transition into engine idle-stop mode if the hood is completely closed, engine operating temperatures are in safe range, and the climate control system has allowed the engine idle-stop based on interior comfort. In addition, the engine transitions into idle-stop mode when the vehicle is shifted into park and all doors are closed. However, the system will not transition into idle-stop mode if the climate control system is set to defrost.

Fuel Cutoff. Engine idle-stop mode allows aggressive deceleration fuel cutoff (FCO) at all speeds higher than 13 mph. In this operating mode, the electric machine controls coast downshifts without having to fuel the engine. FCO is allowed when the operator no longer depresses the accelerator. Driveline forces maintain engine rotation even though fuel is no longer delivered. Normal vehicle functions remain available and depressing the accelerator restores fuel delivery to the engine.

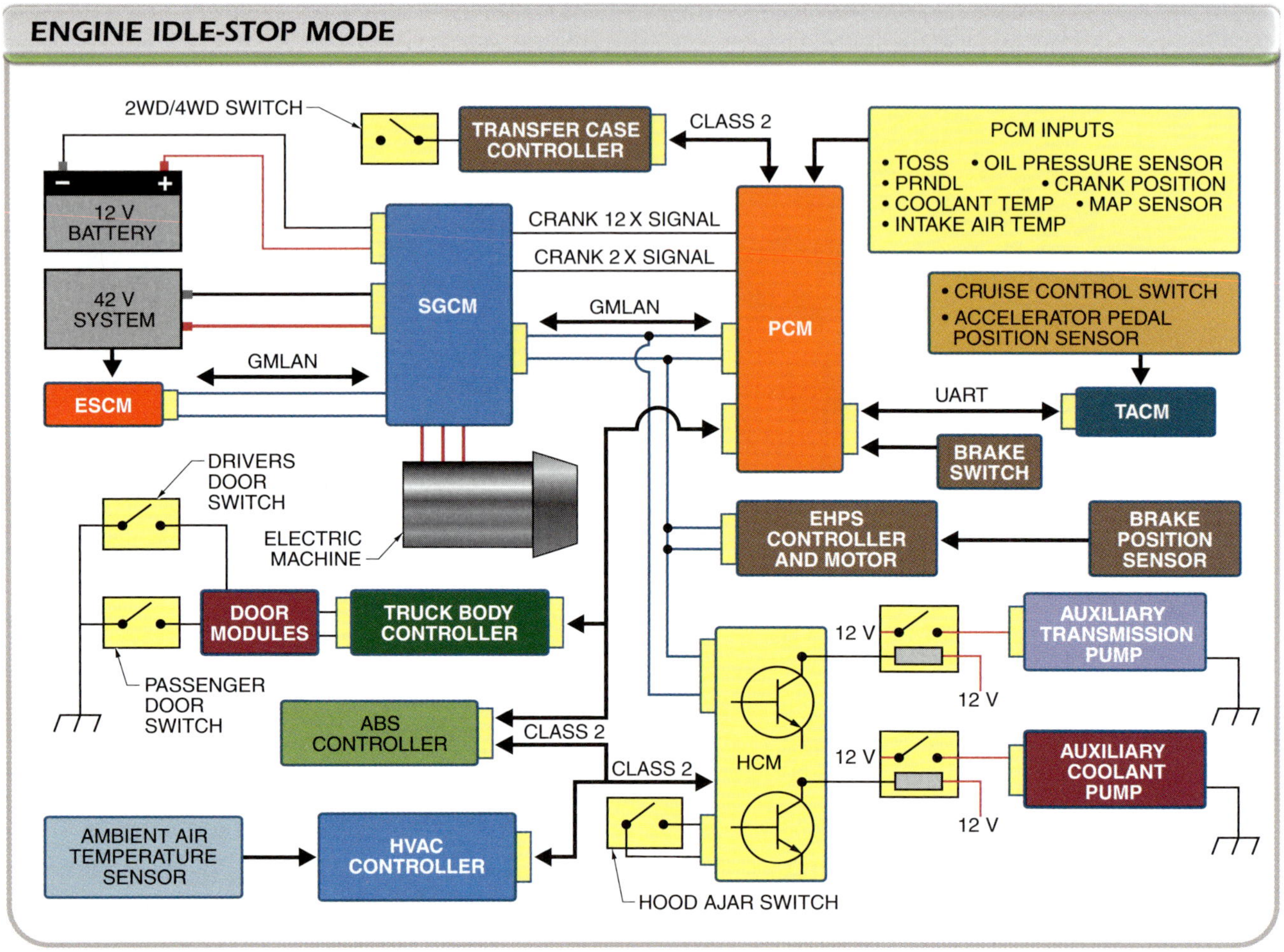

Figure 18-8. Engine idle-stop is an operating mode in which the controller stops the engine when a vehicle is idling.

Engine Auto-Start

Engine idle-stop mode also includes auto-start features. During engine auto-start, the engine is cranked and started during acceleration or when the battery pack is low and the high-voltage system needs to recharge the battery pack. Vehicle auto-start performs the following functions:

- recharges a low 42 V battery-pack SOC
- maintains power steering operation
- maintains cabin comfort when the operator requests the "max" position from the climate control system
- enables the windshield defroster to operate

When the vehicle is operating in idle-stop mode, the engine can auto-start under certain circumstances. When driving the vehicle, the engine will automatically restart when the brake pedal is released or the accelerator pedal pressed.

The vehicle can also auto-start if the vehicle speed exceeds a calibrated specification; the operator selects the D3, D4, or park gear selection; or any door is opened while in park. The reason for auto-start when any door is opened is to alert the operator that the vehicle is in a functioning, operational state (not shutdown) and the vehicle cannot be left unattended. An auto-start can also occur when the operator requests the auxiliary power outlets to operate in a continuous mode.

When parked, the vehicle can continue to provide energy to operate accessories from the 42 V battery pack. The engine will continue operating to maintain 42 V battery-pack SOC while supplying electrical power for the 120 V outlets. In this mode, the engine continues to run if the operator opens the hood with the engine running.

The system must determine when it can execute an automatic restart. If the vehicle is cranked and started after the hood has been opened, the engine idle-stop and auto-start mode are disabled. If the vehicle is cranked and started before the hood is opened, the engine will auto-start but not idle-stop. In addition, auto-start is disabled if the gear selector is in 4Lo in four-wheel-drive models. If the gear selection is not properly set, an engine idle-stop will occur.

Auxiliary Power Outlet System

An *auxiliary power outlet (APO) system* is a system that provides 120 V AC standby power to outlets located in a vehicle cargo bed and a cabin under the passenger seat. **See Figure 18-9.** The hybrid truck system is designed with four 120 V AC outlets that can provide a total of 2.4 kW (20 A at 120 V) for all four electrical outlets.

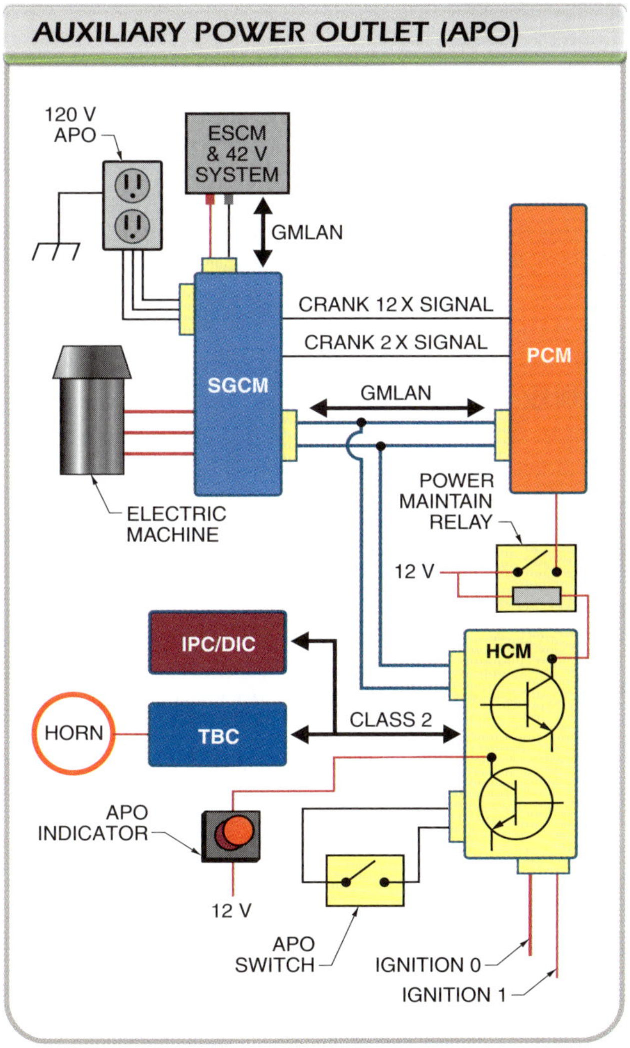

Figure 18-9. The auxiliary power outlet (APO) system provides 120 V AC standby power to outlets located in the vehicle cargo bed and in the cabin under the passenger seat.

The electrical outlets are controlled by a complex software and hardware system. When the APO system is not in use, the electricity generated by the system can be stored in the 42 V battery pack if the battery-pack SOC is less than 100%. APO outlets receive power through a wiring harness and connector system from the SGCM.

The APO is activated by pressing the APO button located to the immediate right of the air conditioning controls on the instrument panel. All APO outlets are designed with a ground fault detection system to prevent shock from accidental contact to ground.

The ground fault detection system can also detect electrical overloads and short circuits. If the APO system is overloaded or if a defective device is connected to any outlet, the ground fault detection system senses the condition and disables the APO system. Additionally, if there is an electrical problem or overload, the APO button light on the instrument panel will flash and the DIC will indicate that there is a problem.

Normal Mode. Normal mode is available only while a vehicle is being driven. Normal mode allows the use of all 120 V outlets. It can be activated by pressing the APO button located on the instrument panel. The APO button indicator lights when it is released to indicate power availability. An operator should use extension cords that are UL approved. If the APO system ceases to supply power and the APO button light is flashing, the APO system has detected an overload and shuts down. The operator must reset the APO system to reactivate it.

The APO system operates in the normal mode until the operator deactivates it by pressing and releasing the APO button or by turning off the ignition key. Additionally, when the APO button is released, the APO button indicator no longer lights. This indicates that the APO system is not operational.

Continuous Mode. Continuous mode is available only while the vehicle is stationary with the engine running. It can provide power with or without an operator in the vehicle, with or without a key in the ignition, or with the vehicle doors closed or locked. Continuous mode can be activated from the normal mode while the engine is running.

The first step is to rotate the ignition to the run position and shift the transmission into park. The APO button must be pressed and held for 2 seconds until the APO button light begins to flash. The operator then has 30 seconds to enter continuous mode by turning the key to the OFF/lock position. The indicator light adjacent to the APO button continues to illuminate to inform the operator that the system is operating in continuous mode.

When the ignition is turned to the OFF/lock position, the key can be either left in the ignition or removed and the desired devices can be plugged into the APO system. The vehicle will operate in continuous mode until the operator depresses the APO button. The APO button light and continuous indicator will no longer illuminate, and the engine will stop running. If the fuel drops below the two-gallon level while in continuous mode, the engine automatically shuts down. A low fuel level message is displayed in the driver information center (DIC), and the horn sounds intermittently.

Communication System

The driver information center (DIC) communicates the status of vehicle functions, including the hybrid system, to the vehicle operator. This communications network for the basic driver information system is comprised of several control modules and electrical components. **See Figure 18-10.** The hybrid electrical architecture uses the Class 2 data bus for nonhybrid functions and the GM local area network (GMLAN) high-speed serial data communication system for all hybrid functions.

The Class 2 data bus for nonhybrid functions is a communication system connecting all controllers that support nonhybrid functions. It primarily links body-related functions such as lighting, driver information center (DIC) displays, and the entertainment system.

The GMLAN high-speed serial data communication system connects the hybrid related control modules. This enables each control module that has direct control in the hybrid system to communicate data relevant to its operation in real-time for concise and precise system control. The GMLAN 500 kbaud transfer and control is nearly 50 times faster than the 10 kbaud Class 2 system.

The hybrid control module provides a gateway between the GMLAN and Class 2 local area networks. If a GMLAN controller transmits a message to the Class 2 bus, the GMLAN controller codes the message and sends it to the HCM. The HCM receives and processes the message and transmits it to the appropriate network.

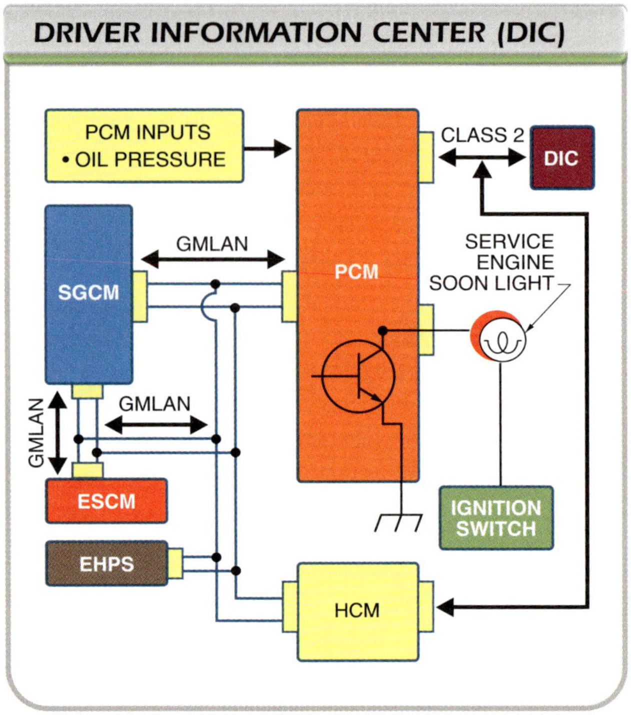

Figure 18-10. The driver information center (DIC) communicates the status of vehicle functions, including the hybrid system, to the operator.

Safety and Security

A right-side impact sensor is used to detect a collision on the side of the vehicle where the 42 V cables are routed from the engine compartment to the ESM. The impact sensor communicates with the airbag sensing diagnostic module (SDM), which has been reconfigured for the hybrid. The SDM airbag deploy message is transmitted to the HCM, and finally to the ESCM in the 42 V ESM. The ESCM activates a battery cutoff switch located inside the ESM to confine the 42 V energy within the ESM assembly.

Theft protection is enabled when the APO system is in continuous mode. The engine will shut off if the vehicle is shifted out of park or if the vehicle is driven at a speed of more than 3 mpg. The electronic throttle control will be disabled if an operator attempts to step on the accelerator pedal when the vehicle is in continuous mode.

Refer to the *Hybrid Electric Vehicle Technology* CD-ROM for a Quick Quiz® on the chapter contents.

POINTS TO REMEMBER

- The hybrid control module (HCM) functions as the master of all vehicle controllers on the data bus in a hybrid vehicle. It is mounted to the left of the engine radiator.
- The powertrain control module (PCM) controls vehicle engine functions and reports data to other controllers on a data bus. The PCM initiates and controls engine cranking and starting.
- The starter-generator control module (SGCM) controls electrical power to a starter-generator and includes a DC-DC buck-boost converter system to charge the 12 V auxiliary battery. The converter can also charge the 36 V battery pack from the 12 V system if necessary.
- The energy storage module (ESM) uses three 12 V advanced lead-acid batteries connected in series to make up a 36 V system that operates at 42 V when the engine is running.
- An electrohydraulic power steering module (EHPSM) is a control module that controls an electrically driven hydraulic motor to provide hydraulic pressure for steering assist. The EHPSM controls power steering assist and power brake assist.
- To conserve fuel, the GM PHT uses engine idle-stop and auto-start, aggressive deceleration fuel cutoff, early torque converter clutch lockup, regenerative braking, and electrohydraulic power steering.
- An auxiliary power outlet (APO) system provides 120 V AC standby power to outlets located in the vehicle cargo bed and the cabin under the passenger seat area.
- A vehicle cabin heating system uses a 12 V auxiliary coolant pump to circulate coolant to the heater core for cabin heating during engine idle-stop mode.
- If the vehicle is in engine idle-stop mode and air conditioning or defrost is requested, the engine will be commanded to crank and run to operate the compressor.

Refer to the *Hybrid Electric Vehicle Technology* CD-ROM for the Review Questions in Microsoft® Word format.

REVIEW QUESTIONS

1. What is the purpose of the ESCM?
2. Describe the EHPS operating modes.
3. How does an HVAC system operate when a vehicle is in engine idle-stop mode?
4. Explain the operation of the PHT engine idle-stop mode.
5. Explain the operation of the APO system.

18 GM FlexPower PHT

HYBRID Electric Vehicle TECHNOLOGY

Trade Test

Name ______________________ Date ______________

_______________ **1.** The GM PHT electric machine is located ___.
A. at the left front of the engine
B. at the right front of the engine
C. between the engine and transmission
D. on the crankshaft pulley

_______________ **2.** An auxiliary transmission oil pump is located ___.
A. to the left of the radiator
B. to the right of the radiator
C. to the left of the electric power steering pump
D. inside the transmission

_______________ **3.** The energy storage control module (ESCM) is located ___.
A. beside the PCM
B. inside the energy storage module
C. beside the SGCM
D. behind the instrument panel

_______________ **4.** The GM PHT contains ___ 120 V AC outlets.
A. two
B. four
C. six
D. eight

_______________ **5.** The auxiliary power outlet (APO) system can operate in ___ different modes.
A. two
B. three
C. four
D. five

_______________ **6.** In engine idle-stop mode, fuel cutoff (FCO) is enabled at speeds higher than ___ mph.
A. 3
B. 13
C. 55
D. 70

_______________ **7.** The GM PHT system contains ___ 12 V batteries in series that provide power for the hybrid components.
A. two
B. three
C. six
D. eight

_______________ **8.** The SGCM generates or controls the 14 V DC, 42 V DC, ___, and 120 V AC power system.
A. 16 V DC
B. 24 V AC
C. 28 V AC
D. 48 V DC

_______________ **9.** The SGCM provides control for engine cranking, torque, and speed control; ___; and the generation of electric power.
A. fuel control input
B. torque smoothing and active damping
C. spark advance input control
D. ABS system control calibrations

_______________ **10.** The ___ controls the jump starting function that allows the vehicle 12 V auxiliary battery to supply energy to the 42 V system batteries.
A. hybrid control module (HCM)
B. powertrain control module (PCM)
C. energy storage control module (ESCM)
D. starter-generator control module (SGCM)

_______________ **11.** If the GM PHT is in a normal idle-stop mode to save fuel and the air conditioning is commanded ON, the engine will ___.
A. restart and continue running
B. restart and run for 45 seconds
C. restart and run for 2 minutes
D. remain off

_______________ **12.** When a vehicle is in engine idle-stop mode, the gear selection is in park, and a door is opened, the hybrid controller module commands the vehicle to ___.
A. start the engine
B. disable the engine
C. start the engine after 30 seconds
D. permit any gear selection

_______________ **13.** If the hood is opened when a vehicle is in engine idle-stop mode and the gear selector is in park, the hybrid controller will command the vehicle to ___.
A. start the engine
B. disable the engine
C. start the engine after 30 seconds
D. start only with Scan Tool

HYBRID
Electric Vehicle
TECHNOLOGY

Belted Alternator Starter (BAS) System

19

A belted alternator starter (BAS) system has an electric machine connected to the engine and driveline by a ribbed drive belt. The electric machine torque adds engine torque or provides regenerative energy throughout the entire vehicle drive cycle. Although the BAS hybrid system does not provide an electric-only propulsion mode, it still has many of the same operating modes as a strong hybrid system. The costs of these features are significantly lower compared to a strong hybrid system.

CHAPTER OBJECTIVES

- Identify BAS components and locations.
- Describe the operation of a drive belt system.
- Explain how a BAS transmission operates during engine idle-stop mode.
- Describe the function of a power electronics box/starter-generator control module (PEB/SGCM).
- Describe the function of a motor-generator unit (MGU).
- Describe the function of an energy storage system (ESS).
- Define the BAS control system operating modes.
- Describe the operation of a BAS hybrid brake delay valve (BDV) and a hill start valve (HSV).

KEY TERMS

belted alternator starter (BAS): A hybrid propulsion system with an electric machine that is connected to an engine and driveline by a ribbed drive belt. The electric machine torque adds engine torque or provides regenerative energy throughout the entire vehicle drive cycle.

power electronics box/starter-generator control module (PEB/SGCM): A power electronics component that controls electrical energy to a motor-generator unit (MGU), controls MGU operation, and converts AC to DC or DC to AC.

motor-generator unit (MGU): A BAS system component, similar to a traditional automotive generator (alternator), that operates as an electric motor or generator during vehicle drive cycles and is coupled to an engine through a drive belt system.

enhanced Lundell electric machine: An electric machine that uses both a traditional generator/alternator synchronous design and permanent magnets on the rotor, resulting in increased efficiency and output. The key elements to the enhanced Lundell rotor are the magnets that are mounted and laser-welded between rotor claws.

energy storage system (ESS): A system made up of battery modules, current sensors, temperature sensors, contactors, cables, connectors, and a thermal system, which stores and provides electrical energy to an electric propulsion system.

battery disconnect unit (BDU): An electronic control unit in a BAS hybrid system that controls energy storage system (ESS) input and output circuits for the contactor, current sensor, microswitches, battery-module voltage sensors, temperature sensors, fuses, and vehicle diagnostics.

brake delay valve (BDV): A brake valve used to minimize vehicle roll by delaying the release of the friction brake system until a vehicle engine has been cranked and started after a hybrid idle-stop.

GM BAS COMPONENTS

In late 2006, General Motors (GM) Corporation began marketing the first ever production belted alternator starter (BAS) system in the Saturn VUE. A *belted alternator starter (BAS)* is a hybrid propulsion system with an electric machine that is connected to an engine and driveline by a ribbed drive belt.

This system was subsequently released in other GM vehicles. The BAS system is categorized as a mild hybrid system (36/42 V). Its construction and system operation resemble a strong hybrid system, but it has a lower systems cost. The electric machine torque adds engine torque or provides regenerative energy throughout the entire vehicle drive cycle.

There are few visible differences between a BAS hybrid and nonhybrid vehicle model. The vehicle electronics control module (ECM) and the bus electrical center (fuse box) are mounted under a beauty cover underhood. **See Figure 19-1.**

One visible difference between a BAS hybrid vehicle and nonhybrid vehicle is the caution labels that are located on the radiator support. The labels indicate that the vehicle has two battery locations. The 12 V auxiliary battery is located in the traditional underhood location. The 36 V battery is located behind the rear seat under the load floor. The location of these battery systems is important to first responders and technicians to allow the electrical system to be disabled in the event of a collision or repair.

BAS BEAUTY COVER

Figure 19-1. The vehicle electronics control module (ECM) and the bus electrical center (fuse box) are mounted under a beauty cover underhood.

The BAS system is comprised of a 2.4 L, four-cylinder ECOTEC engine, motor-generator unit (MGU), starter-generator control module (SGCM), and energy storage system (ESS). Other components include the automatic auxiliary transmission pump module, hybrid battery pack, hybrid and engine controllers with control software, MGU to SGCM 3ϕ power cables, and dual belt tensioner assembly.

Instrumentation System

The instrumentation system for the BAS has been designed to provide the vehicle operator instantaneous feedback as to the precise operation of the system. The vehicle operator is also provided with an "eco" indicator. This indicator illuminates during the most efficient vehicle operating modes and provides a visual feedback indication to the operator that the vehicle is in this mode. **See Figure 19-2.**

INSTRUMENTATION

Figure 19-2. An instrumentation system for the BAS has been designed to provide the vehicle operator instantaneous feedback as to the precise operation of the system.

Internal Combustion Engine

The BAS hybrid internal combustion engine system consists of a 2.4 L four-cylinder double overhead camshaft design with intake and exhaust cam phasers. The cam phasers control the intake and exhaust valves by varying the camshaft timing, which allows good control of torque and power. Control of the intake and exhaust valves optimizes efficiency of the engine regardless of engine speed or load.

The basic engine specifications include a displacement of 2.396 L, compression ratio of 10.4:1, maximum power of 127 kW (170 HP) at 6600 rpm, maximum torque of 220 Nm at 4400 rpm, regular fuel, and a traditional 12 V cranking motor. The engine system also contains a unique drive belt system designed specifically for the hybrid system.

Drive Belt System

The BAS belt drive system permits positive torque electric machine belt-tension control when operating in electric propulsion assist mode. It allows generator negative torque belt-tension control when the MGU is generating electrical power. The belt is made of high-strength aramid cord with seven ribs.

The BAS combines a hydraulic strut tensioner and a friction-damped rotary tensioner on a common pivoting arm. **See Figure 19-3.** The drive belt control system uses the bidirectional system to control the positive and negative dynamic torque loads with two tensioning element systems. This system ensures sufficient tension under all operating modes while reducing overall belt tension and bearing loads.

DRIVE BELT SYSTEM

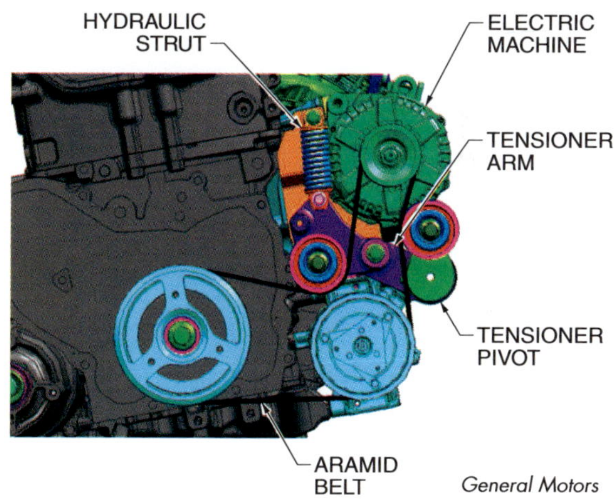

Figure 19-3. The BAS combines a hydraulic strut tensioner and a friction-damped rotary tensioner on a common pivoting arm.

During conventional engine operation, engine torque reduces belt tension in the span between the engine crankshaft pulley and the electric machine (slack side), causing the tensioner to extend to maintain belt tension. During engine restart, electric machine torque tightens the span between the engine crankshaft pulley and the electric machine.

The slack side is then moved to the span between the electric machine and the air conditioning compressor. This causes the rotary tensioner to move toward the belt to control belt tension. Damping the hydraulic strut restricts movement of the pulley under the hydraulic strut during engine restart.

A special hydraulic strut compression tool, EN-48079, is required to remove the drive belt or drive belt tensioner components. Once the strut is compressed, the belt can easily be removed for belt replacement or for completing other repairs. **See Figure 19-4.**

STRUT COMPRESSION TOOL

Figure 19-4. A special hydraulic strut compression tool, EN-48079, is required to remove the drive belt or drive belt tensioner components.

4T45-E Transmission

The 4T45-E transmission used in the BAS hybrid system has unique operating modes. A BAS operating mode that affects transmission operation is the engine idle-stop mode. In this mode, the engine can stop when the vehicle is moving slowly.

In a traditional vehicle system, the engine would not stop as part of the normal operating mode conditions. If a traditional engine were to stop, the engine-driven transmission hydraulic pump would also stop, leading to the loss of hydraulic pressure. Therefore, a BAS needs an auxiliary transmission pump system to assist the engine-driven hydraulic pump in maintaining transmission pressure during normal operating modes. **See Figure 19-5.**

TRANSMISSION

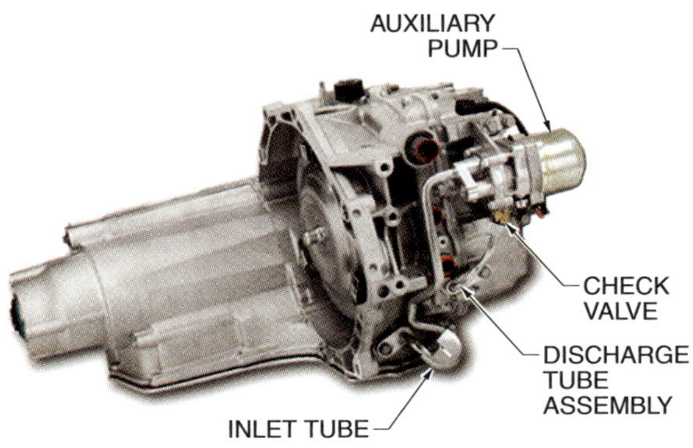

Figure 19-5. A BAS needs an auxiliary transmission pump system to assist the engine-driven hydraulic pump in maintaining transmission pressure during the normal operating modes.

There are other modifications to this transmission that allow the vehicle to operate as a hybrid vehicle. For example, the valve body channel plate, spacer plate, and valve body have been modified. There is an additional solenoid for controlling downshifts during regenerative braking and a unique torque converter and final drive assembly. The transmission gear ratios are 2.960 in first gear, 1.626 in second gear, 1.000 in third gear, 0.681 in fourth gear, and 3.63 in final drive. The torque converter diameter is 245 mm.

The transmission has special service procedures that must be used by a technician when performing service to the auxiliary transmission pump. The auxiliary transmission pump must be primed after repair or replacement. The GM serial data tool (Tech2) has a special output control function that is used to prime the transmission pump instead of requiring a special tool and procedure.

The auxiliary transmission pump specifications call for it to operate under pulse width modulated (PWM) control. It should have a normal operating pressure of 30 psi to 45 psi. The minimum operating pressure is 20 psi. Falling below this pressure could generate a diagnostic trouble code (DTC).

The operating temperature is the normal operating temperature of the transmission during a drive cycle. A varying voltage is applied to the auxiliary transmission pump, depending on the operating temperature. An applied voltage of 12 V means that the pump delivers 140 psi at 65°F (18°C). When the transmission is cold, 6 V is applied to the pump so that it delivers 50% duty for the slower speed and pressure. When the transmission is hot, 9.5 V is applied to the pump to provide more component lubrication and cooling.

Power Electronics Box/Starter-Generator Control Module

A *power electronics box/starter-generator control module (PEB/SGCM)* is a power electronics component that controls electrical energy to a motor-generator unit (MGU), controls MGU operation, and converts AC to DC or DC to AC. The PEB/SGCM is designed to function as both a power inverter and auxiliary power module for the BAS hybrid system.

The power inverter provides 36 V power to the MGU to crank the engine during hybrid operation. It also controls the 42 V electrical power that is generated by the MGU. As part of this control, the power inverter rectifies the 42 V of AC generated by the MGU to DC to charge the energy storage system battery modules and indirectly (through the DC-DC converter) charges the 12 V auxiliary battery. When the vehicle is not operating, with the engine and MGU stopped, the energy storage system is approximately 36 V. When the vehicle and MGU are operating, the system voltage is approximately 42 V.

The PEB/SGCM must control electrical power for all modes of the BAS hybrid operation. This is a single unit that is comprised of numerous internal power electronics and subsystem elements. The PEB/SGCM includes the same elements contained in all power inverter and DC-DC converter systems. The power module section controls the electrical power for the MGU motor, MGU generator operation, and electrical power for vehicle systems. It uses 400 A MOSFET six-pack power semiconductors and a 1500 W DC-DC converter system.

The BAS drive belt system provides positive torque tension control for propulsion and negative torque tension control for regenerative braking.

The PEB/SGCM is approximately 12″ long and 7″ wide. The relatively small power electronics package permits the PEB/SGCM to be mounted in almost any underhood area. The PEB/SGCM is located on the front left side of the engine compartment between the radiator support and auxiliary battery.

Contained within the PEB/SGCM is the auxiliary power module (APM). The APM replaces the conventional alternator (generator) and is used to convert the 36 V of DC to 12 V of DC. It is responsible for supplying electrical power to vehicle systems and the auxiliary battery. Functionally, the APM is a DC-DC converter.

The PEB/SGCM controls the MGU motoring and generating functions. It also controls the transmission auxiliary oil pump, hill-hold solenoids, auxiliary coolant pump, and PEB/SGCM coolant pump. The pumps and solenoids are 12 V PWM-driven components.

All 36 V and 12 V electrical connecting points, except for the auxiliary transmission pump controller, are located on the side of the PEB/SGCM. **See Figure 19-6.** This section of the unit contains 3ϕ cable connections for the MGU, a 12 V positive output terminal, a 175 A, 12 V circuit fuse, a 36 V positive output terminal, and a ground stud for both the 36 V and 12 V systems.

PEB/SGCM

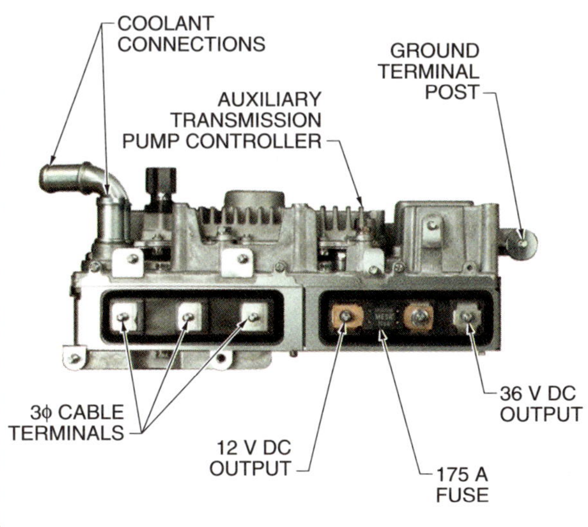

Figure 19-6. All 36 V and 12 V electrical connecting points, except for the auxiliary transmission pump controller, are located on the side of the PEB/SGCM.

The 12 V and 36 V systems use the vehicle body and undercarriage as an electrical return path to the PEB/SGCM. The PEB/SGCM input/output controller connections are located on the end of the unit. The connectors are easily accessible in the underhood area for voltmeter, ohmmeter, ammeter, and oscilloscope measurements.

The auxiliary transmission pump controller is mounted directly on the PEB/SGCM unit. **See Figure 19-7.** This controller is used to control the speed of the auxiliary transmission pump. The pump maintains hydraulic pressure in the transmission when the vehicle and engine are stopped during auto-stop mode.

AUXILIARY TRANSMISSION PUMP CONTROLLER

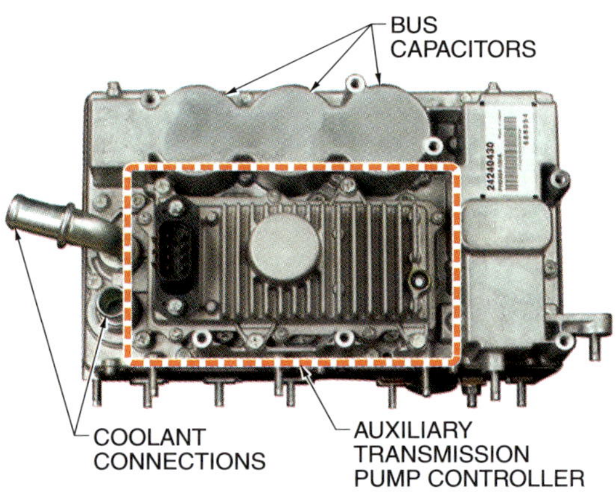

Figure 19-7. The auxiliary transmission pump controller is mounted directly on the PEB/SGCM unit.

Hydraulic pressure must be maintained to ensure that the transmission clutches can lock. The pressure also permits the engine and MGU to apply torque to the driveline when the vehicle transitions from auto-stop mode to auto-start mode. The pump can also be serviced separately from the PEB/SGCM. It should be removed from a failed PEB/SGCM unit and fastened to a new PEB/SGCM unit.

A liquid-cooled, strong hybrid power electronic system uses a separate cooling system with a pump and heat exchanger. The PEB/SGCM for the BAS uses liquid cooling from the engine cooling system loop to cool the power electronics. This simplifies the cooling system and reduces the overall cost of the BAS hybrid system.

The PEB/SGCM has two cooling fittings located on one end of the unit. The fitting with the angled, long neck is the coolant outlet hose, and the short, straight connector is the coolant inlet. When the PEB/SGCM unit is mounted in the vehicle, the coolant hoses can be easily reached for service.

Motor-Generator Unit Assembly

The BAS hybrid system is the first production hybrid system powered by a drive belt rather than an electric machine housed within a transaxle or transmission. The *motor-generator unit (MGU)* is a BAS system component, similar to a traditional automotive generator (alternator), that operates as an electric motor or generator during vehicle drive cycles and is coupled to an engine through a drive belt system. The MGU is located in the front lower right of the underhood compartment between the engine and lower radiator support.

The MGU electric machine couples negative and positive torque through the drive belt as both a 36 V cranking motor and a generator. **See Figure 19-8.** It provides engine electric power assist, engine restarting after engine idle-stop, and torque smoothing during engine transition from cranking to propulsion. The MGU also supplies energy to charge both the 36 V battery system and the 12 V auxiliary battery.

MGU ASSEMBLY

General Motors

Figure 19-8. The MGU electric machine couples its negative and positive torque through the drive belt as both a 36 V cranking motor and as a generator.

MGU Power and Torque. The BAS MGU is designed to generate about three times more power than a traditional generator. However, its physical size is not much larger than a traditional generator because the MGU does not contain voltage regulation circuitry, rectifier diodes, or any special sensing circuits. **See Figure 19-9.**

MGU SIZE

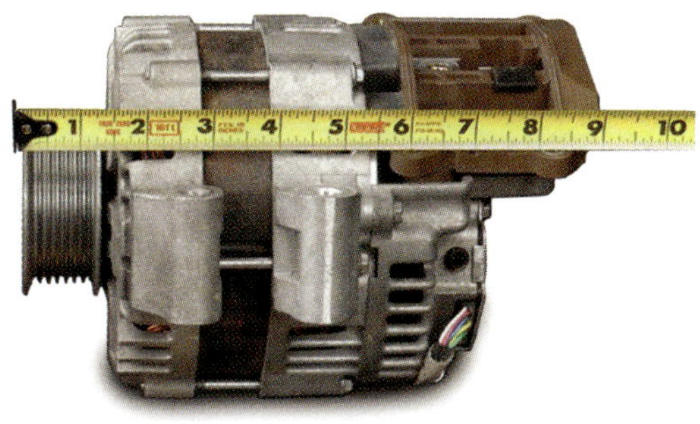

Figure 19-9. The MGU delivers more torque for its size than a traditional automotive generator.

The voltage regulation circuitry is designed into the PEB/SGCM system. The MGU is connected to the PEB/SGCM by three separate power cables. The 3ϕ power cables on the vehicle are covered with blue insulation to indicate that they are intermediate voltage circuits (14 V to 60 V).

The MGU is designed with significantly higher levels of operating torque than a traditional automotive generator. This is necessary for the BAS hybrid system because the MGU assists in cranking the engine and adds torque during acceleration. MGU maximum torque levels are designed at very low MGU (and engine) rpm levels. This design is intentional because low-speed operation occurs where maximum torque is required for engine cranking and electric propulsion assist to the engine.

Low-speed operation occurs many times during the typical drive cycle. Therefore, an MGU must operate at high efficiencies to reduce fuel consumption. The MGU requires an electrical generating power greater than 5 kW and an overall mass less than 12 kg (26 lbs), including the 3ϕ cable assembly. It includes 16 total poles, peak motoring torque greater than 60 Nm (44 lb-ft), peak mechanical motor power greater than 4 kW, and a permanent magnet enhanced Lundell electric machine.

Enhanced Lundell Electric Machines. The modified rotor is the most unique component of MGU design. The BAS MGU rotor has a typical Lundell synchronous AC induction machine design and a wound field rotor with claw poles. It uses 16 neodymium-iron-boron (Ne-Fe-B) permanent bar magnets to increase its overall efficiency at low speeds. This type of MGU is categorized as an enhanced Lundell design because of the magnets. **See Figure 19-10.**

ENHANCED LUNDELL ELECTRIC MACHINES

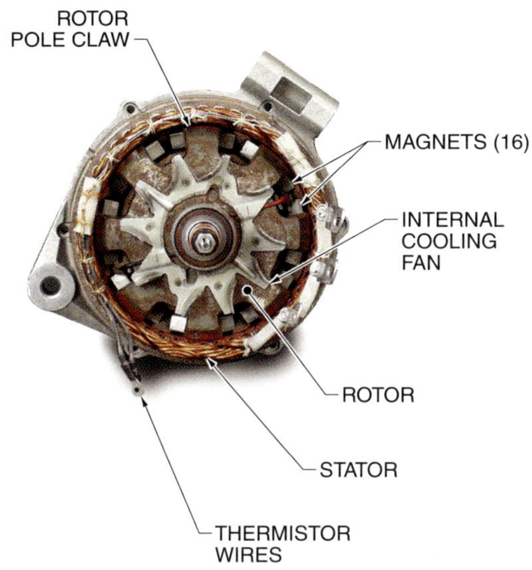

Figure 19-10. The BAS MGU rotor has a typical Lundell design with a wound field rotor and claw poles.

An *enhanced Lundell electric machine* is an electric machine that uses both a traditional generator/alternator synchronous design and permanent magnets on the rotor, resulting in increased efficiency and output. In an enhanced Lundell rotor, the Ne-Fe-B magnets are mounted and laser-welded between the rotor claws. Placing the magnets between the rotor claws increases magnetic strength by reducing air gaps between the claws.

A traditional Lundell electric machine has air gaps between the claws, resulting in a weaker magnetic flux used to create electric current. A BAS MGU can generate maximum electrical power at nearly 85% efficiency. Traditional automotive alternator/generator efficiency is about 50% to 60% efficient at maximum electrical power levels.

MGU Sensing System. The MGU electrical connections for control and sensing systems are very simple in comparison to those of strong hybrid systems found in the Toyota Prius, Lexus RX400h, or the GM Yukon/Tahoe. **See Figure 19-11.** The MGU uses a 3ϕ wye stator configuration. Three current sensors contained within the PEB/SGCM sense the electric current from each of these phases.

MGU EXTERNAL CONNECTIONS

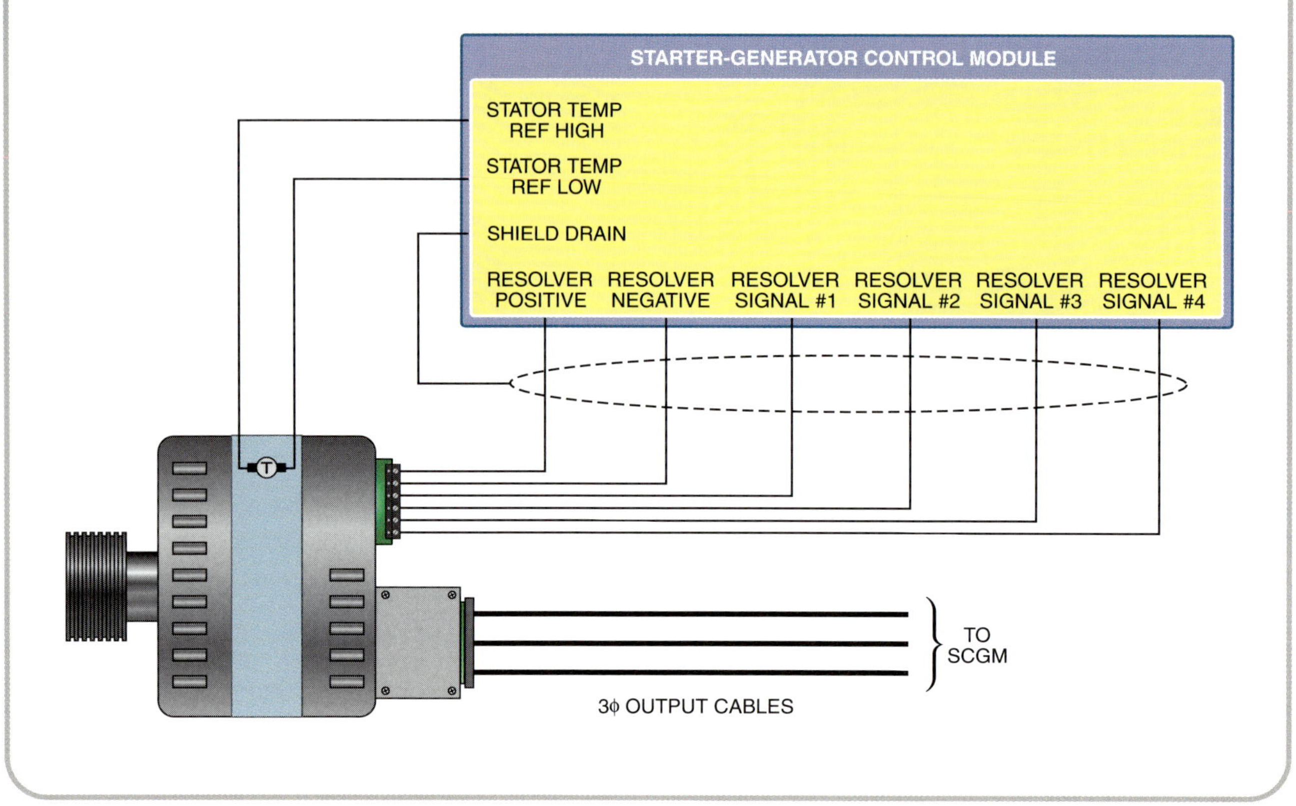

Figure 19-11. MGU electrical connections for control and sensing systems are very simple.

Three shield/drain wires are used to prevent the strong magnetic fields from the MGU output cables from interfering with adjunct electrical circuits. To mitigate corrosion, the connection block at the PEB/SGCM includes a condensation drain as an exit point for any moisture that collects in the connection block. The cable connection point at the MGU also contains a condensation drain.

The MGU uses a traditional slip ring and brush design. The two brushes transfer electric current from the PEB/SGCM to one slip ring connected to one end of the rotor field coil. From there, the current flows through the coil to the other slip ring. The current then returns back to the PEB/SGCM to form a series circuit.

In the MGU circuit, electric current is generated from the PEB/SGCM by a PWM signal through one slip ring. A ground circuit is then provided by the other slip ring through a wire back to the PEB/SGCM and through its framework to the vehicle chassis.

The electric machine uses high current and generates high torque at low speeds in order to launch the vehicle with electric assist. Traditional vehicles do not use electric launch assist, so the application of the MGU and how its power is controlled in a BAS hybrid vehicle is different. Typical alternators need higher speeds to deliver elevated current output. The MGU delivers its highest current at slower generator speeds. **See Figure 19-12.**

GENERATOR SPEED AND CURRENT

Generator Speed	Current
400	3.4
600	3.3
800	3.0
1000	2.8
2000	1.7
3000	1.3
7000	0.9

Figure 19-12. The MGU delivers its highest current at slower generator speeds.

A resolver, located on the rear of the rotor shaft, is used for rotor rpm and position sensing. Resolver signals are transmitted to the PEB/SGCM through a connector that is shared with the stator temperature sensor. As with strong hybrid sensing systems, the output of the resolver is two sine waves that are compared to each other and to a reference sine wave. The hybrid controller calculates the exact rotor position based on the position and amplitude of the resolver output signals.

Energy Storage System

An *energy storage system (ESS)* is a system made up of battery modules, current sensors, temperature sensors, contactors, cables, connectors, and a thermal system, which stores and provides electrical energy to an electric propulsion system. The stored energy comes from captured regenerative braking energy and energy generated by the MGU.

The ESS provides energy to the MGU for propulsion related modes. It is located in the rear of the vehicle under the load floor on Saturn VUE models and can be viewed when the load floor is removed. In passenger car BAS hybrid models, the ESS is located in the trunk area just behind the rear seat upright support.

The ESS is designed with a 36 V NiMH battery pack that weighs less than 55 lb (25 kg). It can deliver 10 kW at 60% SOC and 95°F (35°C) for 10 seconds. It has physical dimensions of 39″ L × 5″ W × 5″ H. The external ESS connections are very similar to that of strong hybrid ESS units. The external connections are encompassed in the battery carrier assembly, battery disconnect unit, cooling fan, and 36 V harness assembly. **See Figure 19-13.**

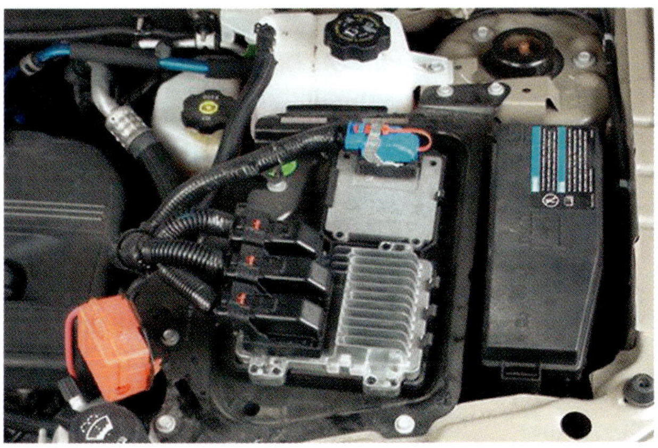

The vehicle ECM and the bus electrical center (fuse box) are the only components mounted under the beauty cover.

ESS COMPONENTS

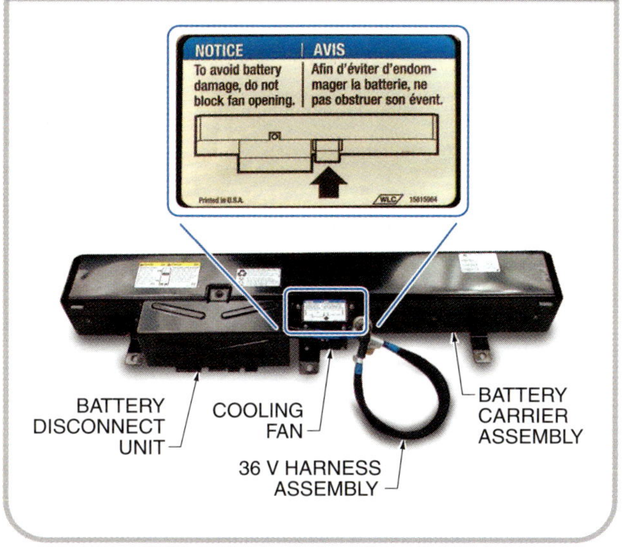

Figure 19-13. The external connections of an ESS include the battery carrier assembly, battery disconnect unit, cooling fan, and 36 V harness assembly.

The ESS has two power cables that connect it to the PEB/SGCM, which is connected to the MGU. The positive cable connects the ESS to the PEB/SGCM and uses a fingerproof terminal design. The cable from the negative terminal connects the ESS to the vehicle undercarriage. The cable ground connection is located on the right side of the vehicle between the ESS carrier and the rear seat.

The BAS hybrid system uses more than one battery for the various vehicle operations. The battery locations are listed on labels located on top of the battery carrier and on the radiator support underhood. These labels are located in areas that can be easily identified by first responders so that they can locate the battery to shut off electrical power after a collision or other incident. The labels clearly state that the vehicle electrical system uses a 36 V BAS system battery and a 12 V auxiliary battery. **See Figure 19-14.**

A 36 V system uses the vehicle body or chassis as ground reference. The BAS hybrid system is not a floating electrical system like a strong hybrid system. The vehicle body or chassis can be used for the ground reference for the 36 V system because the operating level of 36 V is less than the threshold of 60 V. A voltage of 60 V is the point at which systems are considered to be high voltage.

BATTERY LOCATION LABEL

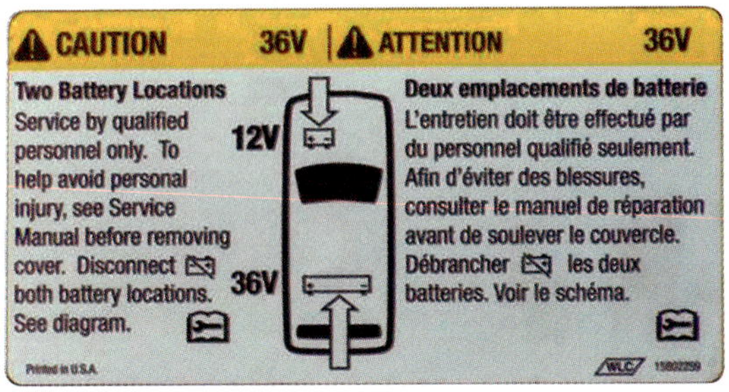

Figure 19-14. The vehicle battery locations are listed on labels located on top of the battery carrier and on the radiator support underhood.

A system voltage less than 60 V has no system safety design requirements. It has fewer safety implications for first responders, service technicians, or collision technicians if there is an ESS voltage leak to the chassis. Although the BAS system does not operate over the 60 V limit, GM recommends the use of personal protective equipment, such as gloves for high voltage, when performing repairs on the system.

ESS Construction. Internal elements of the ESS carrier include three 12 V NiMH battery modules, battery module interconnect cables, and the associated battery module temperature sensors. Each of the three battery modules is made up of two 6 V module cassettes that are electrically connected to achieve a total of 12 V nominal. The three battery modules are connected in series for a total of 36 V nominal.

Since the BAS system does not provide perceptible electric-only propulsion assist, the battery module specific power and electrical power specification rating are much less compared to a strong hybrid ESS. Each 12 V battery module can deliver 8.5 Ah, 110 Wh, 2.5 kW of power, and 1110 W/kg specific power. They are designed with passive cooling, a mass of 2.4 kg (5.3 lbs), and physical dimensions of 12.8″ L × 4.5″ W × 3.4″ H each.

The 12 V battery also provides energy for the first cranking event of a drive cycle or when there is a failure of the 36 V battery system. The 36 V battery pack provides energy for the hybrid propulsion modes, including engine auto-start.

Each 12 V battery module contains two serviceable thermistors (one for each 6 V cassette) to sense module temperature. **See Figure 19-15.** The ESS controller uses the thermistor values as one of several inputs to control regenerative braking and propulsion system energy, the battery module cooling system, and temperature-compensated battery module SOC. Each module is then analyzed. This method of measuring temperature is similar to the operation of an automotive coolant or air temperature sensor circuit.

SENSOR CONNECTIONS

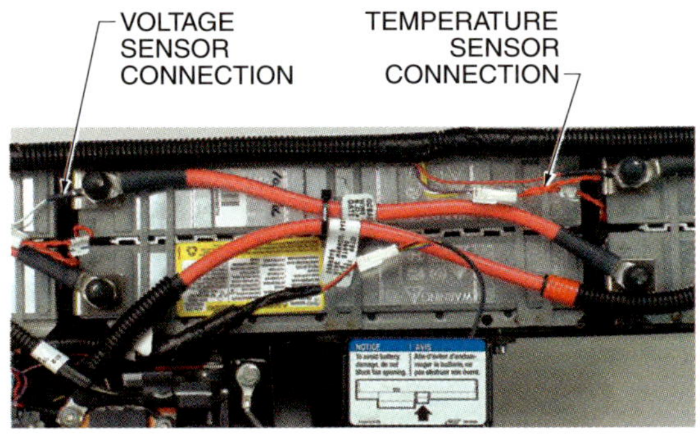

Figure 19-15. Each 12 V battery module contains two serviceable thermistors (one for each 6 V cassette) to sense module temperature.

A current sensor is used to measure electric current that is transferred to and used by the MGU. The ESS controller also uses the terminal voltage values as one of several inputs to determine ESS SOC and to analyze each module. Each module has a wire that is connected to one end of a battery module terminal. The other end is connected to a resistor so that the ESS controller can measure the battery-module terminal voltage.

ESS Control System. The ESS control system is located in the battery disconnect unit (BDU), which is fastened to the battery carrier assembly. The *battery disconnect unit (BDU)* is an electronic control unit in a BAS hybrid system that controls energy storage system (ESS) input and output circuits for the contactor, current sensor, microswitches, battery-module voltage sensors, temperature sensors, fuses, and vehicle diagnostics. **See Figure 19-16.**

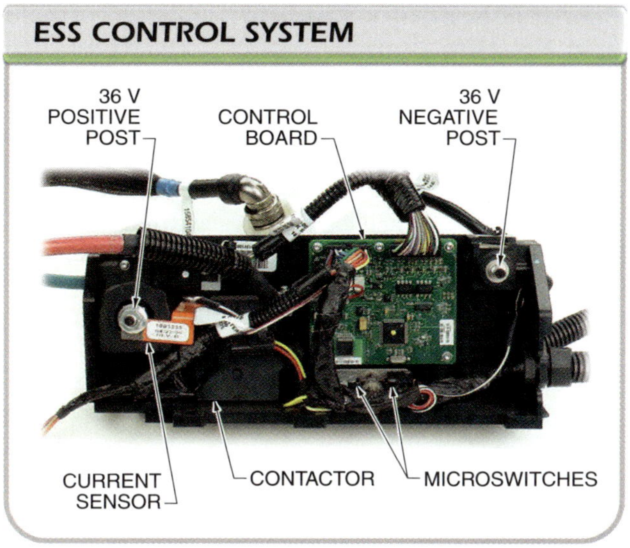

Figure 19-16. The BDU controls the ESS input and output circuits for the contactor, current sensor, microswitches, battery-module voltage sensors, temperature sensors, fuses, and vehicle diagnostics.

The primary function of the BDU is to control the battery-pack SOC and contactor system and to sense electric current. The BDU contains microswitches on the access door to ensure the safety of anyone working on the system. It also contains a 36 V contactor, fuse, current sensor, and the BDU controller.

The 36 V contactor is an intermediate-voltage, continuous duty relay controlled by the ESS controller. It is used to electrically connect or disconnect the ESS from the PEB/SGCM. The 200 A, 58 V fuse is located on the negative cable stud.

The current sensor is a Hall-effect sensor used to measure the electric current that enters and exits the battery pack. The current sensor detects this current on the positive ESS cable and transmits the amplitude of the electric current signal in the form of a voltage signal to the BDU controller. The ESS controller uses the current sensor values as one of several inputs to calculate battery-module SOC and to assist in analyzing the entire battery-module series string.

The BDU controller contains all input/output connections to the temperature and current sensors, battery terminal voltages, BDU access door microswitches, and contactor position. The battery-pack controller provides power through the power terminal to the main contactor.

When the vehicle is powered ON, the main contactor is commanded closed by the battery-pack controller. The ground is provided by the battery controller through the disconnect switch control input. The battery-pack controller senses the BDU access door status through the link voltage and low side PWM signal. **See Figure 19-17.**

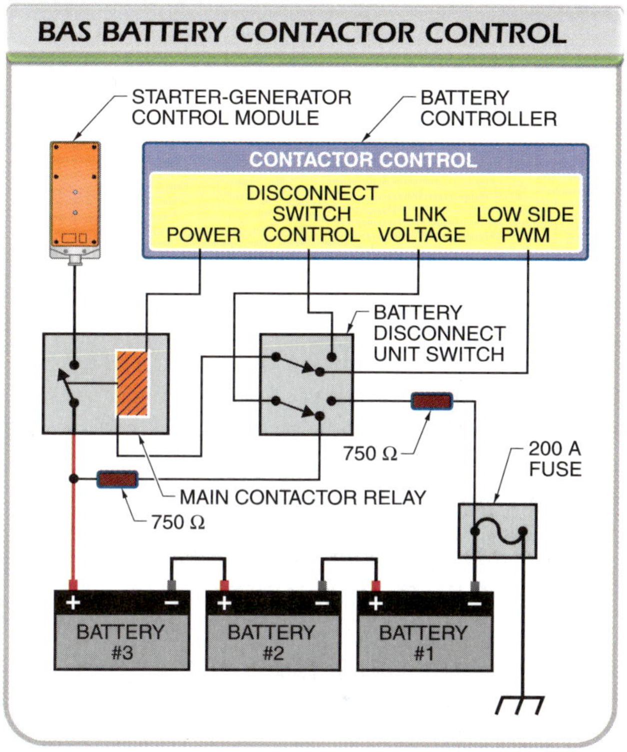

Figure 19-17. The battery-pack controller senses BDU access door status through the link voltage and low side PWM signal.

The BDU uses a battery disconnect switch (BDS). The BDU access door is monitored by two microswitches. The function of the BDS is to sense the status of the BDU access door microswitches. Voltage sensed by the BDU changes when the BDU access door is opened and the microswitches change position.

When the switches move, power is applied to one side of the main contactor coil. The other side of the main contactor is connected to the low side PWM. The link voltage terminal of the BDU is connected to the positive side of the 36 V battery module string. This places 36 V on the link voltage terminal. The change in the voltage status of these terminals causes the main contactor to be commanded to the open position. The PEB/SGCM controller permits the bus capacitors to quickly discharge.

If the BDU spring-loaded plunger is depressed, the access door can be opened while the system is ON. The control system senses this breach and commands the main contactor to the open position. This disconnects the PEB/SGCM from the ESS energy supply and causes a loss of propulsion.

ESS Thermal System. The function of the BAS ESS thermal system is to maintain battery module temperature at a level that permits maximum energy transfer to the PEB/SGCM and the MGU. The ESS thermal system is a passive cooling system. It forces cooler cabin air into the ESS carrier. Therefore, this ESS does not use liquid coolant to cool battery modules.

The cooling fan is fastened to the carrier and located adjacent to the BDU. **See Figure 19-18.** It is a low-power, 12 V, permanent magnet DC unit that is designed to be very quiet when moving air in and out of the ESS carrier. The inlet point for cooling fan air is at the slotted area located on the bottom of the ESS carrier. It is a simple air duct snorkel system that is hidden in the load floor area to reduce any fan noise while routing the exhaust air to the wheel well area where it will ultimately exit the vehicle.

ESS THERMAL SYSTEM

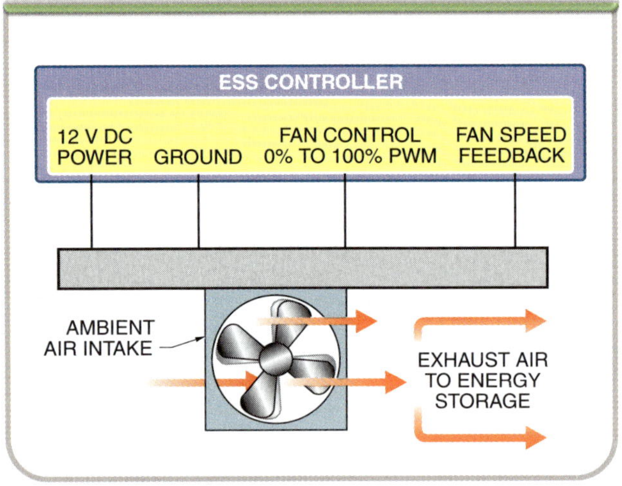

Figure 19-18. The ESS controller provides the PWM control signal for the cooling fan based on battery-module temperature.

The cooling fan electrical control circuit is a very simple design consisting of 12 V power, ground, PWM fan control, and a fan speed feedback circuit. The ESS controller provides a PWM control signal, and the rotating speed of the controller is based on battery module temperature.

Power Electronics Cooling and Heating Pumps

The BAS system uses two auxiliary cooling pumps as part of the hybrid operating system. The heater coolant pump is located in the engine compartment between the engine and radiator support. It is electrically controlled by the PEB/SGCM and located between the engine and bulkhead at the lower left rear of the engine compartment. **See Figure 19-19.**

COOLING AND HEATING PUMPS

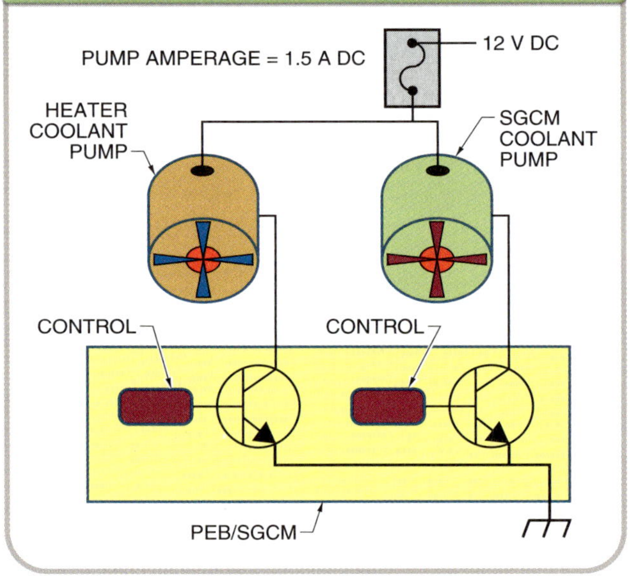

Figure 19-19. The BAS system uses two auxiliary cooling pumps as part of the hybrid operating system.

The heater coolant pump circulates coolant through the heater core when the vehicle is in hybrid auto-stop mode. This particular operating mode assists in maintaining cabin heat during cold weather conditions.

The PEB/SGCM coolant pump is also electrically controlled by the PEB/SGCM and is located on the front lower right of the engine compartment near the air conditioning compressor. Routing of the coolant system connects to the vehicle cooling system and is in parallel with the PEB/SGCM.

The primary function of the PEB/SGCM coolant pump is to assist the engine-driven coolant pump to ensure that the PEB/SGCM receives adequate cooling when the vehicle is in hybrid auto-stop mode. However, the pump will also operate when the engine is operating during normal driving cycles. Therefore, the pump operates any time the ignition system is ON.

GM BAS OPERATION

The simple design and lower cost of the BAS are primarily due to the use of an MGU that is coupled to the engine belt-drive system and not contained within a transaxle/transmission or housed between a transaxle/transmission and the engine. The MGU occupies the same location as a traditional generator (alternator). The battery pack is located in the rear of the vehicle under the load floor. The remaining system components are housed in the engine compartment. **See Figure 19-20.**

BIAS UNDERHOOD COMPONENTS

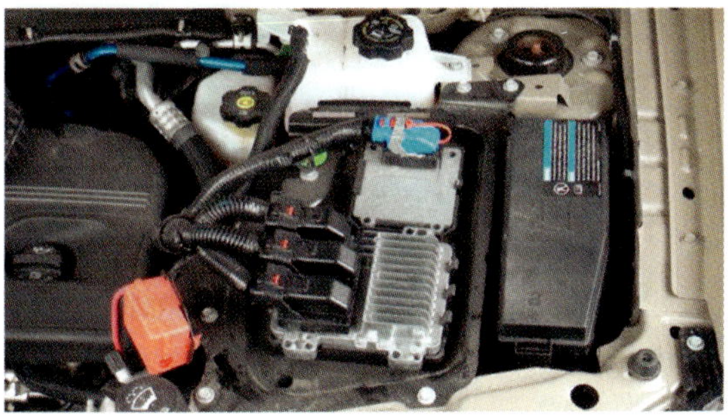

Figure 19-20. Many of the BAS components are located underhood.

One of the prime considerations for GM when designing the BAS system was to reduce system cost while improving fuel consumption by using engine idle-stop and auto-start functions. Operating modes that use the electric machine to increase overall system efficiency were specifically targeted. The BAS system is approximately 20% the cost of a strong hybrid system and can deliver most of the same benefits. The engine idle-stop, auto-start, and special operating modes are designed to increase fuel economy and reduce emissions while providing an equal or enhanced driving experience.

The BAS control system provides several modes to control all functions of the hybrid system. The control system is designed to operate with engine idle-stop, auto-start, 36 V and 12 V battery system charging, early fuel cutoff, regenerative braking, and electric power propulsion assist.

Jump starting for a BAS vehicle is confined to the 12 V auxiliary battery. Traditional jump starting procedures are used to jump start the 12 V battery. If there are any operational problems with the 36 V battery system, the vehicle must be serviced by a qualified technician.

Fuel Economy

The BAS system provides significant fuel economy improvements for both city and highway driving modes. Improvements up to approximately 6 mpg or 24% in the city, and 6 mpg or 19% on the highway can be expected. The BAS hybrid system delivers significant acceleration, city, highway, and combined city/highway improvements over a standard vehicle.

The control system is the focus of the BAS hybrid system and receives the majority of the credit for increased fuel economy and efficiency. The BAS system uses the hybrid control architecture that can be found in all strong hybrid systems except for a high-voltage battery pack.

Low rolling-resistance tires have led to a 3% improvement. Regenerative braking captures vehicle kinetic energy during vehicle coast down for a 1% improvement. Deceleration fuel cutoff has reduced fuel consumption for a 3.5% improvement. Engine idle-off mode, when the vehicle is stopped, results in a 4% improvement.

Efficient mechanical and hydraulic strategies in the engine, transmission, and driveline have resulted in a 4.5% improvement. Vehicle aerodynamics is responsible for a 0.5% improvement. Vehicle control systems account for nearly 80% of city drive cycle improvements and almost 60% of highway drive cycle improvements.

Operating Modes

The BAS system can appear to be a simple idle-stop and auto-start system. However, a complex control system is needed to achieve fuel economy improvements. The BAS system augments the engine with much of the same functionality that a hybrid transaxle or transmission system offers in a strong high-voltage system. However, the BAS hybrid system electric machine is not contained within a transaxle/transmission.

The control system has nearly identical operating modes as a strong hybrid system. It ensures that both the engine and 36 V electric system are combined to achieve high levels of efficiency and drivability. **See Figure 19-21.**

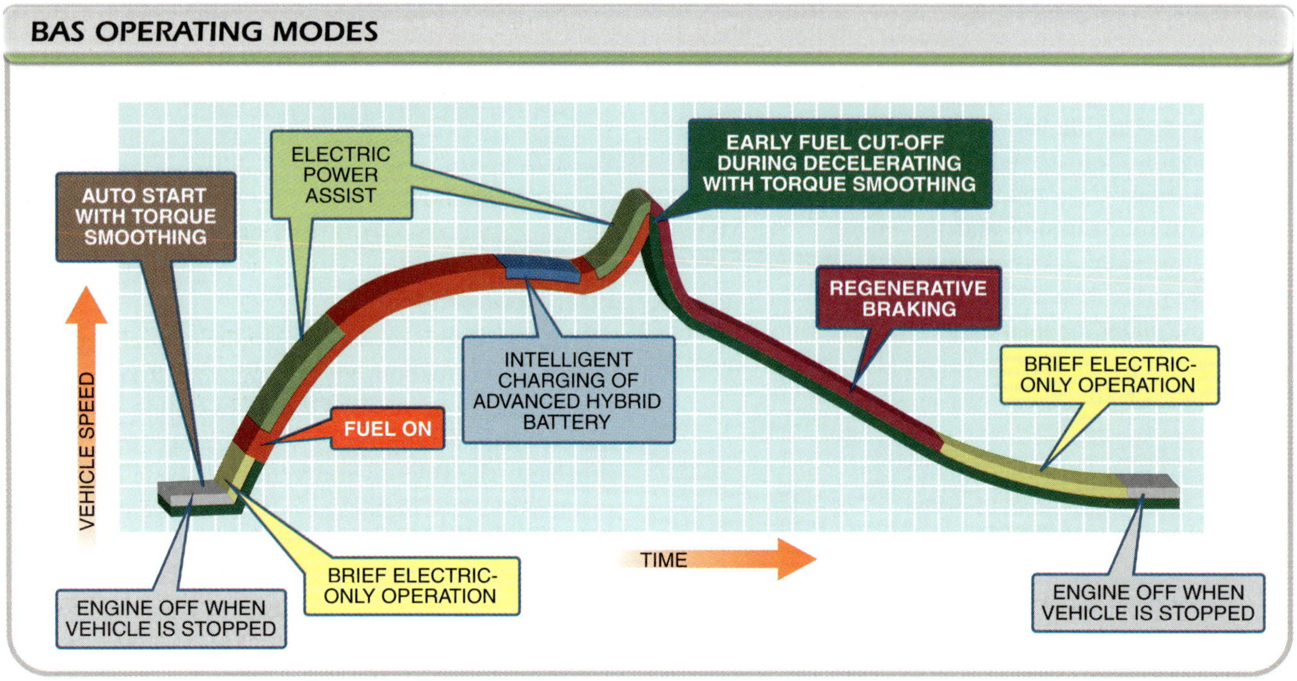

Figure 19-21. There are several modes contained within a drive cycle.

There are many vehicle control transitions during hybrid operation. These transitions are continual and determined by the engine load and speed of the vehicle. There are several modes contained within the drive cycle. Engine idle-off mode is used when the vehicle is stopped. Downshift torque smoothing makes for a smooth transition between gas and electric while decelerating. Other modes include the following:

- auto-start with torque smoothing
- intelligent charging of the advanced hybrid battery
- electric propulsion power assist
- early fuel cutoff with torque smoothing
- regenerative braking
- brief electric only operation

The primary difference between strong hybrid systems and the BAS hybrid is electric-only propulsion. With most strong hybrid systems, there is an electric-only propulsion mode that permits the vehicle to be driven with the electric machine only with no assistance from the engine. Depending on the control system, a strong hybrid vehicle can be propelled with electric-only propulsion to a maximum of 35 mph under a light throttle condition.

In the case of the BAS hybrid, there is never a case when the vehicle gets significant propulsion from the electric machine with no assistance from the engine. However, the BAS hybrid has operating modes that include electric machine assistance to the engine or a special mode. In the special mode, the engine does not receive fuel (although the engine rotates), and the electric machine provides input into the driveline for propulsion control.

Engine Cranking. The BAS system uses a traditional 12 V cranking motor only for the first crank cycle of any ignition cycle. There are three reasons for using the 12 V cranking motor on the first key cycle. First, the availability of energy from the 36 V battery pack is significantly reduced at low temperatures. Second, regular usage of the 12 V motor keeps it functional. Third, the 12 V cranking motor provides a method to crank the engine if the 36 V electrical or mechanical system fails.

Engine Auto-Start. The BAS system provides hybrid auto-start during all hybrid operating modes by using the MGU to crank an engine to approximately 800 rpm. The ECM will then send fuel to the engine in order for it to start. During the transition from cranking and starting to propulsion, the MGU assists the engine by smoothing torque pulsations.

Electric-Only Operation During Acceleration. During the short time frame in which a BAS vehicle operator depresses the throttle pedal and the engine starts, there will be a brief moment of electric-only propulsion. This operating mode is used to accelerate the vehicle electrically while cranking and starting the engine. While the MGU is cranking the engine and providing electric propulsion assist, the ECM is commanding the engine to be fueled. This results in a smooth transition between electric propulsion and engine propulsion.

Electric Power Assist. After the engine has completed a hybrid auto-start, there is a period of vehicle acceleration that will include torque blending between the MGU and engine. Torque blending permits the MGU to provide the majority of the propulsion torque at the beginning of the acceleration period. The engine torque will be dominant at the end of this period to provide more torque for acceleration.

MGU torque is used at the beginning of the acceleration period because the MGU is highly efficient. At low speed, engine efficiency can be extremely low. The engine provides almost 100% of the torque after the initial acceleration period has elapsed, when its efficiency is much higher. Essentially, any time the operator requests acceleration, MGU torque will be used to augment engine torque.

Intelligent Battery Charging. After a vehicle has accelerated and the operator no longer requests high levels of torque from the engine, the BAS system enters the intelligent charging area of the operating cycle. Intelligent charging permits the hybrid battery to be recharged by the MGU during propulsion mode.

The MGU is no longer required to provide propulsion torque when the drive cycle has crested and leveled. In this part of the drive cycle, the engine supplies all of the propulsion torque. It also provides torque to rotate the MGU in a regenerative (generator) mode to charge the hybrid battery modules.

Early Fuel Cutoff. The deceleration period is next in the sequence of operating modes after the acceleration or cruising period. During deceleration mode, fuel to the engine is cut off. This reduces fuel consumption and emissions. With respect to vehicle drivability, this transition period could result in driveline operating harshness with vibrations, shudder, and clunking.

However, the MGU is used to smooth any torque ripple in the system caused by the sudden reduction of engine torque. This is done by providing positive or negative torque input to the engine crankshaft and ultimately through the vehicle driveline. This results in a smooth transition by using the MGU to smooth the severe driveline rpm variations due to transient engine torque operation.

Regenerative Braking. During vehicle deceleration, the BAS hybrid system transitions the MGU to a pure generating mode in order to capture the kinetic energy of the vehicle. This energy is then used to recharge the battery-pack modules. Essentially, the engine drives the MGU to recharge the battery modules. Using the MGU as a generator results in negative torque being placed on the engine and driveline. This acts as a brake and results in a slower vehicle speed.

The BAS hybrid ESS must work with the braking system to blend friction and MGU negative torque. This provides a proper feel for the operator while maintaining safe braking operation. The relative amount of friction and electric braking depends on the battery-pack SOC and temperature. During regenerative braking, the engine continues to operate with minimal fueling.

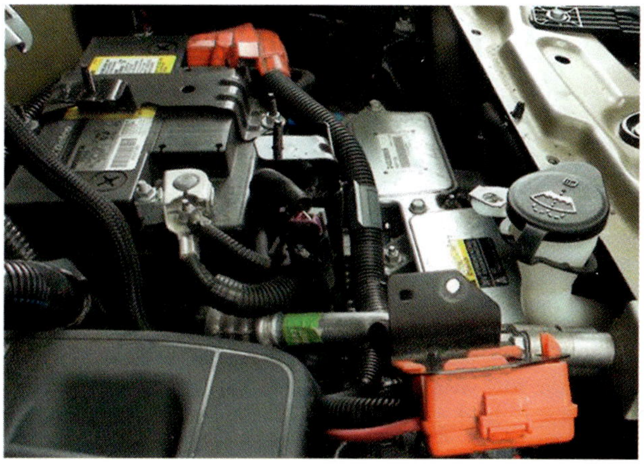

The 12 V battery provides electrical energy to low-voltage vehicle systems.

Electric-Only Operation for Deceleration. When the BAS hybrid system is decelerating, there is a period just prior to the vehicle coasting to a stop when there will be a brief electric-only mode. This mode excludes engine operation and propels the vehicle with electric-only propulsion from the MGU. As with any slow-speed mode, it is more fuel-efficient to move the vehicle with electric propulsion at low vehicle speed than with the engine. This also reduces engine emissions.

Engine Idle-Stop. The BAS engine idle-stop mode stops the engine any time vehicle speed is zero in order to reduce fuel consumption. However, there are many systems requirements that must be met before the hybrid controller allows the engine to stop during a drive cycle.

To allow an engine idle-stop, vehicle speed must have exceeded 4 mph at least once during the drive cycle. The ambient temperature must be greater than 5°F (–15°C). The vehicle gear selector must be placed in drive because engine idle-stop is not enabled if the transmission is in park (P), intermediate (I), low (L), or reverse (R). In addition, the hood of the vehicle must be closed. If the hood is open, the instrument panel indicator will illuminate the "Hood Ajar" warning.

There are many system temperature requirements that must be met before the hybrid controller permits an engine idle-stop during a drive cycle, which include the following:

- hybrid battery temperature must fall between 50°F (10°C) and 122°F (50°C)
- transmission fluid temperature must fall between 77°F (25°C) and 230°F (110°C)
- engine coolant temperature must fall between 140°F (60°C) and 250°F (121°C) for ambients less than 54°F (12°C)
- engine coolant temperature must fall between 180°F (82°C) and 250°F (121°C) for ambients greater than 54°F (12°C)

The vehicle core system also has several requirements that must be met before the hybrid controller permits an engine idle-stop during a drive cycle. These requirements include the following:

- the HVAC compressor system must not be requested when the engine is ON
- the grade must be less than 10%
- there must be a 5″ Hg minimum brake vacuum
- the hybrid battery voltage SOC must be greater than or equal to 70%
- the ESS battery power must be greater than or equal to 6200 W
- system status of the 12 V battery must be an acceptable level
- there can be no wheel slip while the ABS/traction control command is inactive
- the electronically variable pressure regulator (EVPR) cannot be performing a small leak test

Control System Operation

The BAS control system is made up of several elements. **See Figure 19-22.** For the vehicle to operate efficiently, it is necessary that these elements work together.

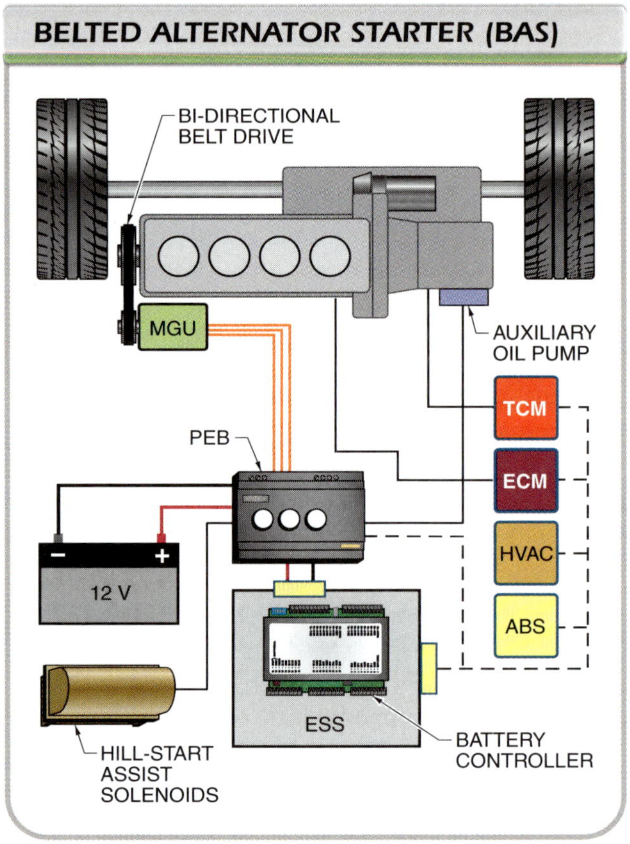

Figure 19-22. The BAS control system is made up of several elements.

The MGU permits smooth engine transitions during engine idle-stop and auto-start modes. It allows the use of coast and deceleration fuel cutoff. Regenerative braking is activated during coast and deceleration to recapture a portion of the vehicle kinetic energy to recharge the hybrid batteries.

The MGU also recharges the battery-pack system if the 36 V battery-pack SOC is low. If the vehicle is not coasting, decelerating, or stopped, and the 36 V battery-pack SOC is low, the hybrid control system selects the optimum engine operating mode to charge the batteries.

Additional Control System Functions. The BAS control system includes additional control system functions. The MGU, electronics control module (ECM), energy storage system (ESS), transmission control module (TCM), antilock brake system (ABS), and HVAC system must work together with a high degree of system integration. This ensures maximum fuel economy and enhanced vehicle drivability.

The MGU electric assist features an electric boost where the MGU supplements engine power for several seconds during engine restart events and acceleration. This reduces fuel usage while the vehicle is cruising. During city cruising, electric assist can account for about 10% of vehicle propulsion power.

The electric machine provides creep operation mode when decelerating to a stop or when transitioning from engine idle-stop to propulsion. This mode provides improved driveline smoothness during fuel ON and OFF transitions and during driver accelerator pedal tip-in maneuvers.

The ECM transmits MGU control commands to the power electronics box (PEB) in the form of torque, speed, or voltage control commands. The PEB then broadcasts the MGU electric machine speed, delivered torque, temperatures, and diagnostics. The ESS control module monitors key battery operating states for optimization. In addition, the ESS control module monitors the cooling fan and battery disconnect control functions. The ECM and TCM continually interact to coordinate gear shift and torque-converter clutch operations to maximize fuel economy.

Brake System

The BAS hybrid brake system is made up of a conventional hydraulic braking system augmented by regenerative braking, the brake delay valve (BDV), and the hill start valve (HSV). The *brake delay valve (BDV)* is a brake valve used to minimize vehicle roll by delaying the release of the friction brake system until a vehicle engine has been cranked and started after a hybrid idle-stop. The brake delay valve is used only when a vehicle is in the engine idle-stop mode and has come to a complete stop.

The PEB/SGCM commands the two BDV solenoids ON, using PWM during each hybrid engine idle-stop. The solenoids in the BDV are then activated in hybrid engine idle-stop and brake pressure is retained by the brake system. The brake pressure signal is transmitted to the SGCM. The PEB/SGCM controls the ON and OFF status of the two PWM solenoids, while the brake pressure release rate depends on the engine start rpm.

When the operator releases pressure on the brake pedal, the brake pressure sensor will signal the ECM. The ECM signal is then transmitted to the PEB/SGCM via the GMLAN. The PEB/SGCM will then command the two linear PWM solenoids to the open or OFF position.

The HSV is the second of the two special brake functions on the BAS hybrid vehicle. The HSV system is designed to enhance drivability on moderate to steep grades. Brake system pressure is captured by brake line solenoids to keep the vehicle from rolling back during the transition from a hybrid engine idle-stop to engine start.

Air Conditioning Controls

The BAS A/C system uses a belt-driven compressor. The A/C instrument panel button is designed as a three-position momentary switch with an OFF, economy A/C, and normal A/C setting. This system provides the appropriate balance between fuel economy and comfort by permitting the operator to select economy or normal A/C modes.

Economy mode favors fuel economy by maximizing hybrid engine idle-stop entry and duration. In economy mode, the ambient air temperature, engine temperature, etc. must be taken into account. Normal A/C mode favors maximum passenger comfort and defogging performance. The hybrid system will not transition into engine idle-stop with the A/C ON. **See Figure 19-23.**

AIR CONDITIONING CONTROLS

Figure 19-23. The A/C instrument panel button is designed as a three-position momentary switch with an OFF, economy A/C, and A/C setting.

Refer to the *Hybrid Electric Vehicle Technology* CD-ROM for a Quick Quiz® on the chapter contents.

POINTS TO REMEMBER

- The 12 V cranking motor is used only on the first engine crank of each operational cycle.
- The MGU is used for every engine crank and start cycle during hybrid operating modes.
- A special hydraulic strut compression tool, EN-48079, is required to remove the drive belt or drive belt tensioner components.
- The auxiliary oil pump on the automatic transmission must be primed through the use of a special scan tool control feature after transmission service.
- The MGU enhanced Lundell design offers significantly higher efficiency than a traditional generator (alternator).
- A resolver contained within the MGU is used to sense rotor position and rpm.
- The PEB/SGCM provides 3ϕ AC to DC and DC to AC conversion.
- The PEB/SGCM provides 12 V of electrical power to replace the traditional generator/alternator and 36 V to control electrical power transmitted to and from the ESS.
- The auxiliary transmission pump controller is mounted on the PEB/SGCM and is serviced as a separate component.
- The BAS system is not a floating ground system, and the 36 V system uses the vehicle body/undercarriage as its ground reference.
- Personal protection equipment (PPE), such as gloves for high voltage and special test equipment, is not required for a BAS hybrid, but the use of PPE is recommended.
- A BDU access door microswitch is used as a safety feature to sense the door position.
- Door position is used to signal when to open the contactor and discharge the PEB/SGCM bus capacitors.
- Battery module temperature sensors are serviceable components.
- Both the heater coolant pump and PEB/SGCM coolant pump are controlled by the PEB/SGCM controller.
- Brake delay and hill start valves are antirollback systems that assist the vehicle in remaining stationary on an incline during engine idle-stop mode.
- The engine idle-stop function is disabled if the operator commands the A/C system ON. The engine will remain ON when the vehicle is stopped.

Refer to the *Hybrid Electric Vehicle Technology* CD-ROM for the Review Questions in Microsoft® Word format.

REVIEW QUESTIONS

1. What modifications were made to the 4T45-E transmission so that it could be used in a BAS system?
2. Explain why the BAS ESS system can use a vehicle body/chassis and not operate as a strong hybrid floating system.
3. Explain the function of the BAS belt-tensioning system.
4. What is an enhanced Lundell electric machine?
5. At what point in a drive cycle does the MGU provide maximum torque?

19 Belted Alternator Starter (BAS) System

Trade Test

Name ______________________ Date ______________

______ **1.** The PEB/SGCM is located ___.
A. under the driver seat
B. between the engine and radiator support
C. beneath the rear load floor
D. under the passenger front seat

______ **2.** The ___ system of the BAS hybrid system is the largest contributor to improving city fuel economy.
A. regenerative braking
B. powertrain hardware
C. vehicle control
D. aerodynamics control

______ **3.** ___ charging is the BAS hybrid drive-cycle mode that permits the battery system to be charged during vehicle propulsion mode (not regenerative braking).
A. Intelligent
B. Opportunity
C. Battery
D. Prime SOC

______ **4.** The drive belt system of a BAS controls ___ torque.
A. lateral and vertical
B. endothermic and exothermic
C. fore and aft
D. positive and negative

______ **5.** During engine auto-start, the electric machine tightens the span between the engine crankshaft pulley and moves the slack side between the ___.
A. electric machine and pivot arm
B. electric machine and A/C compressor
C. A/C compressor and strut arm
D. pivot arm and hydraulic strut

______ **6.** The BAS hybrid automatic transmission uses the auxiliary hydraulic pump during ___ mode.
A. wide open throttle
B. regenerative braking
C. acceleration
D. engine idle-stop

_______________ **7.** The ___ system controller controls the brake delay and hill start valves.
A. ECM
B. PEB/SGCM
C. ABS
D. body

_______________ **8.** The enhanced Lundell design of the MGU rotor features ___.
A. more rotor coil windings
B. permanent magnets
C. increased rotor length
D. coolant fittings

_______________ **9.** The BAS air conditioning system uses a ___ compressor.
A. belt-driven
B. combination belt and electric
C. low-voltage electric
D. high-voltage electric

_______________ **10.** The blue covering over the 3ϕ MGU and ESS cables indicates that the voltage level is ___.
A. less than 24 V
B. between 12 V and 36 V
C. less than 60 V
D. between 60 V and 120 V

_______________ **11.** The PEB/SGCM has a 12 V fuse with a current rating of ___ A.
A. 12
B. 57
C. 100
D. 175

_______________ **12.** When a PEB/SGCM is replaced, the ___ should be removed from the failed unit and installed on the replacement unit.
A. coolant fittings
B. auxiliary transmission pump controller
C. 175 A fuse
D. ground post

_______________ **13.** The ESS contains ___ contactor(s).
A. one
B. two
C. three
D. four

_______________ **14.** The ESS contains ___ 12 V NiMH batteries.
A. two
B. four
C. six
D. eight

__________________ **15.** The negative terminal of the ESS is connected to the ___.
A. PEB/SGCM
B. MGU
C. vehicle alternator
D. vehicle undercarriage

__________________ **16.** The control unit that controls the ESS contactor system is the ___.
A. PEB/SGCM
B. BDU
C. ECM
D. TCM

__________________ **17.** The ESS battery disconnect switch contains ___ microswitches to sense the BDU access door.
A. two
B. four
C. six
D. eight

__________________ **18.** Each battery module has ___ temperature sensors.
A. two
B. four
C. six
D. eight

__________________ **19.** The BAS hybrid system contains ___ different valves to control vehicle position during engine idle-stop.
A. two
B. four
C. six
D. eight

__________________ **20.** The engine is cranked the first time of each key cycle by the ___.
A. 12 V cranking motor
B. MGU
C. expanding compression method
D. ECM

__________________ **21.** Cranking and starting an engine in the hybrid auto-start mode drive cycle is accomplished by the ___.
A. 12 V cranking motor
B. MGU
C. expanding compression method
D. ECM

__________________ **22.** If a vehicle gear selector is placed in park, engine idle-stop mode ___.
A. is enabled when A/C is OFF
B. is enabled when A/C is ON
C. is enabled when speed is less than 4 mph
D. cannot be enabled

_____________________ **23.** If the vehicle operator commands the A/C ON, engine idle-stop mode ___.
A. is enabled after 30 sec
B. is enabled after 1 min
C. is enabled when engine temperature is greater than 80°F
D. cannot be enabled

_____________________ **24.** If the ESS controller senses that the battery module SOC is greater than 70%, it ___.
A. does not permit A/C compressor operation
B. permits hybrid engine idle-stop
C. does not permit ABS brake operation greater than 5 mph
D. permits engine closed loop operation

_____________________ **25.** The body controller senses when the vehicle hood is open and ___.
A. limits the vehicle to 10 mph
B. places the vehicle in engine idle-stop
C. disables engine idle-stop
D. disables 12 V cranking

_____________________ **26.** The BAS hybrid control system accounts for ___% of the total highway drive-cycle fuel economy improvement.
A. 10
B. 25
C. 40
D. 60

_____________________ **27.** A BAS MGU can generate maximum electrical power at nearly ___% efficiency.
A. 65
B. 75
C. 85
D. 95

_____________________ **28.** The BAS PEB/SGCM will rectify ___ V of AC to DC in order to charge the auxiliary battery and energy storage system batteries.
A. 14
B. 24
C. 36
D. 42

_____________________ **29.** The BAS PEB/SGCM controls the MGU by using ___ MOSFET power semiconductor transistors.
A. two
B. four
C. six
D. eight

_____________________ **30.** The BAS DC-DC converter produces ___ W of electrical power.
A. 1000
B. 1500
C. 2000
D. 2500

Answers

Chapter 1—Introduction to HEV Technology

Review Questions

1. A hybrid electric vehicle (HEV) is a vehicle that uses more than one propulsion technology to provide power to the vehicle drive system.

3. With series power flow, the engine is not directly connected to the transmission, transaxle, or driveline. With parallel power flow, the electric machine torque is added in parallel to the engine torque and cannot be used by itself to drive the vehicle. With series/parallel power flow, the electric transaxle provides propulsion when the engine is not operating (series mode) or permits engine torque to be added to the electric transaxle and then transmitted through the drive axles (parallel mode).

Trade Test

1. C
3. A
5. A
7. D

Chapter 2—High-Voltage Electrical Safety

Review Questions

1. The "one hand rule" involves working with only one hand on an electrical circuit in order to avoid accidentally becoming part of a circuit and getting shocked.

3. High-voltage components are rated CAT I, CAT II, CAT III, and CAT IV. Low CAT ratings are for lower voltages and high CAT ratings are for higher voltages. Vehicle manufacturers often recommend tools with a CAT III–1000 V rating.

5. An insulation tester applies high voltage to a circuit and measures the resistance between two test points. If the resistance is lower than expected or specified, the component may have failed.

Trade Test

1. A
3. C
5. D
7. C
9. B

Chapter 3—High-Voltage Vehicle Safety Systems

Review Questions

1. An active safety system functions automatically to provide protection. A passive safety system needs to be manually activated to provide protection.

3. Some of the components that could fail include the battery-pack system, power inverter, DC-DC converter, power steering system, and electric machines.

5. A megohmmeter tests the integrity of the resistance between high-voltage components and the chassis of the vehicle. A technician connects one lead of the megohmmeter to the high-voltage component and the other end to the chassis of the vehicle. If the system has adequate isolation resistance between the high-voltage component and vehicle chassis, then the megohmmeter reading will be a very high, or even infinite (∞). If the system does not have adequate isolation resistance, the megohmmeter reading will be low.

7. A serial interlock loop (SIL) system is a series-circuit system that connects each high-voltage device. The controller senses if any of the high-voltage connectors or the cover of a high-voltage component are disconnected. A local interlock loop (LIP) is a high-voltage safety system that uses discrete switches with the high-voltage bus to signal a controller that a system has been breached.

Trade Test

1. C
3. A
5. D
7. A
9. D

Chapter 4—Hybrid Engines

Review Questions

1. A delayed intake valve closing reduces the compression ratio. When the reduced compression ratio is combined with an increased stroke and/or reduced combustion chamber volume, it permits the expansion ratio to be greater than the compression ratio. This improves fuel economy because the compression ratio in a spark-ignition engine is limited by the octane rating of the fuel. A high expansion ratio delivers a longer power stroke, allowing more expansion of the combustion gases and reducing the amount of heat wasted in the exhaust.
3. Efficiency increases with increasing compression ratio. A supercharger uses less energy than the piston in a naturally aspirated engine. The piston receives the equivalent compression as it would in a standard Otto-cycle engine while requiring only about 70% of the work for compression.
5. Hybrid control systems adjust the power contribution of engine and electric propulsion systems to develop the optimum efficiency for the entire system. The Atkinson cycle has high efficiency at high compression ratios, but it has low power output at low and intermediate loads and high power output at high loads. Electric traction systems have high efficiency at low and intermediate loads and lower efficiency at high loads, with constant torque in all ranges.

Trade Test

1. C
3. A
5. D
7. A

Chapter 5—AC Induction Electric Machines

Review Questions

1. Slip is the difference in rotation speed between the magnetic fields in the stator and the rotor. In an IM, the stator magnetic field always rotates faster than the rotor magnetic field. The slip is positive and the rotor speed lags behind the stator field speed. When an IM is operating as a generator, the slip is negative and the stator field lags behind the rotor.
3. If the stator field is reversed, the rotor also turns in the reverse direction. An IM can rotate as fast in the negative (reverse) direction as it can in the positive (forward) direction. This is a significant advantage in hybrid vehicle design because a transmission is not necessary to reverse a hybrid vehicle. The IM rotation direction can be changed by using a simple software algorithm and a line of software code.
5. The maximum speed of an IM is based on the number of poles per phase and the maximum frequency of the stator field. The formula used to calculate IM rpm is as follows:

$$rpm = \frac{(120)(400)}{4}$$

where
rpm = IM rotational speed, in rpm
Hz = stator field frequency, in Hz
P = number of poles per phase in the stator

For example, when the maximum stator field frequency is 400 Hz with 4 poles per phase, the maximum IM rotational speed of 12,000 rpm is calculated as follows:

$$rpm = \frac{(120)(400)}{4}$$

$$rpm = \frac{48{,}000}{4}$$

$$rpm = \mathbf{12{,}000}$$

Trade Test

1. C
3. A
5. A
7. C
9. A
11. D
13. A
15. B

Chapter 6—Permanent Magnet Electric Machines

Review Questions

1. The rotor design is unique in PM/BLDC motors because permanent magnets are used instead of the rotor bars that are used in other designs. The rotor magnetic field is created by the magnets instead of mutual induction from the stator field. The magnet placement on the rotor determines the timing of interactions with the stator field.
3. Field weakening is the point at which the rotor output voltage (back EMF) exceeds the voltage provided by the stator-pole magnetic fields.
5. The field strength of the rotor is fixed and the field strength of the stator is variable. Therefore, the field strength of the stator is used to control the torque. By increasing the electrical current flow from the power inverter to the stator, the magnetic field strength increases and produces more torque.

Trade Test

1. D
3. D
5. C
7. B

Chapter 7—Power Inverter Systems

Review Questions

1. The polarity of a sine wave that is impressed upon a rotor changes twice as often as the stator field. When the generated voltages and current polarities in the stator are reversed during an interval in the sine wave cycle, the flux field is reversed approximately twice as often in the rotor bars. This means that the rotor frequency is approximately twice the frequency of the stator field.

3. Overmodulation is a technique used to increase the available voltage to a motor by using waveforms other than sine waves. Different waveforms deliver different average voltages to the electric machine. An advantage of overmodulation is that a waveform can be selected in order to maintain the bus voltage. This results in an increase in horsepower and torque and allows higher operating speeds.

Trade Test

1. C
3. A
5. C
7. A
9. A
11. C
13. A
15. B
17. B
19. C
21. C
23. B

Chapter 8—Electric Propulsion Sensing Systems

Review Questions

1. Electric machine speed and position sensing is typically accomplished with resolvers, encoders, or hall-effect sensors.

3. Three-phase current is alternating current. At some point in the alternating cycle, the voltage and current are zero. Therefore, systems that use current sensors typically cannot operate on only one sensor. In this case, the electric propulsion system will not function.

5. The number of reluctor lobes determines the resolution of the signal generated by the resolver sensor and transmitted to the power inverter controller. More reluctor lobes mean a higher resolution signal because smaller rotations are detectable.

Trade Test

1. D
3. C
5. B
7. A
9. D

Chapter 9—DC-DC Converter Systems

Review Questions

1. Boost is a DC-DC converter term that means voltage is stepped up, or increased, from a low voltage to a high voltage. A boost converter steps up output voltage relative to the input voltage.

3. A DC-DC converter typically receives its electrical power from the hybrid high-voltage battery-pack (about 144 V DC to 320 V DC). The battery-pack electrical power is typically first routed through the power inverter as a "pass through" and then to the DC-DC converter.

5. The purpose of an output filter circuit is to smooth out the pulsating DC power from the rectifier circuit. Pulsating DC waveforms can cause electrical noise in an electrical system.

Trade Test

1. C
3. A
5. D
7. A

Chapter 10—Transaxles, Gears, and Cooling Systems

Review Questions

1. The transaxle damper assembly mechanically links the engine to the input shaft of the transmission. It functions as a torsional dampening device to smooth torsional oscillations of the engine by using a system of springs and discs. It is also used to dampen torque fluctuations.

3. Motor-generator one (MG1) is an electric machine contained within the Toyota Prius electric transaxle that determines transaxle gear ratio, generates electrical power for the drive motor and battery-pack systems, and functions as the engine cranking motor. MG1 is mechanically connected to the sun gear of the planetary gear set and determines the gear ratio of the gear set. It splits, or blends, the torque of the engine and MG2 assists the battery pack in providing additional electrical power to the drive motor (MG2) during high load conditions. It recharges the battery pack when the vehicle is in park and starts the engine by functioning as a cranking motor.

5. By connecting the planetary sun gear to MG1, infinite gear ratio combinations and a continuous blending of engine and MG2 torque to the final drive gears can be applied to provide an electronically variable transmission system. Blending engine and MG2 torque is accomplished by varying the planetary gear ratio, which varies the torque and speed of MG1. As the hybrid controller changes MG1 torque and/or speed, the torque and/or speed of the engine and MG2 change(s).

Trade Test

1. B
3. A
5. A
7. A

Chapter 11—Energy Management Hardware Systems

Review Questions

1. Shielding is a braided mesh of aluminum or copper that encases a cable or wire to reduce or eliminate magnetic field emissions, or to ensure that magnetic field emissions do not breach. It is especially important to protect low-voltage microcontroller/microprocessor circuits.
3. A precharge circuit is a circuit with a resistor and contactor that is used to protect high-voltage bus capacitors by limiting electrical current into discharged bus capacitors when a high-voltage system is initialized as the vehicle is turned ON. The purpose of the precharge system is to ensure the safe operation of the high-voltage bus capacitors and the contactor/relay contacts. It limits the amount of electrical current that flows through the contactors/relays and the high-voltage bus capacitors to minimize the possibility of capacitor damage.
5. Battery packs can be designed with one airflow sensor and four or more thermistor temperature sensors for critical modules. The critical modules operate at temperatures greater than other modules because of their placement. Therefore, monitoring the temperatures of these specific battery modules is essential for controlling the health and performance of the battery pack.

Trade Test

1. B
3. A
5. A
7. B
9. D

Chapter 12—Battery Construction and Technologies

Review Questions

1. The state of charge (SOC) is a measure of the remaining capacity of a battery relative to the capacity at full charge, usually given as a percentage. A state of charge of 30% indicates that 30% of the battery's maximum power remains in the battery and 70% has been used.
3. A 2C rate means that a battery is discharged at its rate multiplied by two. An 80 Ah battery can deliver 160 A for ½ hour, so its 2C rate is 160 A.
5. Power density is the amount of power a battery can deliver per unit volume at a specified state of charge, usually expressed in watts per liter (W/L). Specific power is the amount of power a battery can deliver per unit mass at a specified state of charge, usually expressed in watts per kilogram (W/kg). Specific power is a measure of the instantaneous power a battery can deliver based on its mass.

Trade Test

1. B
3. B
5. B
7. A

Chapter 13—Nickel-Metal Hydride Technology

Review Questions

1. The Peukert effect is a characteristic of batteries where the battery capacity depends on the discharge current. For a battery with a high Peukert effect, a higher discharge rate reduces the total power that can be removed from the battery compared to a battery with a low Peukert effect.
3. The discharge curves for NiMH are very flat throughout the majority of the charge and discharge cycles. Flat voltage output means that dynamic terminal voltage is maintained throughout the battery discharge until the end point. At the end point, the terminal voltage drops severely. In other words, the terminal voltage does not change much during discharge and it cannot be used as a reliable indicator of SOC.
5. The self-discharge rate is strongly affected by the temperature at which the batteries are stored. The self-discharge rate is higher for batteries stored at higher temperatures than for batteries stored at lower temperatures.

Trade Test

1. A
3. D

5. D
7. A

Chapter 14—Battery Pack and Energy Management Systems

Review Questions

1. A prismatic battery is a battery module design in which the shape of the module is rectangular instead of cylindrical.
3. The electric current is converted to waste heat. Power loss due to resistance is calculated as follows:

   $P = I^2R$

   where

   P = power (in W)

   I = current (in A)

   R = resistance (in Ω)
5. A series circuit is a circuit with only one current path for electric current flow. A parallel circuit is a circuit with two or more electrical circuit branches for electric current flow.

Trade Test

1. D
3. B
5. B
7. B

Chapter 15—Hybrid Vehicle Braking Systems

Review Questions

1. A hill-hold system is a vehicle propulsion system that uses the electric motor to prevent vehicle rollback down a hill.
3. If the battery-pack controller sends a signal to the hybrid controller that it has the capacity to absorb energy (low SOC), then the hybrid controller sends a message to the power inverter controller to send regen braking energy to the battery pack. The power inverter controller provides energy until the battery-pack controller signals that the battery has reached its target SOC. The power inverter controller then reduces or disallows regen braking and does not allow energy to be sent to the battery pack. When regen is disallowed, the hydraulic braking system provides all the braking power.
5. A brake system power supply is a backup power supply system. It is used to ensure that the electric motors develop brake pressure and have the electrical power to stop the vehicle in the event that the primary vehicle electrical system fails.

Trade Test

1. A
3. A
5. A
7. B
9. A

Chapter 16—Hybrid Climate Control Systems

Review Questions

1. Electric scroll compressors can be about 40% smaller and 50% lighter than conventional belt-driven A/C compressors.
3. A V/Hz motor controller changes the power output to maintain a desired ratio of the voltage to frequency. The ratio is calculated to ensure motor speed stability is maintained throughout the entire speed range of the compressor motor.
5. The first stage is used at startup. The control system uses both compressors to get maximum cooling capacity. In Stage 2, the control system uses the mechanical compressor only. This is used when a stable vehicle cabin temperature is reached. It is also used when the fan blower speed is set on high. In Stage 3, the control system uses the electric motor operation only. This is used when a stable cabin temperature is reached.

Trade Test

1. D
3. A
5. B
7. A
9. C
11. D
13. A
15. D

Chapter 17—Hybrid Vehicle Jump Start and Jump Start Assist Programs

Review Questions

1. Toyota does not provide procedures to charge or jump start the high-voltage battery. If the 12 V battery is adequately charged, but the high-voltage battery pack is depleted, the engine will not start. The vehicle must be towed to a Toyota dealership for service.
3. Although Honda systems are mechanized with high-voltage battery systems, there are no special procedures to jump start their products. Honda Hybrids use a 12 V cranking motor to provide engine cranking in the event of a high-voltage battery pack malfunction or when ambient temperatures are outside the high-voltage battery pack safe operating envelope.

5. The PHT jump starting procedure provides the option of jump starting the vehicle through the 12 V auxiliary battery, the 36 V battery pack, or both. There are two methods for jump starting the vehicle through the 12 V battery. The first method requires the use of a donor vehicle, traditional jump starting methods, and jump start cables. The second method requires a technician equipped with a Tech 2 serial data tool, or scan tool. Within the Tech 2 menus, the "42 V to 12 V Command" function permits the technician to command the system to use the 36 V battery system to charge the 12 V battery. If the 36 V battery has been depleted to the point that it will not crank the engine, the 12 V auxiliary battery can be commanded to transfer energy to the 36 V battery pack.

Trade Test

1. B
3. B
5. B
7. A

Chapter 18—GM FlexPower PHT

Review Questions

1. The energy storage control module (ESCM) is a control module that controls the functions within the energy storage module (ESM) and reports data to other controllers on the data bus. The primary function of the ESM is to provide electrical power for the electrohydraulic power steering and the electric machine. It also communicates the battery-pack status, including each battery module state of charge (SOC), current, and engine idle-stop available time. In addition, it communicates battery-pack constraints, such as the minimum SOC levels and the available charge and discharge power.
3. The HVAC system recirculates cabin air when the engine goes into engine idle-stop mode because the compressor does not operate in this mode. The HVAC blower motor speed is reduced during engine idle-stop mode to maintain a quiet cabin.
5. The auxiliary power outlet (APO) system is a system that provides 120 V AC standby power to outlets located in the vehicle cargo bed and in the cabin under the passenger seat. The APO is activated by pressing the APO button located to the immediate right of the air conditioning controls on the instrument panel. Normal mode is available only while the vehicle is being driven. It allows the use of all 120 V outlets. Normal mode can be activated by pressing the APO button located on the instrument panel.

   Continuous mode is available only while the vehicle is stationary with the engine running. It can provide this power with or without an operator in the vehicle, with or without a key in the ignition, or with the vehicle doors either closed or locked. Continuous mode can be activated from the normal mode while the engine is running.

Trade Test

1. C
3. B
5. A
7. B
9. B
11. D
13. A
15. C

Chapter 19—Belted Alternator Starter (BAS) System

Review Questions

1. A BAS needs an auxiliary transmission pump system to assist the engine-driven hydraulic pump in maintaining transmission pressure during the normal operating modes. The valve body channel plate, spacer plate, and valve body have also been modified. There is an additional solenoid for controlling downshifts during regenerative braking and a unique torque converter and final drive assembly.
3. The BAS combines a hydraulic strut tensioner and a friction damped rotary tensioner on a common pivoting arm. The drive belt control system uses the bidirectional system to control the positive and negative dynamic torque loads with two tensioning element systems. This system ensures sufficient tension under all operating modes while reducing overall belt tension and bearing loads.
5. The electric machine uses high current and generates high torque at low speeds in order to accelerate the vehicle with electric assist.

Trade Test

1. B
3. A
5. B
7. B
9. A
11. D
13. A
15. D
17. A
19. A
21. B
23. D
25. C
27. C
29. C
31. C

Appendix

AC/DC FORMULAS

To Find	DC	AC		
		1ϕ, 115 or 220V	1ϕ, 208, 230, or 240 V	3ϕ—All Voltages
I, HP known	$\frac{HP \times 746}{E \times E_{ff}}$	$\frac{HP \times 746}{E \times E_{ff} \times PF}$	$\frac{HP \times 746}{E \times E_{ff} \times PF}$	$\frac{HP \times 746}{1.73 \times E \times E_{ff} \times PF}$
I, kW known	$\frac{kW \times 1000}{E}$	$\frac{kW \times 1000}{E \times PF}$	$\frac{kW \times 1000}{E \times PF}$	$\frac{kW \times 1000}{1.73 \times E \times PF}$
I, kVA known		$\frac{kVA \times 1000}{E}$	$\frac{kVA \times 1000}{E}$	$\frac{kVA \times 1000}{1.763 \times E}$
kW	$\frac{I \times E}{1000}$	$\frac{I \times E \times PF}{1000}$	$\frac{I \times E \times PF}{1000}$	$\frac{I \times E \times 1.73 \times PF}{1000}$
kVA		$\frac{I \times E}{1000}$	$\frac{I \times E}{1000}$	$\frac{I \times E \times 1.73}{1000}$
HP (output)	$\frac{I \times E \times E_{ff}}{746}$	$\frac{I \times E \times E_{ff} \times PF}{746}$	$\frac{I \times E \times E_{ff} \times PF}{746}$	$\frac{I \times E \times 1.73 \times E_{ff} \times PF}{746}$

E_{ff} = efficiency

METRIC TO ENGLISH EQUIVALENTS

	Unit	British Equivalent		
LENGTH	kilometer	0.62 mi		
	hectometer	109.36 yd		
	dekameter	32.81″		
	meter	39.37″		
	decimeter	3.94″		
	centimeter	0.39″		
	millimeter	0.039″		
AREA	square kilometer	0.3861 sq mi		
	hectacre	2.47 A		
	acre	119.60 sq yd		
	square centimeter	0.155 sq in.		
VOLUME	cubic centimeter	0.061 cu in.		
	cubic decimeter	61.023 cu in.		
	cubic meter	1.307 cu yd		
CAPACITY		*cubic*	*dry*	*liquid*
	kiloliter	1.31 cu yd		
	hectoliter	3.53 cu ft	2.84 bu	
	dekaliter	0.35 cu ft	1.14 pk	2.64 gal.
	liter	61.02 cu in.	0.908 qt	1.057 qt
	cubic decimeter	61.02 cu in.	0.908 qt	1.057 qt
	deciliter	6.1 cu in.	0.18 pt	0.21 pt
	centiliter	0.61 cu in.		338 fl oz
	milliliter	0.061 cu in.		0.27 fl dr
MASS AND WEIGHT	metric ton	1.102 t		
	kilogram	2.2046 lb		
	hectogram	3.527 oz		
	dekagram	0.353 oz		
	gram	0.353 oz		
	decigram	1.543 gr		
	centigram	0.154 gr		
	milligram	0.015 gr		

ENGLISH TO METRIC EQUIVALENTS

		Unit	Metric Equivalent
LENGTH		mile	1.609 km
		rod	5.029 m
		yard	0.9144 m
		foot	30.48 cm
		inch	2.54 cm
AREA		square mile	2.590 km^2
		acre	.405 hectacre, 4047 m^2
		square rod	25.293 m^2
		square yard	0.836 m^2
		square foot	0.093 m^2
		square inch	6.452 cm^2
VOLUME		cubic yard	0.765 m^3
		cubic foot	0.028 m^3
		cubic inch	16.387 cm^3
CAPACITY	*U.S. liquid measure*	gallon	3.785 l
		quart	0.946 l
		pint	0.473 l
		gill	118.294 ml
		fluidounce	29.573 ml
		fluidram	3.697 ml
		minim	0.61610 ml
	U.S. dry measure	bushel	35.239 l
		peck	8.810 l
		quart	1.101 l
		pint	0.551 l
	British imperial liquid and dry measure	bushel	0.036 m^3
		peck	0.0091 m^3
		gallon	4.546 l
		quart	1.136 l
		pint	568.26 cm^3
		gill	142.066 cm^3
		fluidounce	28.412 cm^3
		fluidram	3.5516 cm^3
		minim	0.59194 cm^3
MASS AND WEIGHT	*avoirdupois*	short ton	0.907 t
		long ton	1.016 t
		pound	0.454 kg
		ounce	28.350 g
		dram	1.772 g
		grain	0.0648 g
	troy	pound	0.373 kg
		ounce	31.103 g
		pennyweight	1.555 g
		grain	0.648 g
	apothecaries'	pound	0.373 kg
		ounce	31.103 g
		dram	3.888 g
		scruple	1.296 g
		grain	0.648 g

Ohm's Law

Ohm's law is the relationship between the voltage, current, and resistance in an electrical circuit. Ohm's law states that current in a circuit is proportional to the voltage and inversely proportional to the resistance.

Power Formula

The power formula is the relationship between the voltage, current, and power in an electrical circuit. The power formula states that the power in a circuit is proportional to the voltage and the current.

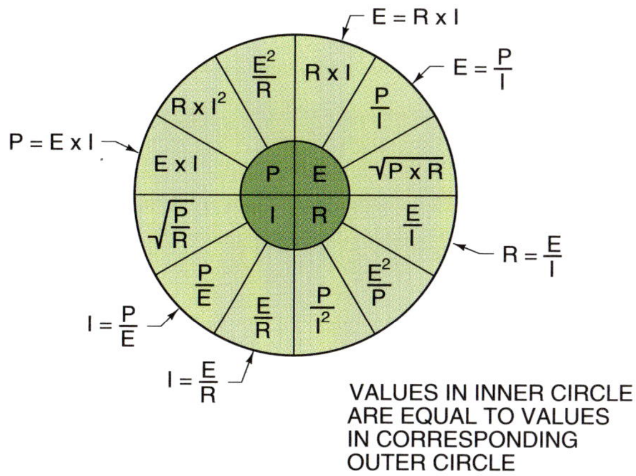

OHM'S LAW AND POWER FORMULA

AC/DC FORMULAS

Phase	To Find	Use Formula	Example		
			Given	Find	Solution
1φ	I	$I = \frac{VA}{V}$	32,000 VA, 240 V	I	$I = \frac{VA}{V}$ $I = \frac{32,000\ VA}{240\ V}$ $I = \mathbf{133\ A}$
1φ	VA	$VA = I \times V$	100 A, 240 V	VA	$VA = I \times V$ $VA = 100\ A \times 240\ V$ $VA = \mathbf{24,000\ VA}$
1φ	V	$V = \frac{VA}{I}$	42,000 VA, 350 A	V	$V = \frac{VA}{I}$ $V = \frac{42,000\ VA}{350\ A}$ $V = \mathbf{120\ V}$
3φ	I	$I = \frac{VA}{V \times \sqrt{3}}$	72,000 VA, 208 V	I	$I = \frac{VA}{V \times \sqrt{3}}$ $I = \frac{72,000\ VA}{360\ V}$ $I = \mathbf{200\ A}$
3φ	VA	$VA = I \times V \times \sqrt{3}$	2 A, 240 V	VA	$VA = I \times V \times \sqrt{3}$ $VA = 2 \times 416$ $VA = \mathbf{832\ VA}$

SELECTED ELECTRICAL SYMBOLS

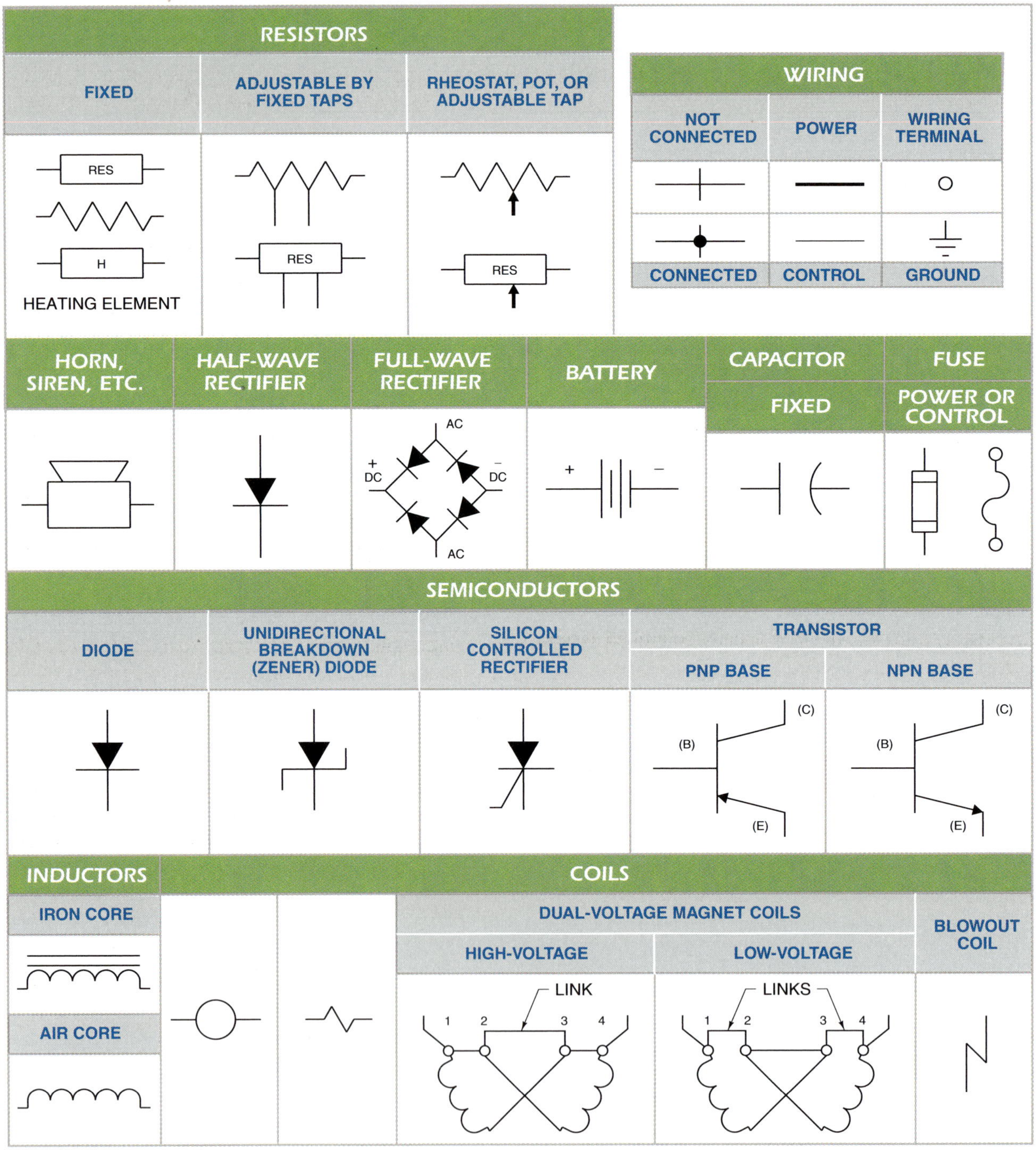

Glossary

A

absorbent glass mat (AGM) technology: Battery-cell technology that uses electrolyte stored in a mat placed between battery plates instead of plates immersed in a liquid electrolyte solution.

active bus-discharge circuit: A bus-discharge circuit that uses electronic circuits to quickly discharge the bus capacitors in a high-voltage system, usually in less than 5 seconds.

active cooling system: A forced-air cooling system that uses an air conditioner to cool a component or system.

active safety system: An electronically controlled system used to provide protection by isolating the battery system and discharging the bus capacitors.

amp-hour (Ah): A unit of measure calculated by multiplying the amount of electrical current a battery will deliver (in amps) by the time in which it will deliver this current (in hours) until the battery terminal voltage reaches a specified level.

Atkinson cycle: A four-stroke engine cycle in which the intake valve is held open longer than normal in order to allow the expansion ratio to exceed the effective compression ratio.

auxiliary power outlet (APO) system: A system that provides 120 V AC standby power to outlets located in a vehicle cargo bed and a cabin under the passenger seat.

B

back electromotive force (EMF): The voltage generated by an electric machine during regenerative braking mode that results in electric current generation.

battery disconnect unit (BDU): An electronic control unit in a BAS hybrid system that controls energy storage system (ESS) input and output circuits for the contactor, current sensor, microswitches, battery-module voltage sensors, temperature sensors, fuses, and vehicle diagnostics.

belted alternator starter (BAS): A hybrid propulsion system with an electric machine that is connected to an engine and driveline by a ribbed drive belt. The electric machine torque adds engine torque or provides regenerative energy throughout the entire vehicle drive cycle.

boost converter: A special DC-DC converter that is used within a power inverter to increase voltage from the battery pack to the electric machine.

boost: A term used to describe DC-DC converter voltage that is stepped up, or increased.

brake delay valve (BDV): A brake valve used to minimize vehicle roll by delaying the release of the friction brake system until a vehicle engine has been cranked and started after a hybrid idle-stop.

brake system power supply: A backup system used to ensure that an electric motor with brake pressure can stop a vehicle if the primary system fails.

buck: A term used to describe DC-DC converter voltage that is stepped down, or reduced.

C

capacitor: An electrical device that stores energy in the form of an electrostatic field.

capacity rate (C-rate): The rate at which the capacity of a battery is discharged until its terminal voltage reaches a specified level.

category (CAT) rating: A standard rating provided by the International Electrotechnical Commission (IEC) that outlines safety categories for electrical meters.

center-tapped inductor: An inductor coil with a secondary winding that has an electrical point in the center allowing it to be used as two electrical circuits.

choke: An inductor used with an electric filtering system to smooth out varying electrical current.

Class 0 glove: An electrical-safety glove designed to protect the wearer from electricity up to a maximum of 1000 V.

coefficient of performance (COP): The ratio of output power to input power in a compressor.

contactor: A large relay used to transfer electric current.

creep-aid system: A type of vehicle braking system that uses a series of solenoid valves and hydraulic pressure on the rear and, in certain modes, the front brakes to prevent a vehicle from rolling back on an incline.

current sensor: A sensor that provides a motor controller feedback information on phase current.

cylindrical battery: A battery module that is cylindrical in shape.

D

DC-DC converter: An electronic device that converts high-voltage DC power to low-voltage DC power, or vice versa.

device under test (DUT): A device that is being tested or analyzed by an instrument or test equipment.

differential scope probe: An instrument that is used to isolate an oscilloscope from a circuit.

E

efficiency (η): The measure of how well a vehicle uses fuel and is calculated based on the ratio of output power to input power.

electrical power: The amount of work done by electric current.

electric machine: A machine with induction or permanent magnet technology that provides torque for electric traction or electric energy during regenerative braking.

electric machine control waveform: A generated, periodic voltage level used to control the voltage level to an electric machine so that the torque and speed can be controlled.

electrohydraulic power steering module (EHPSM): A control module that controls the electrically driven hydraulic motor to provide hydraulic pressure for steering and braking assist.

electromagnetic interference (EMI): The magnetic field emanating from wire or cable that can cause operational disruptions with other electrical circuits.

encoder: A sensor that provides digital, square-wave outputs that the electric machine motor controller counts to determine the approximate rotor position and speed.

energy density: The amount of energy a battery stores per unit volume at a specified discharge rate, usually expressed in watt-hours per liter (Wh/L).

energy storage control module (ESCM): A control module that controls the functions within an energy storage module (ESM) and reports data to other controllers on a data bus.

energy storage system (ESS): A system made up of battery modules, current sensors, temperature sensors, contactors, cables, connectors, and a thermal system, which stores and provides electrical energy to an electric propulsion system.

engine auto-start mode: An operating mode in which an engine is cranked and started during acceleration or a low battery pack is recharged by a high-voltage system.

engine idle-stop mode: An operating mode in which a controller stops or shuts off an engine when a vehicle comes to a stop.

enhanced Lundell electric machine: An electric machine that uses both a traditional generator/alternator synchronous design and permanent magnets on the rotor, resulting in increased efficiency and output. The key elements to the enhanced Lundell rotor are the magnets that are mounted and laser-welded between rotor claws.

F

field weakening: The point at which the rotor output voltage (back EMF) exceeds the voltage provided by the stator-pole magnetic fields.

flooded battery: A lead-acid battery in which all plates in the battery cells are immersed in a liquid electrolyte solution of sulfuric acid and water.

flyback diode: A diode that is connected in the reverse bias position across the emitter and collector of an IGBT in order to steer reactive current during regenerative braking mode.

forward-switching converter: A converter used to transform an input voltage into a regulated output voltage.

full-wave rectifier: A diode system of at least two rectifier diodes that allows the conversion of positive and negative components of an AC sine wave into a DC output.

G

grounded switch interlock system: A modified LIP system that uses switches located within its components to provide a ground for the hybrid-controller sense circuit.

H

heating, ventilating, and air conditioning (HVAC) control module: A control module that controls air conditioning and heating functions and reports data to other controllers on a data bus.

high voltage: A level of system voltage that is greater than 60 V DC or 30 V AC rms.

high-voltage safety system: A system designed to protect the vehicle operator, technicians, and first responder (fire/rescue) personnel from injury caused by the high voltage present in electric propulsion vehicles.

hill-hold system: A type of vehicle propulsion system that uses an electric motor to prevent vehicle rollback down a hill.

hybrid control module (HCM): A control module that functions as the master of all vehicle controllers on the data bus in a hybrid vehicle.

hybrid electric vehicle (HEV): A vehicle that uses more than one propulsion technology to provide power to the vehicle drive system.

I

inductor: An electrical device that stores energy in the form of a magnetic field.

insulation tester: A diagnostic instrument that is used to detect a reduction in resistance between high-voltage circuits, components, or systems and the vehicle chassis.

interlock system: A system used to detect when any high-voltage component or circuit has been breached.

intermediate voltage: A level of system voltage that is greater than 30 V DC and less than or equal to 60 V DC, or that is greater than 15 V AC rms and less than or equal to 30 V AC rms.

ion: An atom or molecule that is either positively or negatively charged.

isolation fault: A condition that results when the resistance between a high-voltage system and the vehicle chassis has degraded to a point that may no longer be safe.

J

jump start assist: A method by which a 12 V auxiliary battery and an onboard vehicle DC-DC boost converter are used to recover (charge) a high-voltage battery pack enough to crank an engine using a high-voltage electric machine.

L

local interlock loop (LIP) system: A high-voltage safety system that uses discrete switches with the high-voltage bus to signal a controller that a system has been breached.

low voltage: A level of system voltage that is less than or equal to 30 V DC, or less than or equal to 15 V AC rms.

M

mild hybrid system: A hybrid electric vehicle system that is typically characterized by system voltages at 36 V or less, may provide electric propulsion assist to the engine, and may have a limited electric-only drive mode.

Miller cycle: A modified Atkinson cycle in which a supercharger is added to an engine in order to increase the compression ratio.

motor-generator one (MG1): An electric machine contained within the Toyota Prius electric transaxle that determines transaxle gear ratio, generates electrical power for the drive motor and battery-pack systems, and functions as the engine cranking motor.

motor-generator two (MG2): An electric machine contained within the Toyota Prius electric transaxle that serves as the propulsion motor and operates in conjunction with the engine system to provide vehicle propulsion and reverse propulsion mode.

motor-generator unit (MGU): A BAS system component, similar to a traditional automotive generator (alternator), that operates as an electric motor or generator during vehicle drive cycles and is coupled to an engine through a drive belt system.

O

Otto cycle: A four-stroke engine cycle used in traditional automotive engine applications in which the amount of time that the intake and exhaust valves open and close is nearly equal.

overmodulation: A technique used to increase the available voltage to a motor by using waveforms other than sine waves.

P

parallel circuit: A circuit with two or more electrical circuit branches for electric current flow.

parallel power flow system: An engine system connected directly to the transmission, transaxle, or driveline with supplemental electric machines connected in parallel.

passive bus-discharge circuit: A bus-discharge circuit that uses resistors to slowly discharge the high-voltage bus capacitors in a high-voltage system, usually in about 5 minutes.

passive cooling system: A forced-air cooling system that uses ambient air to cool a component or system.

passive safety system: A nonelectronically controlled system permanently connected to the high-voltage system that is used to provide protection by discharging the bus capacitors.

pedal position sensor (PPS): A sensor located on the accelerator pedal assembly that provides an input to an electric machine motor controller so that it can sense the desired acceleration rate and level of power desired by the vehicle operator.

Peukert effect: The characteristic of a battery in which the battery capacity depends on the discharge current.

polyalkylene glycol (PAG) oil: A compressor oil that provides a low level of electrical isolation and has a high water saturation value.

polyolester (POE) oil: A compressor oil that provides a high level of electrical isolation to minimize isolation fault conditions.

potassium hydroxide (KOH) electrolyte: An electrolytic solution of potassium hydroxide (KOH) and water.

power density: The amount of power a battery can deliver per unit volume at a specified state of charge, usually expressed in watts per liter (W/L).

power electronics box/starter-generator control module (PEB/SGCM): A power electronics component that controls electrical energy to a motor-generator unit (MGU), controls MGU operation, and converts AC to DC or DC to AC.

power inverter: An electronic component that changes direct current power to alternating current power or alternating current power to direct current power.

powertrain control module (PCM): A control module that controls vehicle engine functions and reports data to other controllers on a data bus.

precharge circuit: A circuit with a resistor and contactor that is used to protect high-voltage bus capacitors by limiting electrical current into discharged bus capacitors when a high-voltage system is initialized.

prismatic battery: A battery module that is rectangular in shape.

proximity sensor: A sensor that provides a digital, square-wave output to the electric machine motor controller by sensing the position of the commutation plate to determine the approximate rotor position and speed.

pumping loss: Energy loss due to restrictions in the engine that inhibit air or gases from flowing or being pumped through the engine intake, engine, or engine exhaust systems.

R

reactive power: Electrical power that is generated during regenerative braking mode when the magnetic fields of the rotor cut across the stator pole windings to produce electrical power that is transferred through the IGBT flyback diodes and to the battery pack.

reactor: A power inductor within a power inverter system that can boost battery-pack voltage to a higher voltage level.

resolver: A sensor that provides two sine waves to a motor controller to indicate absolute rotor position and speed.

rotor: The rotating part of an electric machine.

S

self-discharge rate: The rate at which a battery will discharge without being connected to an external load and is due to the natural chemical reactions in the battery or micro shorting in a cell plate.

serial interlock loop (SIL) system: A series-circuit, or daisy-chain, system that connects each high-voltage device in a hybrid-vehicle power system.

series circuit: A circuit with only one current path for electric current flow.

series/parallel power flow system: An engine and electric transaxle system that permits the electric transaxle to provide propulsion when the engine is not operating (series mode) or permits engine torque to be added to the electric transaxle and then transmitted through the drive axles (parallel mode).

series power flow system: An engine system connected in series to a generator that provides electrical power to the electric drive motor and the battery pack.

service disconnect switch: A mechanical lever or rotating switch device that is used to remove power from a high-voltage circuit.

shielding: A braided mesh of aluminum or copper that encases a cable or wire to negate magnetic field emissions, or to ensure that magnetic field emissions do not breach a wire or cable.

six-pack motor drive: A power transistor configuration that uses six IGBTs to control the torque and speed of an electric machine.

slip: The difference in rotation speed between the magnetic fields in the stator and the rotor.

specific energy: The amount of energy a battery stores per unit mass at a specified discharge rate, usually expressed in watt-hours per kilogram (Wh/kg).

specific power: The amount of power a battery can deliver per unit mass at a specified state of charge, usually expressed in watts per kilogram (W/kg).

starter-generator control module (SGCM): A control module that controls electrical power to a starter-generator and includes a DC-DC buck-boost converter system to charge a 12 V auxiliary battery. The converter can also charge a 36 V battery pack from the 12 V system if necessary.

starved electrolyte battery: A battery with only a small amount of electrolyte.

state of charge (SOC): A measure of the remaining capacity of a battery relative to the capacity at full charge, usually expressed as a percentage.

stator: The nonrotating part of an electric machine that encloses the rotor.

strong hybrid system: A hybrid electric vehicle system that is typically characterized by system voltages between 144 V and 600 V, an electric-only drive mode, and/or electric propulsion assist to the engine.

synchronous field control: A control system that causes the rotor and stator fields to interact at consistent and repeatable angles through the 360° circle in which the rotor revolves.

T

thermal system: A cooling and heating system used to control the temperature of a battery pack.

throttle position sensor (TPS): A sensor located on the engine throttle body that is used to monitor the position of the throttle.

torque ripple: Rough torque delivered by an electric machine caused by the use of waveforms that have aggressive changes in voltage, such as six-step and rectangular waveforms.

transaxle damper assembly: A device that permits electric machine power and torque to be transmitted from an engine to an electronically variable transmission.

V

valve regulated lead-acid (VRLA) battery: A battery that has a pressure regulation valve located inside to control the internal pressure.

volts per hertz (V/Hz) motor control: An electric motor control strategy in which the frequency and voltage applied to a motor is changed to provide a corresponding change in motor speed and torque levels.

Index

Page numbers in italic refer to figures.

N

O

P

R

S

T

V

W

Z

USING THE HYBRID ELECTRIC VEHICLE TECHNOLOGY INTERACTIVE CD-ROM

Before removing the Interactive CD-ROM from the protective sleeve, please note that the book cannot be returned for refund or credit if the CD-ROM sleeve seal is broken.

System Requirements

To use this Windows®-compatible CD-ROM, your computer must meet the following minimum system requirements:

- Microsoft® Windows Vista™, Windows 2000®, or Windows NT® operating system
- Intel® 1.3 GHz processor (or equivalent)
- 128MB of available RAM (256MB recommended)
- 335MB of available hard disc space
- 1024 x 768 monitor resolution
- CD-ROM drive (or equivalent optical drive)
- Sound output capability and speakers
- Microsoft® Internet Explorer® 6.0 or Firefox® 2.0 web browser
- Active Internet connection required for Internet links

Opening Files

Insert the Interactive CD-ROM into the computer CD-ROM drive. Within a few seconds, the home screen will be displayed allowing access to all features of the CD-ROM. Information about the usage of the CD-ROM can be accessed by clicking on Using This Interactive CD-ROM. The Quick Quizzes®, Illustrated Glossary, Media Clips, Flash Cards, Review Questions, and ATPeResources.com can be accessed by clicking on the appropriate button on the home screen. Clicking on the American Tech web site button (www.go2atp.com) accesses information on related educational products. Unauthorized reproduction of the material on this CD-ROM is strictly prohibited.

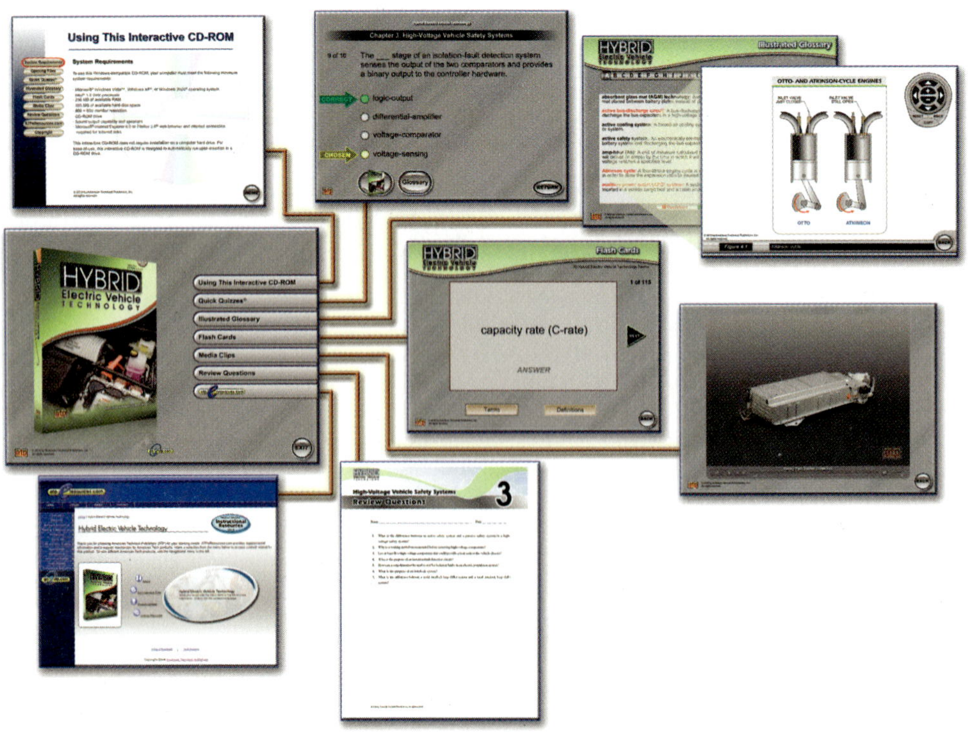